Lecture Notes in Economics and Mathematical Systems

425

Founding Editors:

M. Beckmann
H. P. Künzi

Editorial Board:

H. Albach, M. Beckmann, G. Feichtinger, W. Hildenbrand, W. Krelle
H. P. Künzi, K. Ritter, U. Schittko, P. Schönfeld, R. Selten

Managing Editors:

Prof. Dr. G. Fandel
Fachbereich Wirtschaftswissenschaften
Fernuniversität Hagen
Feithstr. 140/AVZ II, D-58097 Hagen, Germany

Prof. Dr. W. Trockel
Institut für Mathematische Wirtschaftsforschung (IMW)
Universität Bielefeld
Universitätsstr. 25, D-33615 Bielefeld, Germany

Kurt Marti   Peter Kall (Eds.)

Stochastic Programming

Numerical Techniques
and Engineering Applications

Proceedings of the 2nd GAMM/IFIP-Workshop
on "Stochastic Optimization: Numerical Methods
and Technical Applications" held at the
Federal Armed Forces University Munich,
Neubiberg/München, Germany, June 15-17, 1993

Springer

Editors

Prof. Dr. Kurt Marti
Univerisität der Bundeswehr München
Fk. LRT
D-85577 Neubiberg/München, Germany

Prof. Dr. Peter Kall
Universität Zürich
Institut für Operations Research
Moussonstr. 15
CH-8044 Zürich

ISBN 978-3-540-58996-9   ISBN 978-3-642-88272-2 (eBook)
DOI 10.1007/978-3-642-88272-2

CIP data applied for

This work is subject to copyright. All rights are reserved, whether the whole or part of the material is concerned, specifically the rights of translation, reprinting, re-use of illustrations, recitation, broadcasting, reproduction on microfilms or in any other way, and storage in data banks. Duplication of this publication or parts thereof is permitted only under the provisions of the German Copyright Law of September 9, 1965, in its current version, and permission for use must always be obtained from Springer-Verlag. Violations are liable for prosecution under the German Copyright Law.

© Springer-Verlag Berlin Heidelberg 1995

Originally published by Springer-Verlag Berlin Heidelberg New York in 1995.

Typesetting: Camera ready by author
SPIN: 10486737   42/3142-543210 - Printed on acid-free paper

PREFACE

This volume includes a selection of papers presented at the
2nd GAMM/IFIP-Workshop on "Stochastic Optimization: Numerical
Methods and Technical Applications", held at the Federal Armed
Forces University Munich, June 15-17, 1993.

In order to obtain more reliable optimal solutions (e.g. optimal designs) of
concrete technical problems, the often known stochastic variations of many
technical parameters/coefficients (e.g. material coefficients, tolerances, external
loads, manufacturing errors, technological coefficients) have to be taken into
account already in the planning phase (offline)!

Hence, the aim of this Workshop was to bring together scientists from
Stochastic Programming, Numerical Optimization and from those Engineering
areas, where optimization models are common tools, as e.g. in Optimal Structural
Design, Path Planning for Robots, Target Costing.

The following International Programme Committee was formed:

P. Abel (Germany)

H.A. Eschenauer (Germany)

P. Kall (Switzerland)

K. Marti (Germany, Chairman)

J. Mayer (Hungary/Switzerland)

F. Pfeiffer (Germany)

G.I. Schueller (Austria).

The area covered by the lectures during the workshop was rather broad, although
the number of participants had to be small for technical reasons. New theoretical
insight into several branches of reliability-oriented optimization of stochastic
technical systems was given, new computational approaches for stochastic
programs were discussed, and - in the main part - applications of stochastic
optimization methods to Optimal Path Planning of Robots, Optimal Reliability of
Mechanical Structures, Optimal Design and Target Costing were given.

In order to guarantee a high scientific level of the present Proceedings, all papers were refereed. Hence, we express our gratitude to all referees and to all contributors for delivering the final version of their papers in due time.

We gratefully acknowledge the financial support of GAMM (Gesellschaft für Angewandte Mathematik und Mechanik), IFIP (International Federation For Information Processing) and Federal Armed Forces University Munich; especially, we thank the commander of the student division for the kind accomodational support.

Finally, we thank Springer-Verlag for including the Proceedings in the Springer Lecture Notes Series.

München/Zürich

December 1994   K.Marti, P. Kall

TABLE OF CONTENTS

III. ENGINEERING APPLICATIONS

TYPES OF ASYMPTOTIC APPROXIMA-
TIONS FOR NORMAL PROBABILITY
INTEGRALS

K. Breitung [1]

[1]Schellingstr. 21, D-80799 München, Germany.

Abstract. For many problems in reliability and optimization it is necessary to calculate the probabilities of large deviations of normal random vectors. Using the structure of the normal probability density it is possible to derive simple asymptotic approximations for such integrals. Here three types of such approximations are described: approximations for the logarithm of the probabilities, approximations for the probabilities and asymptotic expansions for them.

Keywords. Asymptotic analysis, asymptotic expansions, Gaussian distribution, Laplace method, large deviations, normal distribution, random vectors

1 Introduction

In many problems of probability theory, mathematical statistics and their applications probabilities of large deviations are of interest. Here an overview of different approximation methods is given.

We consider a Gaussian random vector $\boldsymbol{X} = (X_1, \ldots, X_n)$ with a standard normal distribution, i.e. its mean vector is the zero vector $\boldsymbol{0} = (0, \ldots, 0)$, its covariance matrix is the n-dimensional unity matrix $\boldsymbol{I}_n$ and its density is given by

$$\varphi_n(\boldsymbol{x}) = (2\pi)^{-n/2} \exp\left(-\frac{1}{2}|\boldsymbol{x}|^2\right) \tag{1}$$

with $|\boldsymbol{x}|$ the euclidean norm of the vector $\boldsymbol{x}$. The probability content of a set $A \subset \mathbb{R}^n$ for this probability measure is denoted by $\Phi_n(A)$. In the following we will study the asymptotic behaviour of the large deviation probabilities given by

$$\Phi_n(\beta A) = \int\limits_{\beta A} \varphi_n(\boldsymbol{x}) \, d\boldsymbol{x}$$

as $\beta \to \infty$. We assume that A is a closed set and $A^c = \mathbb{R}^n \setminus A$ – the complement of A – is a neighborhood of the origin in $\mathbb{R}^n$. This means that A^c is an absorbing set and therefore $\Phi_n(\beta A) = 1 - \Phi_n(\beta A^c) \to 0$ as $\beta \to \infty$.

Such results can be derived by making a transformation $x \to y = \beta^{-1}x$, which gives

$$\Phi_n(\beta A) = \beta^n \int_A \varphi_n(\beta x) \, dx. \tag{2}$$

Here now the integration domain is fixed and only the integrand depends on the parameter β. Such integrals are called Laplace type integrals and their asymptotic behavior is studied in a number of papers and books (for example [2], chap. 8). The first asymptotic approximations for the case of normal integrals were derived in [3]. They give a connection between the differential geometry of the boundary surface ∂A and the probability content of βA as $\beta \to \infty$.

Another approach to this problem is to make a further transformation to spherical coordinates, i.e. $x \to (\tau, u)$, where $\tau = |x|^{-1}x$ and $u = |x|$. This gives then

$$\Phi_n(\beta A) = \beta^n \int_0^\infty \int_{S_n(1)} 1_{u\tau \in A} u^{n-1} \varphi_n((\beta u)\tau) \, d\tau du \tag{3}$$

Here $S_n(1)$ is the n-dimensional unit sphere. Since the distance of A to the origin is unity and the standard normal density has a rotational symmetry, we get

$$= \beta^n \int_1^\infty \left[\int_{S_n(1) \cap u^{-1}A} d\tau \right] u^{n-1} (2\pi)^{-n/2} \exp(-(\beta u)^2/2) du \tag{4}$$

If we denote by $F(A; u)$ the probability that for the uniform probability distribution on $S_n(1)$ a random point is in $u^{-1}A$, i.e.

$$F(A; u) = \omega_n \left(u^{-1}A \cap S_n(1) \right) \tag{5}$$

we can write this as

$$\Phi_n(\beta A) = \frac{\beta^n \omega_n}{(2\pi)^{n/2}} \int_0^\infty F(A; u) u^{n-1} \exp\left(-(\beta u)^2/2\right) \, du. \tag{6}$$

with $\omega_n = 2\pi^{n/2}/\Gamma(n/2)$ the surface area of the n-dimensional unit sphere $S_n(1) = \{x \in \mathbb{R}^n; |x| = 1\}$. The representation in equation (6) was studied in [1], [15] and [16].

We will study in the following three types of approximations for $\Phi_n(\beta A)$:

1. approximations for $\log(\Phi_n(\beta A))$,

2. approximations for $\Phi_n(\beta A)$,

3. expansions for $\Phi_n(\beta A)$ in the sense of Poincaré.

All these results show that asymptotically the probabilities depend on the structure of the boundary ∂A in the neighborhood of the set $M = \{x; x \in A \cap S_n(1)\}$, i.e. the set of the points in A with minimal distance to the origin, where the density is maximal.

2 Basic concepts of asymptotic approximations and expansions

In asymptotic analysis one of the main problems is to find simple approximations for functions in an asymptotic sense. For two given functions $f(x)$ and $g(x)$, we write $f(x) \sim g(x)$, $x \to x_0$ iff $\lim_{x \to x_0} g(x)/f(x) = 1$. This means that

$$\lim_{x \to x_0} \frac{f(x) - g(x)}{f(x)} = 0. \tag{7}$$

Therefore the relative approximation error for approximating $f(x)$ by $g(x)$ approaches zero as $x \to x_0$. This relations says nothing about the error for a fixed $x \neq x_0$, only that in the limit the error goes to zero. It is an important problem in asymptotic analysis to find explicit error bounds, but in many cases this is not possible. Then only by studying suitable examples one can get an idea about the error.

An asymptotic scale is a sequence $\{\psi_n(x)\}$ of continuous functions such that

$$\psi_{n+1}(x) = o(\psi_n(x)), \quad x \to x_0 \tag{8}$$

An asymptotic expansion of a function $f(x)$, as $x \to x_0$, with respect to the the scale $\{\psi_n\}$ is given by the formal series $\sum_{n=0}^{\infty} a_n \psi_n(x)$ if for all $k = 0, 1, \ldots$ always

$$f(x) = \sum_{m=0}^{k} a_m \psi_m(x) + O(\psi_{k+1}(x)), \quad x \to x_0. \tag{9}$$

Such an expansion is said to be of Poincaré type. The expansion does not need to converge. If an expansion up to k terms is made, the following term $a_{k+1}\psi_{k+1}(x)$ gives the order of magnitude of the approximation error. Textbooks about asymptotic analysis are for example [2] and [17].

An important result, which relates the asymptotic expansion of a function near zero with the asymptotic expansion of its Laplace transform near infinity, is Watson's lemma. We consider a locally integrable function $f : (0, \infty) \to \mathbb{R}$, which is bounded on finite intervals. Further we assume that $f(t) = O(e^{at})$, $t \to \infty$ for a real number a. For $t \to 0+$ let the function $f(t)$ have the following expansion

$$f(t) \sim \sum_{m=0}^{\infty} c_m t^{a_m},$$

where the a_m are complex numbers such that $\mathrm{Re}(a_m)$ increases monotonically to $+\infty$ as $m \to \infty$ and $\mathrm{Re}(a_0) > -1$. Then Watson's lemma says that the Laplace transform of $f(t)$ has the following asymptotic expansion

$$\int_0^{\infty} e^{-\lambda t} f(t)dt \sim \sum_{m=0}^{\infty} \frac{c_m \Gamma(a_m + 1)}{\lambda^{a_m + 1}}, \quad \lambda \to \infty. \tag{10}$$

A proof is given for example in [2], p. 103-4.

3 Approximations for the logarithms

This case is the simplest. We consider the logarithm $\log(\Phi_n(\beta A))$. This can written as

$$\log(\Phi_n(\beta A)) = -\frac{\beta^2}{2} - \frac{n}{2}\log(2\pi) + \log\left[\int_A \exp\left(\frac{\beta^2}{2}(|\boldsymbol{x}|^2 - 1)\right)d\boldsymbol{x}\right] \quad (11)$$

Introducing spherical coordinates gives then

$$\log(\Phi_n(\beta A)) = \quad (12)$$

$$-\frac{\beta^2}{2} - \frac{n}{2}\log(2\pi) + \log\left[\omega_n\int_1^\infty F(A; 1 + u)u^{n-1}\exp\left(\frac{\beta^2(u^2-1)}{2}\right)du\right].$$

Since $F(A; 1 + u)$ is a probability its value $F(A; 1 + \delta)$ lies between zero and one. The results of the following paragraph show that the last two terms are asymptotically negligible in comparison with the first. Therefore we get

$$\log(\Phi_n(\beta A)) \sim -\frac{\beta^2}{2}, \quad \beta \to \infty. \quad (13)$$

In structural reliability the distance β of the domain βA to the origin is called reliability index. To use this distance as measure of the reliability was proposed in [11] and [12]. It gives an rough estimate for the order of magnitude of the probability content of βA.

4 Approximations for the probabilities

To obtain an approximation for the probability $\Phi_n(\beta A)$ as $\beta \to \infty$, we use the Laplace method for multivariate integrals. We assume for simplicity that the set $M = \{y \in A; |y| = 1\}$ is a simply connected submanifold of ∂A. If it consists of several disjunct sets, the contributions from the various sets have to be added to obtain the approximation. If the set M is an one-point set, i.e. $M = \{\boldsymbol{x}_0\}$ we define the dimension as zero.

We get

$$\Phi_n(\beta A) \sim C_A \cdot \bar{Q}_{k+1}(\beta), \quad \beta \to \infty. \quad (14)$$

Here $\bar{Q}_{k+1}(\beta)$ is the complimentary cumulative distribution function of the χ-distribution with $k + 1$ degrees of freedom and C_A is a constant depending on the shape of A.

In the case of an one-point set we have $\bar{Q}_1(\beta) = 2\Phi(-\beta)$. Then the probability content of βA is asymptotically proportional to $\Phi(-\beta)$. Here the constant C_A is given by

$$C_A = \frac{1}{2}\prod_{i=1}^{n-1}(1 - \kappa_i)^{-1/2} \quad (15)$$

if $M = \{x_0\}$ with the κ_i's being the main curvatures of ∂A at x_0.

If the dimension of M is not zero, then the constant C_A is obtained by an integration over M

$$C_A = \frac{\Gamma((k+1)/2))}{2\pi^{(k+1)/2}} \int\limits_M |\det(C(y))|^{-1/2} \, ds_M(y) \tag{16}$$

with $M \subset S_n(1)$ a k-dimensional manifold with $1 \le k \le n-1$. Here $ds_M(y)$ denotes surface integration over M. Here the matrix $C(y)$ is defined by

$$C(y) = P(y)H(y)P(y) + (I_n - P(y)). \tag{17}$$

Here

$$H(y) = I_n - |\nabla g(y)|^{-1} G(y). \tag{18}$$

$P(y)$ is the projection matrix onto the linear subspace spanned by the orthogonal complement of the the subspace spanned by $\nabla g(x)$ and the tangential space $T_M(y)$. $G(y)$ is the Hessian of g at y.

For the special case of an one-point set the result is proved in [3] and [9]. A proof for the general result in equation (16) can be found in [8]. These results show the influence of the local differential geometry of the boundary of A on the asymptotic probability. The dimension of M determines the order of magnitude of the probability and then the integral over M gives the exact coefficient. Some special cases are considered in [9]

5 Asymptotic expansions for the probabilities

Here we will derive an asymptotic expansion of the probabilities with respect to the asymptotic scale $\{\Phi(-\beta)\beta^{-i}\}$ as $\beta \to \infty$. The exact derivation for this result is given in [10]. The probability can be written in the form

$$\Phi_n(\beta A) = \frac{\beta^n \omega_n}{(2\pi)^{n/2}} \int\limits_1^\infty F(A; u) u^{n-1} \exp\left(-(\beta u)^2/2\right) du \tag{19}$$

Making the substitution $u \to z = u^2 - 1$ gives then with $du/dz = 1/(2\sqrt{1+z})$

$$\Phi_n(\beta A) \tag{20}$$

$$= \frac{\omega_n}{(2\pi)^{n/2}} \frac{\beta^n}{2} e^{-\beta^2/2} \int\limits_0^\infty \left[F(A; (1+z)^{1/2})(1+z)^{(n-2)/2}\right] \exp\left(-\frac{\beta^2}{2} \cdot z\right) dz$$

For the function $F(A; (1+z)^{1/2})(1+z)^{(n-2)/2}$ we find under some regularity conditions, given in [10], the following asymptotic expansion as $z \to 0+$

$$F(A; (1+z)^{1/2})(1+z)^{(n-2)/2} \sim z^{(n-2)/2} \sum_{i=1}^\infty \gamma_i z^{i/2}. \tag{21}$$

Then $\Phi_n(\beta A)$ can be written as a Laplace transform and then we get using Watson's lemma that asymptotically

$$\Phi_n(\beta A) \sim \Phi(-\beta) \sum_{i=0}^{\infty} a_i \beta^{-i}, \ \beta \to \infty. \tag{22}$$

with respect to the asymptotic scale $\{\Phi(-\beta)\beta^{-i}\}$. The coefficients are given by

$$a_{i-1} = \gamma_i \frac{\Gamma((n+i)/2)}{\Gamma(n/2)} \pi^{1/2} 2^{(i+1)/2} \tag{23}$$

if i is odd and $a_{i-1} = 0$ if i is even. The first term a_0 is given by $a_0 = \prod_{j=1}^{n-1}(1 - \kappa_j)^{-1/2}$. This expansion shows that the local structure of the domain A near the maximum manifold M determines the asymptotic behavior of the probabilities $\Phi_n(\beta A)$ as $\beta \to \infty$.

6 Summary

Different methods of asymptotic approximations for large deviation probabilities of Gaussian random vectors are compared. Asymptotic approximations for surface integrals over normal densities are given in [4], [5] and [7].

Such methods can be used for non-normal random vectors too. Such approximations are derived in [6], [14] and [13].

References

[1] H. Birndt and W.-D. Richter. Vergleichende Betrachtungen zur Bestimmung des asymptotischen Verhaltens mehrdimensionaler Laplace-Gauß-Integrale. *Zeitschrift für Analysis und ihre Anwendungen*, 4(3):269–276, 1985.

[2] N. Bleistein and R.A. Handelsman. *Asymptotic Expansions of Integrals*. Dover Publications Inc., New York, 1986.

[3] K. Breitung. Asymptotic approximations for multinormal integrals. *Journal of the Engineering Mechanics Division ASCE*, 110(3):357–366, 1984.

[4] K. Breitung. Asymptotic crossing rates for stationary Gaussian vector processes. *Stochastic Processes and Applications*, 29:195–207, 1988.

[5] K. Breitung. The extreme value distribution of non-stationary vector processes. In A. H.-S. Ang, M. Shinozuka, and G.I. Schuëller, editors, *Proceedings of ICOSSAR '89 5th Int'l Conf. on structural safety and reliability*, volume II, pages 1327–1332. American Society of Civil Engineers, 1990.

[6] K. Breitung. Probability approximations by log likelihood maximization. *Journal of the Engineering Mechanics Division ASCE*, 117(3):457–477, 1991.

[7] K. Breitung. Crossing rates of Gaussian vector processes. In *Transactions of the 11th Prague Conference on Information Theory, Statistical Decision Functions and Random Processes 1990*, volume I, pages 303–314, Prague, Czech. Rep., 1992. Academia.

[8] K. Breitung. *Asymptotic Approximations for Probability Integrals*. Springer, New York, 1994. to appear.

[9] K. Breitung and M. Hohenbichler. Asymptotic approximations for multivariate integrals with an application to multinormal probabilities. *Journal of Multivariate Analysis*, 30:80–97, 1989.

[10] K. Breitung and W.-D. Richter. An asymptotic expansion fot large deviation probabilities of Gaussian random vectors. *Journal of Multivariate Analysis*, 1993. submitted.

[11] A.M. Freudenthal. Safety and the probability of structural failure. *Transactions of the ASCE*, 121:1337–1397, 1956.

[12] A.M. Hasofer and N.C. Lind. An exact and invariant first-order reliability format. *Journal of the Engineering Mechanics Division ASCE*, 100(1):111–121, 1974.

[13] M.A. Maes, K. Breitung, and D.J. Dupuis. Asymptotic importance sampling. *Structural Safety*, 1993. to appear.

[14] M.A. Maes, K. Breitung, and P. Geyskens. Asymptotic importance sampling. In Y.K. Lin, editor, *Probabilistic Mechanics and Structural and Geotechnical Reliability, Proceedings of the sixth specialty conference*, pages 96–99. American Society of Civil Engineers, 1992.

[15] W.-D. Richter. Laplace-Gauss integrals, Gaussian measure asymptotic behavior and probabilities of moderate deviations. *Zeitschrift für Analysis und ihre Anwendungen*, 4(3):257–267, 1985.

[16] W.-D. Richter. Remarks on moderate deviations in the multidimensional central limit theorem. *Mathematische Nachrichten*, 122:167–173, 1985.

[17] R. Wong. *Asymptotic Approximations of Integrals*. Academic Press, San Diego, 1989.

Strong Convexity and Directional Derivatives of Marginal Values in Two–Stage Stochastic Programming *

Darinka Dentcheva[1], Werner Römisch[1] and Rüdiger Schultz[2]

[1] Humboldt-Universität Berlin, Institut für Angewandte Mathematik, Unter den Linden 6, D-10099 Berlin

[2] Konrad-Zuse-Zentrum für Informationstechnik Berlin, Heilbronner Str. 10, D-10711 Berlin

Abstract. Two-stage stochastic programs with random right-hand side are considered. Verifiable sufficient conditions for the existence of second-order directional derivatives of marginal values are presented. The central role of the strong convexity of the expected recourse function as well as of a Lipschitz stability result for optimal sets is emphasized.

Keywords. Two-stage stochastic programs, directional derivatives of marginal values, strong convexity, sensitivity analysis

1991 Mathematics Subject Classification: 90C15, 90C31

1 Introduction

Consider the following two-stage stochastic program

(1.1) $\min\{g(x) + Q_\mu(Ax) : x \in C\}$

(1.2) $Q_\mu(\chi) = \int_{I\!\!R^s} \tilde{Q}(z - \chi)\mu(\mathrm{d}z),$

(1.3) $\tilde{Q}(t) = \min\{q^\top y : Wy = t, y \geq 0\}$

where $g : I\!\!R^m \to I\!\!R$ is a convex function, $C \subset I\!\!R^m$ is a non-empty closed convex set and μ is a Borel probability measure on $I\!\!R^s$. Furthermore, $q \in I\!\!R^{\overline{m}}$ and $A \in L(I\!\!R^m, I\!\!R^s), W \in L(I\!\!R^{\overline{m}}, I\!\!R^s)$. To have (1.1.)-(1.3) well-defined we assume

*This research has been supported by the Schwerpunktprogramm " Anwendungsbezogene Optimierung und Steuerung" of the Deutsche Forschungsgemeinschaft

(A1) $W(I\!\!R_+^{\overline{m}}) = I\!\!R^s$ (complete recourse),

(A2) $M_D := \{u \in I\!\!R^s : W^\top u \le q\} \ne \emptyset$ (dual feasibility),

(A3) $\int\limits_{I\!\!R^s} \|z\|\mu(dz) < +\infty$ (finite first moment).

By linear programming duality, (A1) together with (A2) implies that $\tilde{Q}(t) \in I\!\!R$ for all $t \in I\!\!R^s$. Due to (A3) also the integral in (1.2) is finite (cf. [10], [24] and the beginning of Section 2).

The model (1.1)-(1.3) is derived from an optimization problem with uncertain data, where some evidence on the probability distribution of the random data is at hand or has been gained on the basis of statistical information. We have a first-stage decision x to be made *here and now* (i.e. before the realization of z), and a second-stage decision (recourse action) y that has to be fixed after the realization of the random parameters. (1.1) then aims at fixing an x that minimizes the sum of the first-stage costs and the expected second-stage costs caused by the corrective action y. Further details and fundamental properties of (two-stage) stochastic programs can be found in [10], [24].

The present paper contributes to the stability analysis of (1.1) if μ (and hence Q_μ) is subjected to perturbations. We consider (1.1) with convex functions $v : I\!\!R^s \to I\!\!R$ instead of Q_μ and study the optimal (marginal) value φ of (1.1) as a function of v. Resorting to convex perturbations v is motivated by the fact that , given (A1) and (A2), Q_μ is convex for any probability measure μ fulfilling (A3) (cf. [10], [24] and Section 2).

Our investigations focus on second-order directional derivatives of the marginal-value function φ. In [18], [19], [20] such objects are considered for general parametric optimization problems. Lacking smoothness and non-uniqueness of optimal solutions prevent a direct application of the techniques from [19], [20] in the present setting. In contrast to the very general paper [18] we do not utilize a second-order strong stability condition imposed there. Our independent approach uses ideas from [18], [20] and is essentially based on a Lipschitz-stability result for optimal solutions ([16]) and on the strong (strict) convexity of Q_μ ([15],[17]). Accent is placed on ending up with conditions that are verifiable for the problem class (1.1)-(1.3). The issue of first-order directional differentiability of φ in the context of two-stage stochastic programs is essentially settled in [7], [22]. For the reader's convenience we display a central result in this respect. In the general non-linear programming context, first-order directional derivatives of marginal values are addressed in [8], [14], cf. also the references therein.

Our paper is organized as follows: In Section 2 we review improved convexity properties (strict and strong convexity) of Q_μ that were established in [17]. As essential prerequisites, these properties enter Section 3, where we analyze the second-order directional differentiability of Q_μ.

2 Strong Convexity

Given (A1) and (A2), linear programming duality implies

(2.1) $\tilde{Q}(t) = \max\{t^\top u : W^\top u \le q\}.$

Moreover, M_D is also bounded, i.e. it has the vertices $d_i (i = 1, \ldots, l)$. By (2.1) we obtain

$$\tilde{Q}(t) = \max_{i=1,\ldots,l} d_i^\top t,$$

i.e. $\tilde{Q}$ is piecewise linear and convex.

Together with (A3) this implies that Q_μ is a real-valued convex function on $\mathbb{R}^s$. It is natural to ask for stronger properties of Q_μ. Concerning smoothness there are sufficient conditions for Q_μ to be (twice) continuously differentiable: If μ has a density, then Q_μ is continuously differentiable and

(2.2) $$\nabla Q_\mu(\chi) = \sum_{i=1}^{l} (-d_i)\mu(\chi + K_i),$$

where $K_i (i = 1, \ldots, l)$ denotes the normal cone to M_D at d_i (cf. [10], [24]). The function Q_μ is twice continuously differentiable if $\mu \circ B$ has a continuously differentiable distribution function for any nonsingular transformation $B \in L(\mathbb{R}^s, \mathbb{R}^s)$ (for details consult [11], [23] and [15]).

In the present paper we focus on improved convexity properties. Recall that Q_μ is strictly convex if the convexity inequality holds strictly for different arguments; Q_μ is called strongly convex on a convex subset $V \subset \mathbb{R}^s$ if there exists a constant $\kappa > 0$ such that for all $\chi_1, \chi_2 \in V, \lambda \in [0,1]$

$$Q_\mu(\lambda\chi_1 + (1-\lambda)\chi_2) \le \lambda Q_\mu(\chi_1) + (1-\lambda)Q_\mu(\chi_2) - \kappa\lambda(1-\lambda)\|\chi_1 - \chi_2\|^2.$$

The strong convexity of Q_μ on a convex subset V is equivalent to the strong monotonicity of the gradient ∇Q_μ on V and to the positive definiteness of the Hessian $\nabla^2 Q_\mu$ on V, i.e. $\langle \nabla^2 Q_\mu(\chi)h, h \rangle \ge 2\kappa\|h\|^2$ for all $\chi \in V, h \in \mathbb{R}^s$ (cf. [12]).

Let us consider two illustrative examples to provide some initial insight into the situation.

Example 2.1 Let $\tilde{Q}(t) = \min\{y^+ + y^- : y^+ - y^- = t, \ y^+ \ge 0, y^- \ge 0\}$ and $\mu \in \mathcal{P}(\mathbb{R})$ be given by the density

$$\Theta(\tau) = \begin{cases} 4|\tau| & \text{if } -\frac{1}{2} \le \tau \le \frac{1}{2} \\ \\ 0 & \text{otherwise} \end{cases}.$$

Linear programming duality yields

$$\tilde{Q}(t) = \max\{t\,u : -1 \le u \le 1\} = |t|.$$

A straightforward calculation then provides

$$
Q_\mu(\chi) = \begin{cases} \frac{4}{3}|\chi|^3 + \frac{1}{3} & \text{if } -\frac{1}{2} \leq \chi \leq \frac{1}{2} \\ |\chi| & \text{otherwise} \end{cases}.
$$

Two conclusions can be drawn from this representation. Firstly, Q_μ is piecewise linear and, hence, not strictly convex outside the support of μ. Secondly, inside the support of μ the function Q_μ is strictly convex but not strongly convex since the second derivative vanishes at $\chi = 0$.

Example 2.2 *Let $W = \begin{pmatrix} 1 & 0 & -1 & 0 \\ 1 & 1 & 1 & -1 \end{pmatrix}$, $q = (1,-1,1,1)^\mathsf{T}$ and assume that $\mu \in \mathcal{P}(\mathbb{R}^2)$ fulfils (A3) and has a density.*

It is easy to see that (A1) is fulfilled and that $M_D = conv\left((-2,-1)^\mathsf{T},(2,-1)^\mathsf{T}\right)$. However, in view of (2.2) the second component of $\nabla Q_\mu(\chi)$ is identical -1 for all $\chi \in \mathbb{R}^s$. Therefore, ∇Q_μ cannot be strictly monotone and Q_μ is not strictly convex.

The following theorems (proved in [17]) give positive answers with respect to the improved convexity of Q_μ. Roughly speaking, it suffices to eliminate the pathologies encountered above.

Theorem 2.3 *Assume (A1), (A3) and*

(A2)* $int\ M_D \neq \emptyset$,

(A4)   *μ has a density.*

Then Q_μ is strictly convex on any open convex subset of the support of μ.

Theorem 2.4 *Assume (A1), (A2)* , (A3) and*

(A4)*   *there exists an open convex set $V \subset \mathbb{R}^s$, constants $r > 0, \rho > 0$, and a density Θ_μ of μ such that*

$$\Theta_\mu(\tau) \geq r \quad \text{for all}\quad \tau \in \mathbb{R}^s \quad \text{with } dist(\tau, V) \leq \rho.$$

Then Q_μ is strongly convex on V.

In addition to these two theorems it is shown in [17] that under (A1)-(A4) the assumption (A2)* is also necessary for the strict convexity of Q. There are instances where (A2)* becomes especially handy: For simple recourse (i.e. $W = (I, -I)$, where I denotes the identity in $\mathbb{R}^s$) it is equivalent to $q^+ + q^- > 0$ (componentwise), where $q^\mathsf{T} = (q^{+\mathsf{T}}, q^{-\mathsf{T}}), q^+, q^- \in \mathbb{R}^s$; in case $W \in L(\mathbb{R}^{s+1}, \mathbb{R}^s)$ fulfils (A1) and (A2) is valid, (A2)* is equivalent to $q \notin W^\mathsf{T}(\mathbb{R}^s)$ (for details consult [17]).

3 Directional Derivatives of Marginal Values

Consider perturbations

$$(3.1) \qquad \min\{g(x) + v(Ax) : x \in C\}$$

of (1.1), where $v : \mathbb{R}^s \to \mathbb{R}$ is a convex function. We are interested in the directional behaviour of the value function

$$\varphi(v) := \inf\{g(x) + v(Ax) : x \in C\}$$

at Q_μ into convex directions.
Let

$$\psi(v) := \operatorname{argmin}\{g(x) + v(Ax) : x \in C\}.$$

The following Lipschitz result on ψ with respect to the Hausdorff distance d_H will become a fundamental prerequisite.

Theorem 3.1 *Assume (A1)-(A3) and let $\psi(Q_\mu)$ be non-empty, bounded. Let g be convex quadratic, C be convex polyhedral, and Q_μ be strongly convex on some open convex set $V \supset A(\psi(Q_\mu))$.*
Then, for each convex function $v : \mathbb{R}^s \to \mathbb{R}$ there exist constants $L > 0, \delta > 0$ such that

$$d_H(\psi(Q_\mu), \psi(Q_\mu + tv)) \leq Lt$$

whenever $0 < t < \delta$.

Proof: The proof splits into two parts: First one has to show that $\psi(Q_\mu + tv) \neq \emptyset$ for $t > 0$ sufficiently small and then the Lipschitz rate has to be established. Guidelines for both parts are given by results in [16] (Proposition 2.3, Theorem 2.4). However, in [16] perturbations of Q_μ are of the type Q_ν with $\nu \in \mathcal{P}(\mathbb{R}^s)$. Therefore, some preparation is needed for drawing conclusions from [16] in case the perturbations are of the type $Q_\mu + tv$.

The analysis in [16] is based on a subgradient distance d which, in the present setting, reads as follows: Let $U = \operatorname{cl} U_0$, where U_0 is an open convex bounded set such that $\psi(Q_\mu) \subset U_0$ and $A(U) \subset V$, for convex $Q : \mathbb{R}^s \to \mathbb{R}$ we define

$$(3.2) \qquad d(Q, Q_\mu; U) := \sup\{\|z^*\| : z^* \in \partial(Q - Q_\mu)(Ax) : x \in U\}$$

where ∂ denotes the Clarke subdifferential ([5]).
Inserting $Q = Q_\mu + tv$ $(t > 0)$ into the above relation yields

$$d(Q_\mu + tv, Q_\mu; U) := t \cdot \sup\{\|z^*\| : z^* \in \partial v(Ax) : x \in U\}$$

where ∂ specifies to the subdifferential of convex analysis.
Since U is compact, we have

$$L_0 := \sup\{\|z^*\| : z^* \in \partial v(Ax) : x \in U\} < +\infty$$

and

$$d(Q_\mu + tv, Q_\mu; U) = L_0 \cdot t.$$

Re-interpreting Proposition 2.3. and Theorem 2.4 in [16] in terms of the definition (3.2) now provides that $\psi(Q_\mu + tv)$ is non-empty for $t > 0$ sufficiently small and that there exist constants $L > 0, \delta > 0$ such that

$$d_H(\psi(Q_\mu), \psi(Q_\mu + tv)) \leq L \cdot t$$

whenever $0 < t < \delta$. $\qquad\qquad\qquad\qquad\qquad\qquad\qquad\qquad\qquad\qquad$ □

Remark 3.2 Note that the above results (and those to follow) remain valid if v is replaced by $v - Q_\mu$. Both the assumptions on g, C and the strong convexity of Q_μ are indispensable in the above theorem. This is illustrated by several examples in [16]. To give an idea we quote the one justifying the polyhedrality assumption on C: Let $m := 2, s := 1, g(x) \equiv 0, A := (1,0), q := (1,1)^\top, W := (1,-1), C := \{x = (x_1, x_2) \in \mathbb{R}^2 : (x_2)^2 \leq x_1\}$ and μ be the uniform distribution on $\left[-\frac{1}{2}, \frac{1}{2}\right]$. Then $Q_\mu(Ax) = x_1^2 + \frac{1}{4}$ for $0 \leq x_1 \leq \frac{1}{2}$ and $\psi(Q_\mu) = \{(0,0)\}$. Let further $v := Q_{\delta_1} - Q_\mu$ where δ_1 is the measure putting unit mass at $z = 1$. Then

$$d_H(\psi(Q_\mu), \psi(Q_\mu + tv)) \geq \sqrt{\tfrac{t}{2}} \text{ for } t > 0 \text{ sufficiently small.}$$

Another precondition for the subsequent second-order analysis consists in a first-order directional differentiability result for φ as obtained, for instance, in [7], [22].

Theorem 3.3 *Assume (A1)-(A3) and let $\psi(Q_\mu)$ be non-empty, bounded. Then φ is (Gateaux) directionally differentiable at Q_μ in any convex direction v : $\mathbb{R}^s \to \mathbb{R}$ and it holds*

$$\varphi'(Q_\mu; v) := \lim_{t \to 0+} \frac{1}{t}(\varphi(Q_\mu + tv) - \varphi(Q_\mu)) = \min\{v(Ax) : x \in \psi(Q_\mu)\}.$$

Conclusion 3.4 *If Q_μ is strictly convex on some open convex neighbourhood of $A(\psi(Q_\mu))$ (cf. Theorem 3.3) we obtain for all convex $v : \mathbb{R}^s \to \mathbb{R}$*

$$\varphi'(Q_\mu; v) = v(\chi_*) \quad where \quad A(\psi(Q_\mu)) = \{\chi_*\}.$$

Remark 3.5 Provided that C is bounded, techniques from [15] (Proposition 2.1) can be utilized to establish that the marginal-value function φ is locally Lipschitzian (with respect to the uniform distance on $A(C)$) at any convex function v. Then, the directional differentiability of φ in the sense of Hadamard (cf. [21]) is a direct consequence of Proposition 3.5 in [21].

In what follows we explore whether φ has second-order directional derivatives at (certain) Q_μ.

Theorem 3.6 *Assume (A1)-(A3) and let $\psi(Q_\mu)$ be non-empty, bounded. Let g be convex quadratic, C be convex polyhedral, Q_μ be strongly convex on some open convex neighbourhood of $A(\psi(Q_\mu))$ and twice continuously differentiable at χ_* with $A(\psi(Q_\mu)) = \{\chi_*\}$. Then we have for all convex $v : \mathbb{R}^s \to \mathbb{R}$ and any $x \in \psi(Q_\mu)$*

$$(3.3) \qquad \varphi''(Q_\mu; v) = \lim_{t \to 0+} \frac{1}{t^2}(\varphi(Q_\mu + tv) - \varphi(Q_\mu) - t\varphi'(Q_\mu; v))$$

$$= \inf\{\frac{1}{2}\langle Hy, y\rangle + \frac{1}{2}\langle \nabla^2 Q_\mu(\chi_*)Ay, Ay\rangle + v'(\chi_*; Ay) : y \in S(x)\}$$

where $S(x) = \{y \in \mathbb{R}^m : y \in T_C(x), \nabla g(x)y + \nabla Q_\mu(\chi_*)Ay = 0\}$, $T_C(x) = \liminf_{t \to 0+} \frac{1}{t}(C - x)$ *is the tangent cone to* C *at* x *(with set convergence in Kuratowski's sense [1]),* $H := \nabla^2 g(x)$ *and* $v'(\chi_*, .)$ *denotes the directional derivative of* v *at* χ_*.

Observe that, given the function v, *the value of* $\varphi''(Q_\mu; v)$ *is the same for all* $x \in \psi(Q_\mu)$. *Moreover, the infimum in (3.3) is attained.*

Proof: Let $v : \mathbb{R}^s \to \mathbb{R}$ be convex and $x \in \psi(Q_\mu)$. Let $L > 0$ and $\delta > 0$ be as in Theorem 3.1 and $t \in (0, \delta)$. By Theorem 3.1 there exists an $x(t) \in \psi(Q_\mu + tv)$ such that $\|x(t) - x\| \leq Lt$.

It holds

$$\varphi(Q_\mu + tv) - \varphi(Q_\mu) - t\varphi'(Q_\mu; v) =$$

$$= g(x(t)) + Q_\mu(Ax(t)) + tv(Ax(t)) - g(x) - Q_\mu(Ax) - tv(Ax)$$

$$= \nabla g(x)(x(t) - x) + \frac{1}{2}\langle H(x(t) - x), x(t) - x\rangle +$$

$$+ \nabla Q_\mu(Ax)(A(x(t) - x)) + \frac{1}{2}\langle \nabla^2 Q_\mu(Ax)(A(x(t) - x)), A(x(t) - x)\rangle$$

$$+ t(v(Ax(t)) - v(Ax)) + o(\|x(t) - x\|^2)$$

where we have used Theorem 3.3 for the first identity and the twice differentiability of g at x and Q_μ at $Ax = \chi_*$ for the second identity, respectively. Moreover, the above remarks imply that $o(\|x(t) - x\|^2) = o(t^2)$. This provides for all $t \in (0, \delta)$

$$\frac{1}{t^2}(\varphi(Q_\mu + tv) - \varphi(Q_\mu) - t\varphi'(Q_\mu; v)) =$$

$$= \frac{1}{t^2}(\nabla g(x)(x(t) - x) + \langle A^\top \nabla Q_\mu(Ax), x(t) - x\rangle +$$

$$+ \frac{1}{2}\langle H(x(t) - x), x(t) - x\rangle +$$

$$+ \frac{1}{2}\langle A^\top \nabla^2 Q_\mu(\chi_*)A(\frac{1}{t}(x(t) - x)), \frac{1}{t}(x(t) - x)\rangle +$$

$$+ \frac{1}{t}(v((Ax(t)) - v(Ax)) + o(1).$$

The optimality of x implies

$$\langle \nabla g(x) + A^\top \nabla Q_\mu(Ax), x(t) - x \rangle \geq 0.$$

Now take $t_k \xrightarrow[k \to \infty]{} 0+$ in such a way that

$$\liminf_{t \to 0+} \frac{1}{t^2}(\varphi(Q_\mu + tv) - \varphi(Q_\mu) - t\varphi'(Q_\mu; v)) =$$

$$= \lim_{k \to \infty} \frac{1}{t_k^2}(\varphi(Q_\mu + t_k v) - \varphi(Q_\mu) - t_k \varphi'(Q_\mu; v)).$$

By $\|\frac{1}{t_k}(x(t_k) - x)\| \leq L$ for $k \in N$ sufficiently large, there exists a subsequence $\{t_k\}_{k \in N'}$ such that

$$y_k := \frac{1}{t_k}(x(t_k) - x) \xrightarrow[k \in N']{} y.$$

Now $y \in T_C(x)$ and $x(t_k) = x + t_k y_k$ for all $k \in N'$. Theorem 3.3 yields

$$
\begin{aligned}
v(Ax) &= \varphi'(Q_\mu; v) = \lim_{\substack{k \to \infty \\ k \in N'}} \frac{1}{t_k}(\varphi(Q_\mu + t_k v) - \varphi(Q_\mu)) \\
&= \lim_{\substack{k \to \infty \\ k \in N'}} \frac{1}{t_k}(g(x + t_k y_k) + (Q_\mu + t_k v)(A(x + t_k y_k)) - g(x) - Q_\mu(Ax)) \\
&= \nabla g(x) y + \nabla Q_\mu(Ax) Ay + v(Ax).
\end{aligned}
$$

The above relation implies

$$\nabla g(x) y + \nabla Q_\mu(Ax) Ay = 0,$$

thus $y \in S(x)$.
Therefore

$$\lim_{\substack{k \to \infty \\ k \in N'}} \frac{1}{t_k^2}(\varphi(Q_\mu + t_k v) - \varphi(Q_\mu) - t_k \varphi'(Q_\mu; v)) \geq$$

$$\geq \frac{1}{2} \lim_{\substack{k \to \infty \\ k \in N'}} \left(\langle Hy_k, y_k \rangle + \langle A^\top \nabla^2 Q_\mu(\chi_*) Ay_k y_k \rangle + \right.$$

$$\left. + \frac{1}{t_k}(v(Ax + t_k Ay_k) - v(Ax)) \right)$$

$$= \frac{1}{2}\langle Hy, y \rangle + \frac{1}{2}\langle \nabla^2 Q_\mu(\chi_*) Ay, Ay \rangle + v'(\chi_*; Ay)$$

$$\geq \inf_{y \in S(x)} \left\{ \frac{1}{2}\langle Hy, y \rangle + \frac{1}{2}\langle \nabla^2 Q_\mu(\chi_*) Ay, Ay \rangle + v'(\chi_*; Ay) \right\}.$$

Hence

$$\liminf_{t \to 0+} \frac{1}{t^2}(\varphi(Q_\mu + tv) - \varphi(Q_\mu) - t\varphi'(Q_\mu; v)) \geq$$

$$\geq \inf_{y \in S(x)} \left\{ \frac{1}{2} \langle Hy, y \rangle + \frac{1}{2} \langle \nabla^2 Q_\mu(\chi_*) Ay, Ay \rangle + v'(\chi_*; Ay) \right\}.$$

Now we establish the reverse inequality for the limes superior.

To this end, let $y \in S(x)$ be arbitrary, i.e., in particular , $y \in T_C(x)$. The polyhedrality of C now implies that, given a sequence $\{t_k\}$ with $t_k \to 0+$, we have $x + t_k y \in C$ for sufficiently large k. This allows the following estimate

$$\varphi(Q_\mu + t_k v) - \varphi(Q_\mu) - t_k \varphi'(Q_\mu; v) \leq$$

$$\leq g(x + t_k y) + Q_\mu(A(x + t_k y)) + t_k v(A(x + t_k y)) - g(x) - Q_\mu(Ax) - t_k v(Ax)$$

$$= t_k \nabla g(x) y + \frac{1}{2} t_k^2 \langle Hy, y \rangle + t_k \nabla Q_\mu(Ax) Ay$$

$$+ \frac{1}{2} t_k^2 \langle \nabla^2 Q_\mu(Ax) Ay, Ay \rangle + o(t_k^2) + t_k(v(A(x + t_k y)) - v(Ax))$$

$$= \frac{1}{2} t_k^2 \langle Hy, y \rangle + \frac{1}{2} t_k^2 \langle \nabla^2 Q_\mu(Ax) Ay, Ay \rangle + o(t_k^2) +$$

$$+ t_k(v(A(x + t_k y)) - v(Ax)).$$

The last identity is valid since $y \in S(x)$ implies

$$\nabla g(x) y + \nabla Q(Ax) Ay = 0.$$

Now we obtain

$$\limsup_{k \to \infty} \frac{1}{t_k^2} (\varphi(Q_\mu + t_k v) - \varphi(Q_\mu) - t_k \varphi'(Q_\mu; v))$$

$$\leq \frac{1}{2} \langle Hy, y \rangle + \frac{1}{2} \langle \nabla^2 Q_\mu(Ax) Ay, Ay \rangle + v'(Ax; Ay).$$

Since $y \in S(x)$ was arbitrary, (3.3) is established.

To prove that the infimum is actually attained, let us denote

$$h(y) := \frac{1}{2} \langle Hy, y \rangle + \ell(Ay),$$

where

$$\ell(\chi) := \frac{1}{2} \langle \nabla^2 Q_\mu(\chi_*) \chi, \chi \rangle + v'(\chi_*, \chi).$$

We will show that the function h is constant on each common direction of recession of h and $S(x)$. Theorem 27.3 in [13] then states that h attains its infimum over $S(x)$.

Let $u \in \mathbb{R}^m$ be a common direction of recession of h and $S(x)$, i.e. $u \in \mathbb{R}^m$ fulfils

$$y + \lambda u \in S(x) \quad \text{and} \quad h(y + \lambda u) \leq h(y)$$

for all $\lambda \geq 0$ and all $y \in S(x)$.

Since $S(x)$ is a polyhedral cone, $u \in \mathbb{R}^m$ is a direction of recession of $S(x)$ if and only if $u \in S(x)$.
Let $u \in S(x)$ and $\lambda \geq 0$. It holds

$$h(y + \lambda u) = \frac{1}{2}\langle Hy, y\rangle + \lambda\langle Hu, y\rangle + \frac{\lambda^2}{2}\langle Hu, u\rangle + \ell(Ay + \lambda Au).$$

By assumption, the function ℓ is strongly convex on $\mathbb{R}^s$ and, hence, obeys a unique minimizer $\bar{\chi}$ on $AS(x)$. Moreover, we have, with a suitable constant $\alpha > 0$

$$\ell(\chi) \geq \ell(\overline{\chi}) + \alpha\|\chi - \overline{\chi}\|^2 \quad \text{for all} \quad \chi \in AS(x) \quad ([12]).$$

Therefore,

$$\ell(Ay + \lambda Au) \geq \ell(\overline{\chi}) + \alpha(\|Ay - \overline{\chi}\|^2 + 2\lambda\langle Ay - \overline{\chi}, Au\rangle + \lambda^2\|Au\|^2).$$

For $Au \neq 0$ this implies $\ell(Ay + \lambda Au) \underset{\lambda\to\infty}{\longrightarrow} \infty$. Together with $\langle Hu, u\rangle \geq 0$ we obtain $h(y + \lambda u) \underset{\lambda\to\infty}{\longrightarrow} \infty$, i.e. u is no direction of recession of h.
In case $Au = 0$ and $\langle Hu, u\rangle > 0$ we again obtain $h(y + \lambda u) \underset{\lambda\to\infty}{\longrightarrow} \infty$, showing that u is no direction of recession of h.
It remains to check the case where $Au = 0$ and $\langle Hu, u\rangle = 0$. Then we have $Hu = 0$, yielding $h(y + \lambda u) = \frac{1}{2}\langle Hy, y\rangle$ for all $\lambda \geq 0$, i.e. h is constant in direction u and Theorem 27.3 in [13] works. $\qquad\square$

Example 3.7 *Let $m := s := 1, g(x) \equiv 0, A := 1, C := \mathbb{R}$ and select Q_μ as in Example 2.1. Then it holds $\psi(Q_\mu) = \{0\}, \varphi(Q_\mu) = \frac{1}{3}$. With $v(x) := -x \ (x \in \mathbb{R})$ we obtain for all $t \in [0, 1]$*

$$\psi(Q_\mu + tv) = \{x \in \mathbb{R} : Q'_\mu(x) = t\} = \left\{\frac{1}{2}\sqrt{t}\right\},$$

$$\varphi(Q_\mu + tv) = \frac{1}{3}(1 - t^{\frac{3}{2}})$$

and

$$\varphi'(Q_\mu; v) = \min\{v(x) : x \in \psi(Q_\mu)\} = 0.$$

Thus

$$\frac{1}{t^2}(\varphi(Q_\mu + tv) - \varphi(Q_\mu) - t\varphi'(Q_\mu; v)) = -\frac{1}{3}t^{-\frac{1}{2}}$$

for all $t \in [0, 1]$.
Hence, φ has no second-order directional derivative at Q_μ in direction v. Note that there exists a neighbourhood of $\psi(Q_\mu)$ where Q_μ is strictly convex. However, there is no such neighbourhood where Q_μ is strongly convex. This shows that the strong convexity in Theorem 3.6 is indispensable.

The next result shows that the estimate for the upper second-order directional derivative of φ, which constitutes the final part of the proof of Theorem 3.6, remains valid under more general hypotheses. Second-order tangents sets to C (cf. [1], [6]) turn out to be essential in this respect. Higher-order sets of such type are studied in [9] in the context of (higher-order) necessary optimality conditions for abstract mathematical programs.

Proposition 3.8 *Assume* (A1)-(A3) *and let* $\psi(Q_\mu)$ *be non-empty, bounded. Let* Q_μ *be strictly convex on some open convex neighbourhood of* $A(\psi(Q_\mu))$ *and continuously differentiable at* χ_* *with* $A(\psi(Q_\mu)) = \{\chi_*\}$. *Assume that* Q_μ *has a second-order directional derivative at* χ_*, *i.e. there exist*

$$Q_\mu''(\chi_*; h) = \lim_{t \to 0+} \frac{1}{t^2}(Q_\mu(\chi_* + th) - Q_\mu(\chi_*) - t\nabla Q_\mu(\chi_*)h)$$

for all $h \in \mathbb{R}^s$. *Let* g *be twice continuously differentiable. Then we have for all convex* $v : \mathbb{R}^s \to \mathbb{R}$ *and* $x \in \psi(Q_\mu)$

$$\limsup_{t \to 0+} \frac{1}{t^2}(\varphi(Q_\mu + tv) - \varphi(Q_\mu) - t\varphi'(Q_\mu; v))$$

$$\leq \inf_{y \in S(x)} \inf_{z \in T_C^2(x,y)} \left\{ \nabla g(x)z \;+\; \nabla Q_\mu(\chi_*)Az + \frac{1}{2}\langle\nabla^2 g(x)y, y\rangle \right.$$

$$\left. +\; Q_\mu''(\chi_*; Ay) + v'(\chi_*; Ay) \right\}$$

where $S(x)$ *is given as in Theorem 3.6 and* $T_C^2(x, y)$ *is the second-order tangent set to* C *at* $x \in C$ *in direction* y, *i.e.*

$$T_C^2(x, y) = \liminf_{t \to 0+} \frac{1}{t^2}(C - x - ty).$$

Proof:
Let $y \in S(x)$ be arbitrary. If $T_C^2(x, y) = \emptyset$, then the assertion trivially holds. Hence, let $z \in T_C^2(x, y)$. Then, for arbitrary $t_k \to 0+$ there exists a sequence $\{z_k\}$ such that $z_k \to z$ and $x + t_k y + t_k^2 z_k \in C$ for all $k \in \mathbb{N}$. This allows the following estimate

$$\varphi(Q_\mu + t_k v) - \varphi(Q_\mu) - t_k \varphi'(Q_\mu; v)$$

$$\leq g(x + t_k y + t_k^2 z_k) + Q_\mu(A(x + t_k y + t_k^2 z_k)) + t_k v(A(x + t_k y + t_k^2 z_k))$$

$$-g(x) - Q_\mu(Ax) - t_k v(Ax)$$

$$= [g(x + t_k y + t_k^2 z_k) - g(x) - t_k \nabla g(x)y] +$$

$$+[Q_\mu(A(x + t_k y + t_k^2 z_k)) - Q_\mu(Ax) - t_k \nabla Q_\mu(Ax)Ay] +$$

$$+t_k[v(A(x + t_k y + t_k^2 z_k)) - v(Ax)].$$

After dividing by t_k^2 the right-hand side converges to (cf. [3], p. 484)

$$\nabla g(x)z + \frac{1}{2}\langle\nabla^2 g(x)y, y\rangle + \nabla Q_\mu(Ax)Az + Q_\mu''(Ax; Ay) + v'(Ax; Ay).$$

Taking infima on the right-hand side yields the assertion. □

An upper bound similar to the above one is given in [2], Proposition 1.

Remark 3.9 The following condition allows to extend the estimate from Proposition 3.8 to the limes inferior $\liminf\limits_{t\to 0+}$: For each $\varepsilon > 0$ there exist $x \in \psi(Q_\mu), y \in S(x)$ such that for arbitrary $t_k \to 0+$ there exists a sequence $\{z_k\}$ such that $z_k \to z, x+t_k y+t_k^2 z_k \in C$ and $g(x+t_k y+t_k^2 z_k)+(Q_\mu+t_k v)(A(x+t_k y+t_k^2 z_k)) \leq \varphi(Q_\mu + t_k v) + \varepsilon t_k^2$ for $k \in \mathbb{N}$ sufficiently large.

This condition is employed in [18], Theorem 4.1, where it is called *second-order strong stability condition*. Verifying it in the context of two-stage stochastic programs is an open problem.

We finally combine the techniques from Theorem 3.6 and Proposition 3.8. In this way, the additional assumptions on g and C can be dropped. However, more implicit hypotheses on $\psi(Q_\mu)$ have to be verified.

Corollary 3.10 *Assume (A1)-(A3) and let $\psi(Q_\mu)$ be non-empty, bounded. Let Q_μ be strongly convex on some open convex neighbourhood of $A(\psi(Q_\mu))$ and twice continuously differentiable at χ_* with $A(\psi(Q_\mu)) = \{\chi_*\}$. Let g be twice continuously differentiable and $v : \mathbb{R}^s \to \mathbb{R}$ be convex. Assume that $x \in \psi(Q_\mu)$ has the following properties:*

(i) $\qquad d(x, \psi(Q_\mu + tv)) = O(t),$

(ii) $\qquad 0 \in T_C^2(x, y) \quad$ *for all $y \in T_C(x)$.*

Then $\qquad \varphi''(Q_\mu; v) =$

$$= \inf\left\{\frac{1}{2}\langle\nabla^2 g(x)y, y\rangle + \frac{1}{2}\langle\nabla^2 Q_\mu(\chi_*)Ay, Ay\rangle + v'(\chi_*; Ay) : y \in S(x)\right\},$$

where $S(x)$ is given as in Theorem 3.6.

Proof: In view of (i), the same technique as in the proof of Theorem 3.6 applies and one ends up with the right-hand side of the assertion as a lower bound for the $\liminf\limits_{t\to 0+}$. For the $\limsup\limits_{t\to 0+}$, we use Proposition 3.8 and the fact that the necessary optimality condition yields:

$$\nabla g(x)z + \nabla Q_\mu(\chi_*)Az \geq 0 \quad \text{whenever} \quad z \in T_C^2(x, y), y \in T_C(x).$$

(ii) now implies the assertion. □

Acknowledgement:

This paper was partly written during the Workshop " Approximation of Stochastic Optimization Problems" held at the International Institute for Applied Systems Analysis, Laxenburg (Austria) in July 1993. We wish to thank the organizers, Georg Pflug and Andrzej Ruszczyński, for providing excellent working conditions and a stimulating atmosphere.
Further thanks are due to Alexander Shapiro (Georgia Institute of Technology, Atlanta) for valuable comments. Moreover, we are indebted to Alberto Seeger (University of Avignon) for pointing out an incorrectness in an earlier version of this paper.

References

[1] J.-P. Aubin and H. Frankowska, Set-Valued Analysis, Birkhäuser, Boston 1990.

[2] A. Auslender and R. Cominetti, First and second order sensitivity analysis of nonlinear programs under directional constraint qualification conditions, Optimization 21 (1990), 351-363.

[3] A. Ben-Tal and J. Zowe, Directional derivatives in nonsmooth optimization, Journal of Optimization Theory and Applications 47(1985), 483-490.

[4] C. Berge, Espaces Topologiques, Functions Multivoques, Dunod, Paris 1959.

[5] F.H. Clarke, Optimization and Nonsmooth Analysis, Wiley, New York 1983.

[6] R. Cominetti, Metric regularity, tangent sets and second-order optimality conditions, Applied Mathematics and Optimization 21 (1990), 265-287.

[7] J. Dupačová, Stability and sensitivity analysis for stochastic programming, Annals of Operations Research 27 (1990), 115-142.

[8] J.-B. Hiriart-Urruty, Approximate first-order and second-order directional derivatives of a marginal function in convex optimization, Journal of Optimization Theory and Applications 48 (1986), 127-140.

[9] K.-H. Hoffmann and H.J. Kornstaedt, Higher-order necessary conditions in abstract mathematical programming, Journal of Optimization Theory and Applications 26(1978), 533-568.

[10] P. Kall, Stochastic Linear Programming, Springer-Verlag, Berlin 1976.

[11] K. Marti, Approximationen der Entscheidungsprobleme mit linearer Ergebnisfunktion und positiv homogener, subadditiver Verlustfunktion, Zeitschrift für Wahrscheinlichkeitstheorie und verwandte Gebiete 31 (1975), 203-233.

[12] B. Poljak, Existence theorems and convergence of minimizing sequences in extremum problems with restrictions, Soviet Math. Dokl. 7 (1966), 72-75 (Dokl. Akad. Nauk SSSR 166 (1966), 287-290).

[13] R. T. Rockafellar, Convex Analysis, Princeton University Press, Princeton 1970.

[14] R.T. Rockafellar, Directional differentiability of the optimal value function in a nonlinear programming problem, Mathematical Programming Study 21 (1984), 213-226.

[15] W. Römisch and R. Schultz, Stability of solutions for stochastic programs with complete recourse, Mathematics of Operations Research 18 (1993), 590-609.

[16] W. Römisch and R. Schultz, Lipschitz stability for stochastic programs with complete recourse, Schwerpunktprogramm " Anwendungsbezogene Optimierung und Steuerung" der DFG, Report No. 408, 1992 and submitted to SIAM Journal Optimization.

[17] R. Schultz, Strong convexity in stochastic programs with complete recourse, Journal of Computational and Applied Mathematics 56 (1994) (to appear).

[18] A. Seeger, Second order directional derivatives in parametric optimization problems, Mathematics of Operations Research 13 (1988), 124-139.

[19] A. Shapiro, Second-order derivatives of extremal-value functions and optimality conditions for semi-infinite programs, Mathematics of Operations Research 10 (1985), 207-219.

[20] A. Shapiro, Sensitivity analysis of nonlinear programs and differentiability properties of metric projections, SIAM Journal Control and Optimization 26 (1988), 628-645.

[21] A. Shapiro, On concepts of directional differentiability, Journal of Optimization Theory and Applications 66 (1990), 477-487.

[22] A. Shapiro Asymptotic analysis of stochastic programs, Annals of Operations Research 30 (1991), 169-186.

[23] J. Wang, Distribution sensitivity analysis for stochastic programs with complete recourse, Mathematical Programming 31 (1985), 286-297.

[24] R.J.-B. Wets, Stochastic programs with fixed recourse: the equivalent deterministic program, SIAM Review 16 (1974), 309-339.

Computation of Probability Functions and its Derivatives by means of Orthogonal Function Series Expansions

K. Marti

Federal Armed Forces University Munich,

Faculty of Aero-Space Engineering and Technology,

D-85577 Neubiberg/München, Germany

Abstract.

In reliability-oriented optimization of engineering systems, the computation of derivatives of the probability $P(x)=P(y1\leq y(a(\omega),x)\leq y2)$ of systems survival plays an important role. Here $y=y(a,x)$ denotes the response vector depending on the design vector x and the random parameter vector $a=a(\omega)$, and $y1,y2$ are vectors of given bounds for y. Whereas probabilities can be computed e.g. by sampling methods or asymptotic expansion techniques based on Laplace integral representations of certain multiple integrals, efficient techniques for the computation of derivatives of probability functions are still missing. Series representations of $P=P(x)$ and its derivatives result by an orthogonal function series expansion (e.g. by Hermite, Legendre, Laguerre series) of the density $f=f(y|x)$ of $y=y(a(\omega),x)$. The coefficients of these series are defined by certain expectations, and the integrals over the basis functions can be evaluated analytically. Estimation techniques for the coefficients and approximations (including error estimates) of the series by finite sums are given. Moreover, the properties of the resulting new estimators for probability functions and its derivatives are examined.

Keywords. Probability functions, differentiation, density estimation, series expansion, Hermite series.

1. Introduction

One of the main tools in reliability analysis of technical or economic stochastic systems [4],[7],[10], [18] are probability functions of the type

$$P(x) := P(y_{\ell i} < y_i(a(\omega),x) < y_{ui}, \ 1 \leq i \leq m). \tag{1}$$

Here,

$$y_{\ell i} < y_i(a,x) < y_{ui}, \ i = 1,\ldots,m, \tag{1.1}$$

are the basic bahavioral constraints of the system, where

$$y = (y_1,\ldots,y_i,\ldots,y_m)' \tag{2}$$

are the response or output functions under consideration, e.g. displacement, stress variables in technical system, or cost functions, deviations between the demands and the production outputs in a production problem. The response variables y_i are functions

$$y_i = y_i(a,x), \ i = 1,\ldots,m, \tag{2.1}$$

of a decision r-vector

$$x = (x_1,\ldots,x_k,\ldots,x_r)' \in D_o \ (\text{with } D_o \subseteq \mathfrak{R}^n) \tag{2.2}$$

and a parameter ν-vector

$$a = (a_1,\ldots,a_j,\ldots,a_\nu)' \in A_o \ (\text{with } A_o \subseteq \mathfrak{R}^\nu), \tag{2.3}$$

where x_k, $k = 1,\ldots,r$, are the **deterministic (nominal)** design, input variables or deterministic system coefficients, e.g. sizing, geometric variables in technical systems, or factors of production, production inputs in production problems; moreover,

$$a_j = a_j(\omega), \quad j=1,\ldots,\nu, \tag{2.4}$$

are the **random** system parameters or coefficients, e.g. noise, material, load parameters, manufacturing errors in technical systems, or demands, technological coefficients, cost factors in production problems. We assume that the random ν-vector

$$a(\omega) := (a_1(\omega),\ldots,a_\nu(\omega))' \tag{2.5}$$

is defined on a probability space $(\Omega,\mathcal{O}\!\!\!\!\!\;,P)$ with a given distribution P.

Finally,

$$y_\ell = (y_{\ell 1},\ldots,y_{\ell i},\ldots,y_{\ell m})',$$

$$y_u = (y_{u1},\ldots,y_{ui},\ldots,y_{um})' \tag{2.6}$$

are the m-vectors of lower, upper bounds (margins) $y_{\ell i} < y_{ui}$, i=1,...,m, for the response vector y. In some cases [4] also the bounds $y_{\ell i}, y_{ui}$ are random and must be modelled also as random variables

$$y_{\ell i} = y_{\ell i}(\omega), \quad y_{ui} = y_{ui}(\omega), \quad i=1,\ldots,m, \tag{2.6.1}$$

on $(\Omega,\mathcal{O}\!\!\!\!\!\;,P)$. However, since in this case

$$P(x)$$
$$= \int P(y_\ell < y(a(\omega),x) < y_u | y_\ell(\omega) = y_\ell, y_u(\omega) = y_u) \mu(dy_\ell, dy_u), \tag{2.6.2}$$

where μ denotes the distribution of $(y_\ell(\omega), y_u(\omega))$, the results for **deterministic** bounds $y_{\ell i}, y_{ui}$, i=1,...,m, obtained in the following can be transfered easily to the more general case of stochastic bounds.

In most practical cases, the response variables $y_i = y_i(a,x)$, i=1,...,m, are complicated, highly nonlinear functions of the vectors a and x. Hence, the **numerical evaluation** of P(x) and its first and higher order derivatives $\frac{\partial P}{\partial x_k}(x)$,

$$D_{\underline{\ell}}P(x) := \frac{\partial}{\partial x_{\ell_1}} \frac{\partial}{\partial x_{\ell_2}} \cdots \frac{\partial}{\partial x_{\ell_s}}P(x) \qquad (2.7)$$

with $\underline{\ell} := (\ell_1, \ell_2, \ldots, \ell_s)$,

resp., is in general a very difficult task, which was solved up to now only for special situations.

Especially, in reliability analysis and design of mechanical structures very efficient methods were developed for the computation of (low) probabilities of failure $p_f = 1 - P(x)$, see [2],[3],[5],[6].

Furthermore, general first order differentiation formulas for probability functions and parameter-dependent integrals are known from the literature, see [3],[16], [20],[21]. Evaluating these formulas, the following difficulties occur: i) Several second order derivatives of $y_i = y_i(a,x)$, $1 \le i \le m$, with respect to the parameter vector a are needed, ii) the density $f_o = f_o(a)$ of $a(\omega)$ must be known explicitly and iii) a certain multiple integral has to be calculated.

In the following the computation of $P(x)$ and its derivatives is achieved by using an orthogonal function series expansion of the probability density $f(;x)$ and its derivatives with respect to x.

In this way we find **series expansions of P(x) and its derivatives $D_{\underline{\ell}}P(x)$**, where i) the probability density $f_o = f_o(a)$ has not to be known explicitly, ii) the uni- and multivariate case, $m=1$, $m>1$, resp., can be treated by the same basic Fourier series expansion techniques, and iii) a priori informations on the random parameter vector $a(\omega)$ can be incorporated easily.

Note. For sake of simplicity we denote the random vector $y = y(a(\cdot),x)$ and its values $y = y(a(\omega),x)$ by the same symbol $y = y_x$.

2. Integral representations of $P(x)$ and its derivatives

In the following we suppose that for all x under consideration the random m-vector

$$y_x(\omega) := y(a(\omega),x) = (y_1(a(\omega),x),\ldots,y_m(a(\omega),x))' \quad (3)$$

has a probability density

$$f = f(y;x). \tag{3.1}$$

The density $f(y;x)$ exists under weak assumptions:

Lemma 2.1. Let $m \le \nu$. For given x, suppose that the Jacobian $\frac{\partial y}{\partial a}(a,x)$ of the mapping $a \longrightarrow y(a,x)$ exists and has $\mathrm{rank}\frac{\partial y}{\partial a}(a,x) = m$ for all $a \in \mathfrak{R}^\nu$ up to a set of $P_{a(\cdot)}$-measure zero, where $P_{a(\cdot)}$ denotes the probability distribution of $a = a(\omega)$. If the random parameter vector $a = a(\omega)$ has a probability density $f_0 = f_0(a)$, then the density $f = f(y;x)$ exists.

Proof. First a vector $\eta = \eta(a,x) := (y_{m+1}(a,x),\ldots,y_\nu(a, x))'$ of suitable smooth functions $y_{m+1}(a),\ldots,y_\nu(a)$ is added to the given functions $y_1(a,x),\ldots,y_m(a,x)$ such that $\mathrm{rank}\frac{\partial \tilde{y}}{\partial a}(a,x) = \nu$ for all $a \in \mathfrak{R}^\nu$ up to a $P_{a(\cdot)}$-measure zero, where $\tilde{y}(a,x) := (y_1(a,x),\ldots,y_m(a,x),y_{m+1}(a,x),\ldots,y_\nu(a,x))'$, see [11]. Defining then for a given, fixed vector x the 1-1-transformation

$$T_x = T_x(a) := \tilde{y}(a,x), \tag{4}$$

the density $\tilde{f} = \tilde{f}(\tilde{y};x)$ of $\tilde{y} = \tilde{y}(a(\omega),x)$ follows by means of the standard transformation rule for densities

$$\tilde{f}(\tilde{y};x) = f_0(T_x^{-1}(\tilde{y}))\frac{1}{|\det\frac{\partial T_x}{\partial a}(T_x^{-1}(\tilde{y}))|}. \tag{4.1}$$

Finally, as a marginal density of $\bar{f}(\bar{y};x)$, $f(y,x)$ is given by

$$f(y;x) = \int_{-\infty}^{+\infty} \ldots \int_{-\infty}^{+\infty} \bar{f}(\binom{y}{\eta};x)d\eta. \qquad (4.2)$$

Besides the density representation by means of (4)-(4.2), the probability density of $y_x(\omega)=y(a(\omega),x)$ can also be described by using known integral representations of derivatives for probability functions and parameter-dependent integrals, see [3],[16],[21]. Indeed, if $a=a(\omega)$ has a density $f_0=f_0(a)$, then the distribution function

$$F(\eta;x):= P(y(a(\omega),x)\leq\eta), \quad \eta\in\mathbb{R}^m, \qquad (4.3)$$

of $y_x(\omega)$ is a special probability function being represented by the integral

$$F(\eta;x) = \int_{y(a,x)\leq\eta} f_0(a)da. \qquad (4.3.1)$$

Hence, the following result is obtained by applying one of the above mentioned differentiation formulas first to (4.3.1) and then to the resulting integral representation of the density $f(y;x)$:

Lemma 2.2. Let $m=1$ and suppose that $a(\omega)$ has a density $f_0=f_0(a)$. Furthermore, assume that the derivatives and integrals under consideration exist (and are finite, resp.). Then

$$f(\eta;x) = \int_{y(a,x)=\eta} f_0(a)\frac{dS}{||\nabla_a y(a,x)||}$$
$$= \int_{y(a,x)\leq\eta} \text{div}(f_0(a)\frac{\nabla_a y(a,x)}{||\nabla_a y(a,x)||^2})da, \qquad (4.4)$$

where dS denotes the surface element on $S=S_{x,\eta}:=\{a\in\mathbb{R}^\nu: y(a,x)=\eta\}$, and

$$\frac{\partial f}{\partial x_k}(\eta;x) = - \int\limits_{y(a,x)\leq\eta} \text{div}(\text{div}(f_o(a)\frac{\partial y}{\partial x_k}(a,x)$$

$$\times \frac{\nabla_a y(a,x)}{||\nabla_a y(a,x)||^2}) \frac{\nabla_a y(a,x)}{||\nabla_a y(a,x)||^2})da.$$

$$(4.5)$$

Remark 2.1. Higher order derivatives of $f(y;x)$ with respect to x can be obtained in an iterative way, compare (4.4) and (4.5).

The case $m>1$ can be treated by using more general differentiation formulas. On the other hand, very simple representations of approximative density representations can be found by means of **stochastic completion technique**, see also [8],[9]: Here, the distribution function $F(y;x)$ is approximated first by

$$\widetilde{F}(\eta;x):= P(y(a(\omega),x) + \delta(\omega)\leq\eta), \quad \eta\in\mathbb{R}^m, \qquad (4.3a)$$

where the stochastic completion term $\delta=\delta(\omega)$ is a random m-vector being independent of $y_x(\omega)$ and having a smooth probability density $\varphi=\varphi(\delta)$. Since

$$\widetilde{F}(\eta;x) = E\Phi(\eta-\delta(\omega);x) = E\Phi(\eta-y(a(\omega),x)), \qquad (4.6)$$

where $\Phi=\Phi(\delta)$ denotes the distribution function of $\delta(\omega)$, approximations of $f(y;x)$ and its derivatives can be obtained as follows:

Lemma 2.3. Suppose that the expectations under consideration exist and differentiation and expectations may be interchanged. Approximating $f=f(\eta;x)$ by

$$\widetilde{f}(\eta;x):= \frac{\partial^m}{\partial\eta_1\ldots\partial\eta_m}\widetilde{F}(y;x),$$

we have

$$\widetilde{f}(\eta;x):= E\varphi(\eta-y(a(\omega),x)), \qquad (4.7)$$

$$\frac{\partial\widetilde{f}}{\partial x_k}(\eta;x) = -E\sum_{i=1}^{m}\frac{\partial\varphi}{\partial\delta_i}(\eta-y(a(\omega),x))\frac{\partial y_i}{\partial x_k}(a(\omega),x). \qquad (4.8)$$

Furthermore, $\tilde{f}(\eta;x)\longrightarrow f(\eta;x)$ and $\frac{\partial \tilde{f}}{\partial x_k}(\eta;x)\longrightarrow \frac{\partial f}{\partial x_k}(\eta;x)$ as

$\delta(\omega)\longrightarrow 0$ a.s., provided that $f(\cdot;x)$, $\frac{\partial f}{\partial x_k}(\cdot;)$ are contin-

uous.

Remark 2.2. Approximations to higher order partial derivatives of $f=f(y;x)$ with respect to x can be obtained in the same way.

Under the above assumptions the probability function $P=P(x)$ can be represented now by the multiple integral

$$P(x) = \int_{y_\ell}^{y_u} f(y;x)dy \tag{5}$$

having the very simple **fixed** domain of integration

$B = \{y \in \mathbb{R}^m : y_\ell \leq(<) y \leq(<) y_u\}.$

In order to compute the partial derivative $\frac{\partial P}{\partial x_k}(x)$

with respect to x_k at points

$$x = x(t) = (x_1^o,\ldots,x_{k-1}^o,t,x_k^o,\ldots,x_r^o)',\ t\in I, \tag{6}$$

where x_j^o, $j\neq k$, are given fixed values for the components x_j, $j\neq k$, and $I\subset\mathbb{R}$ is a certain finite interval, we need the following first assumption on the density $f=f(y;x)$:

$\frac{\partial f}{\partial x_k}(y;x(t))$ exists for all $t\in I$ and has an inte-

grable majorant $h=h(y)$ on B, i.e. for a nonneg-

ative, integrable function h on B we have a.e. $\left.\rule{0pt}{60pt}\right\}$ (7)

(almost everywhere) on B

$\quad |\frac{\partial f}{\partial x_k}(y;x(t))| \leq h(y)$ for all $t\in I$.

Remark 2.3. If B is a bounded domain, then condition (7) holds e.g. if the function $(y,t) \longrightarrow \frac{\partial f}{\partial x_k}(y,x(t))$ is piecewise continuous on $B \times \bar{I}$, where $\bar{I}$ denotes the closure of I. Indeed, put $h(y) := \sup\{|\frac{\partial f}{\partial x_k}(y,x(t))|: y \in B, t \in \bar{I}\}$.

If (7) holds, then we may [1] interchange differentiation and integration in (5), hence, for all $x \in \{x(t): t \in I\}$ the derivative $\frac{\partial P}{\partial x_k}(x)$ exists and is given by

$$\frac{\partial P}{\partial x_k}(x) = \int_{y_\ell}^{y_u} \frac{\partial f}{\partial x_k}(y;x)\,dy. \tag{8}$$

Iterating the above procedure, under corresponding assumptions we also have, cf. (2.7),

$$D_{\underline{\ell}}P(x) = \int_{y_\ell}^{y_u} D_{\underline{\ell}}f(y;x)\,dy. \tag{8.1}$$

Density estimation and approximation of more general functions by **orthogonal function series expansions** is a well established technique, see e.g. [12-14],[17],[19].

Hence, in the following we consider estimates/approximations of $P = P(x)$ and its derivatives based on **orthogonal function series expansions** of $f(\cdot\,;x)$, $\frac{\partial f}{\partial x_k}(\cdot\,;x), D_{\underline{\ell}}f(\cdot\,;x)$, resp., with respect to the orthonormal systems of

- Hermite ($y_x(\omega)$ has range $\Re^m$, unknown range, resp.)
- trigonometric ($y_x(\omega)$ has bounded range)
- Legendre ($y_x(\omega)$ has bounded range)
- Laguerre ($y_x(\omega)$ is bounded from the left, right, resp.)

functions.

3. Expansions in Hermite functions in case m=1

In the following, let x be an arbitrary, but fixed r-vector. For simplification we consider here first the case of a **scalar** response function $y(a,x) = y_1(a,x)$.

If the random variable $y_x(\omega)=y(a(\omega),x)$ takes values from $-\infty$ to $+\infty$ or if it is difficult to ascertain its range, then we consider expansions in Hermite functions: Defining [12],[15] the Hermite polynomials H_j by

$$H_j(y) := (-1)^j e^{y^2} \frac{d^j}{dy^j} e^{-y^2}, \quad j=0,1,2,\ldots, \tag{9}$$

e.g.

$$H_0(y) = 1$$
$$H_1(y) = 2y$$
$$H_2(y) = 4y^2-2$$
$$H_3(y) = 8y^3-12y \tag{9.1}$$
$$H_4(y) = 16y^4-48y^2+12$$
$$H_5(y) = 32y^5-160y^3+120y$$
$$\ldots,$$

and the normalized Hermite functions φ_j [12],[15] by

$$\varphi_j(y) := (\pi^{1/2} 2^j j!)^{-\frac{1}{2}} e^{-\frac{y^2}{2}} H_j(y), \quad j=0,1,2,\ldots, \tag{9.2}$$

it is known [12],[15],[19] that

$$\varphi_j \in L^1(\mathfrak{R}) \cap L^2(\mathfrak{R}), \quad j=0,1,2,\ldots \tag{9.3.1}$$

and the sequence

$$(\varphi_j) \text{ forms a complete, orthonormal system in } L_2(\mathfrak{R}). \tag{9.3.2}$$

Furthermore, there exists a uniform constant $d_2=\pi^{-\frac{1}{4}} 1.086435\ldots$, see [15], such that

$$|\varphi_j(y)| \leq d_2 \text{ for all } y\in\mathfrak{R} \text{ and each } j=0,1,2,\ldots . \tag{9.3.3}$$

Assuming that

$$f(\cdot\,;x) \in L^2(\mathfrak{R}) \tag{10}$$

and

$$\frac{\partial f}{\partial x_k}(\cdot\,;x) \in L^2(\mathfrak{R}), \tag{10.1}$$

$$D_{\ell} f(\cdot\,;x) \in L^2(\mathfrak{R}), \tag{10.1.1}$$

resp., then these functions can be expanded in the orthogonal series

$$f(y;x) = \sum_{j=0}^{\infty} c_j(x)\varphi_j(y) \tag{11}$$

$$\frac{\partial f}{\partial x_k}(y;x) = \sum_{j=0}^{\infty} c_{kj}(x)\varphi_j(y) \tag{11.1}$$

$$D_{\underline{\ell}}f(y;x) = \sum_{j=0}^{\infty} c_{\underline{\ell}j}(x)\varphi_j(y), \tag{11.1.1}$$

where the (Fourier) coefficients $c_j = c_j(x)$, $c_{kj} = c_{kj}(x)$, $c_{\underline{\ell}j} = c_{\underline{\ell}j}(x)$, resp., are defined [12],[15],[19] by

$$c_j(x) := \int f(y;x)\varphi_j(y)\,dy \tag{12}$$

$$c_{kj}(x) := \int \frac{\partial f}{\partial x_k}(y;x)\varphi_j(y)\,dy \tag{12.1}$$

and

$$c_{\underline{\ell}j}(x) := \int D_{\underline{\ell}}f(y;x)\varphi_j(y)\,dy. \tag{12.1.1}$$

Whereas under the assumption (10),(10.1),(10.1.1) the series in (11), (11.1),(11.1.1), resp., are convergent in the L^2-sense, for the pointwise convergence of the series towards $f(\cdot;x)$, $\frac{\partial f}{\partial x_k}(\cdot;x)$, $D_{\underline{\ell}}f(\cdot;x)$ some additional assumptions [12],[15] are needed:

Lemma 3.1. Let $g = g(y;x)$ denote one of the function $f(\cdot;x)$, $\frac{\partial f}{\partial x_k}(\cdot;x)$, $D_{\underline{\ell}}f(\cdot;x)$, respectively. Suppose that $g(\cdot;x) \in L^1(\mathfrak{R}) \cap L^2(\mathfrak{R})$. If $g(\cdot;x)$ is continuous and of bounded variation in an interval $[y_1,y_2]$, then the series in (11),(11.1),(11.1.1), resp., converges uniformly to $g(\cdot;x)$ in any interval interior to $[y_1,y_2]$.

Since $f = f(y;x)$ is the probability density function of the random variable $y = y(a(\omega),x)$, see (3),(3.1), for $c_j(x)$ we have also the representation

$$c_j(x) = E\varphi_j(y(a(\omega),x)), \quad j = 0,1,2,\ldots \ . \tag{13}$$

Moreover, if there exists an integrable function

$h_k = h_k(y)$ on $\mathfrak{R}$ such that, with $x = x(t)$ given by (6),

$$|\frac{\partial f}{\partial x_k}(y;x(t))\varphi_j(y)| \leq h_k(y), \text{ for all } y \in \mathfrak{R} \text{ and } t \in I, \quad (14)$$

cf. (7) and (9.3.3), then, by interchanging differentiation and expectation in (12), from (12),(12.1) and (13) for $x \in \{x(t):t \in I\}$ we obtain the important relation

$$\frac{\partial}{\partial x_k}c_j(x) = \int \frac{\partial f}{\partial x_k}(y;x)\varphi_j(y)dy = c_{kj}(x). \quad (14.1)$$

Iterating this procedure, i.e. replacing (12) by (12.1) and $f(y;x)$ by $\frac{\partial f}{\partial x_k}(y;x)$ etc., under corresponding modifications of assumption (7) we also find

$$D_\ell E\varphi_j(y(a(\omega),x)) = D_\ell c_j(x) = c_{\ell j}(x). \quad (14.2)$$

Because of (5) and the series representation (11) of $f(\cdot;x)$, for arbitrary finite bounds $y_\ell, y_u, -\infty < y_\ell < y_u < +\infty$, we get

$$|P(x) - \sum_{j=0}^{n} c_j(x) \int_{y_\ell}^{y_u} \varphi_j(y)dy|$$

$$= |\int_{y_\ell}^{y_u} (f(y,x) - \sum_{j=0}^{n} c_j(x)\varphi_j(y))dy|$$

$$\leq \int_{y_\ell}^{y_u} |f(y;x) - \sum_{j=0}^{n} c_j(x)\varphi_j(y)|dy$$

$$\leq (y_u - y_\ell)^{1/2}||f(\cdot;x) - \sum_{j=0}^{n} c_j(x)\varphi_j(\cdot)||_2, \quad (15)$$

where $||\cdot||_2$ denotes the norm of $L^2(\mathfrak{R})$. In the same way from (8),(8.1) and (11.1),(11.1.1), resp., we obtain

$$|\frac{\partial P}{\partial x_k}(x) - \sum_{j=0}^{n} c_{kj}(x) \int_{y_\ell}^{y_u} \varphi_j(y)dy|$$

$$\leq (y_u - y_\ell)^{1/2}||\frac{\partial f}{\partial x_k}(\cdot;x) - \sum_{j=0}^{n} c_{kj}(x)\varphi_j(\cdot)||_2 \quad (15.1)$$

$$|D_{\underline{\ell}}P(x) - \sum_{j=0}^{n} c_{\underline{\ell}j}(x) \int_{y_\ell}^{y_u} \varphi_j(y)dy|$$

$$\leq (y_u-y_\ell)^{1/2}||D_{\underline{\ell}}f(\cdot;x) - \sum_{j=0}^{n} c_{\underline{\ell}j}(x)\varphi_j(\cdot)||_2. \quad (15.2)$$

Since the series in (11),(11.1,(11.1.1), resp., is convergent in the L^2-norm, from (15)-(15.2) we obtain our first result:

Theorem 3.1. Based on the assumptions concerning the existence of the density $f=f(\cdot;x)$ and its derivatives with respect to x mentioned in the above, suppose that $P(x)$, $\frac{\partial P}{\partial x_k}(x)$, $D_{\underline{\ell}}P(x)$, resp., can be represented for a given vector x by (5),(8),(8.1), and assume that the orthogonal series representation (11),(11.1),(11.1.1), resp., holds. Then $P(x)$, $\frac{\partial P}{\partial x_k}(x)$, $D_{\underline{\ell}}P(x)$, resp., can be represented for finite bounds $-\infty<y_\ell<y_u<+\infty$ by the following convergent series:

$$P(x) = \sum_{j=0}^{\infty} c_j(x) \int_{y_\ell}^{y_u} \varphi_j(y)dy, \quad (16)$$

$$\frac{\partial P}{\partial x_k}(x) = \sum_{j=0}^{\infty} c_{kj}(x) \int_{y_\ell}^{y_u} \varphi_j(y)dy, \quad (16.1)$$

$$D_{\underline{\ell}}P(x) = \sum_{j=0}^{\infty} c_{\underline{\ell}j}(x) \int_{y_\ell}^{y_u} \varphi_j(y)dy. \quad (16.2)$$

3.1. The integrals over the basis functions and the coefficients of the orthogonal series

The integrals over the basis functions φ_j, see (9.1), (9.2), read

$$\int_{y_\ell}^{y_u} \varphi_j(y)dy = C_j \sum_{i=0}^{j} h_{ji}J_i \quad (17)$$

where C_j, h_{ji}, $0 \leq i \leq j$, $j \geq 0$, are given, fixed coefficients, and for the definite integrals

$$J_i := \int_{y_\ell}^{y_u} e^{-\frac{y^2}{2}} y^i dy, \quad i=0,1,\ldots, \tag{17.1}$$

by partial integration we find the recursion:

$$J_i = y_\ell^{i-1} e^{-\frac{y_\ell^2}{2}} - y_u^{i-1} e^{-\frac{y_u^2}{2}} + (i-1)J_{i-2}, \quad i \geq 2. \tag{17.2}$$

Hence, the sequence (J_i) can be obtained very easily
from (17.2) and

$$J_0 = \sqrt{2\pi}(\Phi(y_u) - \Phi(y_\ell)), \tag{17.3}$$

$$J_1 = e^{-\frac{y_\ell^2}{2}} - e^{-\frac{y_u^2}{2}}, \tag{17.3.1}$$

where Φ designates the distribution function of the
$N(0,1)$-normal distribution. E.g., for $i=2,3$ we get

$$J_2 = y_\ell e^{-\frac{y_\ell^2}{2}} - y_u e^{-\frac{y_u^2}{2}} + \sqrt{2\pi}(\Phi(y_u) - \Phi(y_\ell)) \tag{17.3.2}$$

$$J_3 = y_\ell^2 e^{-\frac{y_\ell^2}{2}} - y_u^2 e^{-\frac{y_u^2}{2}} + 2(e^{-\frac{y_\ell^2}{2}} - e^{-\frac{y_u^2}{2}}). \tag{17.3.3}$$

Mean value representations for the Fourier coeffi-
cients $c_{kj}(x)$ and $c_{\ell j}(x)$ can be obtained by interchanging differentiation and integration in the mean value
representation (13) for $c_j(x)$.

If in addition to condition (14) there is a function
$\tilde{h}_k = \tilde{h}_k(a)$ on $\aleph^\nu$ such that $\int \tilde{h}_k(a) f_0(a) da < +\infty$ and

$$|\varphi_j'(y(a,x(t))) \frac{\partial y}{\partial x_k}(a,x(t))| \leq \tilde{h}_k(a) \text{ a.s. on } \aleph^\nu \tag{18}$$

for all $t \in I$,

where again $x=x(t)$, $t \in I$, is given by (6), then (13) and
(14.1) yield for $x=x(t)$, $t \in I$,

$$c_{kj}(x) = \frac{\partial}{\partial x_k} E\varphi_j(y(a(\omega),x))$$

$$= E\varphi_j'(y(a(\omega),x)\frac{\partial y}{\partial x_k}(a(\omega),x). \tag{19}$$

Because of

$$\varphi_j'(y) = -y\varphi_j(y) + \sqrt{2j}\varphi_{j-1}(y)$$

$$= \frac{\sqrt{j}}{\sqrt{2}}\varphi_{j-1}(y) - \frac{\sqrt{j+1}}{\sqrt{2}}\varphi_{j+1}(y), \quad j \geq 0, \tag{20}$$

cf. [12], where $\varphi_{-1}(y) \equiv 0$, and (9.3.3) we get

$$|\varphi_j'(y)| \leq \frac{d_2}{\sqrt{2}}(\sqrt{j} + \sqrt{j+1}) \text{ for all } y \in \Re \text{ and } j=0,1,2,\ldots, \tag{20.1}$$

hence, also the derivatives φ_j' of φ_j are bounded on $\Re$.

Furthermore, under corresponding assumptions from (19) and (14.2) we get, see also (2.7),

$$c_{\underline{\ell}j}(x) = ED_{\underline{\ell}}\varphi_j(y(a(\omega),x))$$

$$= \sum_{t=1}^{s} E\varphi_j^{(t)}(y(a(\omega),x))$$

$$\times \{ \sum_{\substack{\lambda_1,\ldots,\lambda_t \text{ partition} \\ \text{of } (\ell_1,\ldots,\ell_s)}} \prod_{j=1}^{t} D_{\lambda_j} y(a(\omega),x)) \}, \tag{21}$$

where (20) yields corresponding representations for the higher derivatives $\varphi_j^{(t)}$ of the Hermite functions φ_j, which show then the boundedness of $\varphi_j^{(t)}$ on $\Re$ for arbitrary values of t.

Based on the mean value representations (13),(19) and (21), the Fourier coefficients can be estimated by sampling methods, e.g. by the sample means

$$\hat{c}_j(x) := \frac{1}{n} \sum_{t=1}^{n} \varphi_j(y(a^t,x)), \tag{22}$$

$$\hat{c}_{kj}(x) := \frac{1}{n} \sum_{t=1}^{n} \varphi_j'(y(a^t,x))\frac{\partial y}{\partial x_k}(a^t,x), \tag{22.1}$$

and

$$\hat{c}_{\underline{\ell}j}(x) := \frac{1}{n} \sum_{t=1}^{n} D_{\underline{\ell}}\varphi_j(y(a^t,x)), \qquad (22.2)$$

where a^t, $t=1,..,n$, denote independent realizations of $a=a(\omega)$.

If the moments

$$\mu_i(x) := E(y(a(\omega),x))^i, \quad i=0,1,2,...$$

of $y_x(\omega)=y(a(\omega),x)$, the moments $\mu_i(x)$ and its derivatives, resp., up to a certain order are available, then approximations to the Fourier coefficients $c_j(x)$, $c_{kj}(x)$, $c_{\underline{\ell}j}(x)$ can be obtained for given x

* by Taylor approximation $e^{-t} = T_N(t)+R_N(t)$ of e^{-t}

  at $t=0$ of a certain order N, or $\qquad (23a)$

* by Taylor approximation of the functions $\qquad (23b)$

$$\psi_j(a) := \varphi_j(y(a,x)),$$
$$\psi_{kj}(a) := \varphi_j'(y(a,x))\frac{\partial y}{\partial x_k}(a,x),$$
$$\psi_{\underline{\ell}j}(a) := D_{\underline{\ell}}\varphi_j(y(a,x)),$$

resp., with respect to a at the mean $\bar{a} := Ea(\omega)$; in practice mainly **second order** approximations (N=2) are used in this case.

In **case (23a)** the Hermite functions $\varphi_j(y)$ occuring in $c_j(x)$, $c_{kj}(x)$, $c_{\underline{\ell}j}(x)$, cf. (13),(19),(21) and (20) are approximated by the polynomial

$$\tilde{\varphi}_j(y) := C_j T_N(\frac{y^2}{2})H_j(y) \qquad (24)$$

of order 2N+j, where the error $\varphi_j-\tilde{\varphi}_j$ can be estimated by

$$|\varphi_j(y)-\tilde{\varphi}_j(y)| \leq \frac{1}{N!}|\varphi_j(y)| \int_0^{\frac{1}{2}y^2} s^N e^s ds. \qquad (24.1)$$

Consequently, if $c_j(x)$, see (13), is approximated by

$$\tilde{c}_j(x) := EC_j T_N(\tfrac{1}{2}y^2(a(\omega),x))H_j(y(a(\omega),x)) \qquad (24.2)$$

$$= C_j \sum_{k=0}^{2N+j} h_{ji}\mu_i(x),$$

where C_j, h_{ji}, $i=0,1,\ldots,2N+j$, are certain coefficients, then

$$|c_j(x) - \tilde{c}_j(x)| \le \frac{d_2}{N!} \int_0^{-\infty} s^N e^s P(|y(a(\omega),x)| > \sqrt{2s})ds,$$

$$j=0,1,\ldots . \qquad (24.3)$$

For $c_{kj}(x)$ the corresponding approximation reads, cf. (20),

$$\tilde{c}_{kj}(x) := \frac{1}{\sqrt{2}}E(\sqrt{j}\ \tilde{\varphi}_{j-1}(y(a(\omega),x)) - \sqrt{j+1}\ \tilde{\varphi}_{j+1}(y(a(\omega),x)))$$

$$\times \frac{\partial y}{\partial x_k}(a(\omega),x), \qquad (24.4)$$

where

$$|c_{kj}(x) - \tilde{c}_{kj}(x)| \le d_2 \frac{\sqrt{j}+\sqrt{j+1}}{\sqrt{2}N!}E|\frac{\partial y}{\partial x_k}(a(\omega),x)|$$

$$\times \int_0^{\frac{1}{2}y^2(a(\omega),x)} s^N e^s ds. \qquad (24.5)$$

Moreover, similar approximations and error bounds can also be derived for $c_{\ell j}(x)$.

In **case (23b)** the Fourier coefficients are approximated (for N=2) by

$$\tilde{c}_j(x) := \varphi_j(y(\bar{a},x)) + \frac{1}{2}E(a(\omega)-\bar{a})^T$$

$$\times(\varphi_j''(y(\bar{a},x))\nabla_a y(\bar{a},x)\nabla_a y(\bar{a},x)^T \qquad (24.6)$$

$$+ \varphi_j'(y(\bar{a},x))\nabla_a^2 y(\bar{a},x))(a(\omega)-\bar{a});$$

the error is bounded then by

$$|c_j(x) - \tilde{c}_j(x)| \le \frac{1}{2} \int_0^1 (1-t)^2 E||\nabla_a^3 \varphi_j(y(\bar{a}+t(a(\omega)-\bar{a})))||$$

$$\times ||a(\omega)-\bar{a}||^3 dt, \qquad (24.7)$$

where

$$||\nabla_a^3\varphi_j(y(a,x))|| \leq |\varphi_j'(y(a,x))|\cdot||\nabla_a^3y(a,x)||$$

$$+ 3|\varphi_j''(y(a,x))|\cdot||\nabla_a^2y(a,x)||\cdot||\nabla_ay(a,x)||$$

$$+ |\varphi_j'''(y(a,x))|\cdot||\nabla_ay(a,x)||^3. \qquad (24.8)$$

Obviously, related approximations and error bounds can be obtained also for $c_{kj}(x)$ and $c_{\ell j}(x)$.

3.2. Estimation/approximation of P(x) and its derivatives

Based on the expansions (11)-(11.1.1), estimates of $f(y;x)$ and its derivatives $\frac{\partial f}{\partial x_k}(y;x)$, $D_{\ell}f(y;x)$, resp., can be defined, cf. [13-15], by

$$\hat{f}(y;x) := \sum_{j=o}^{q(n)} \hat{c}_j(x)\varphi_j(y), \qquad (25)$$

$$\widehat{\frac{\partial f}{\partial x_k}}(y;x) := \sum_{j=o}^{q(n)} \hat{c}_{kj}(x)\varphi_j(y), \qquad (25.1)$$

$$\widehat{D_{\ell}f}(y;x) := \sum_{j=o}^{q(n)} \hat{c}_{\ell j}(x)\varphi_j(y), \qquad (25.2)$$

where $q=q(n)$ is a certain integer and $\hat{c}_j(x)$, $\hat{c}_{kj}(x)$, $\hat{c}_{\ell j}(x)$ are unbiased estimators of the Fourier coefficients $c_j(x)$, $c_{kj}(x)$, $c_{\ell j}(x)$, resp., as defined by (22)-(22.2). Replacing in (5),(8),(8.1) the integrands $f(y;x)$, $\frac{\partial f}{\partial x_k}(y;x)$, $D_{\ell}f(y;x)$, resp., by the above estimates, for $P(x)$ and its derivatives we get the estimators

$$\hat{P}(x) := \int_{y_\ell}^{y_u} \hat{f}(y;x)dy = \sum_{j=o}^{q(n)} \hat{c}_j(x) \int_{y_\ell}^{y_u} \varphi_j(y)dy, \qquad (26)$$

$$\widehat{\frac{\partial P}{\partial x_k}}(x) := \int_{y_\ell}^{y_u} \widehat{\frac{\partial f}{\partial x_k}}(y;x)dy = \sum_{j=o}^{q(n)} \hat{c}_{kj}(x) \int_{y_\ell}^{y_u} \varphi_j(y)dy, \qquad (26.1)$$

$$\hat{D}_{\underline{\ell}}P(x) := \int_{y_{\ell}}^{y_u} \hat{D}_{\underline{\ell}}f(y;x)dy - \sum_{j=o}^{q(n)} \hat{c}_{\underline{\ell}j} \int_{y_{\ell}}^{y_u} \varphi_j(y)dy. \qquad (26.2)$$

Since $\hat{f}(\cdot;x), \frac{\partial \hat{f}}{\partial x_k}(\cdot;x), D_{\underline{\ell}}f(\cdot;x) \in L^2(\Re)$, for the mean square error of (26) we find

$$E(P(x) - \hat{P}(x))^2 = E|\int_{y_{\ell}}^{y_u} (f(y;x) - \hat{f}(y;x))dy|^2 \qquad (27)$$

$$\leq E(\int_{y_{\ell}}^{y_u} |f(y;x) - \hat{f}(y;x)|dy)^2$$

$$\leq (y_u - y_{\ell})E\int_{y_{\ell}}^{y_u} (f(y;x) - \hat{f}(y;x))^2 dy$$

$$\leq (y_u - y_{\ell})MISE_{\hat{f}},$$

where

$$MISE_{\hat{f}} := E||f(\cdot;x) - \hat{f}(\cdot;x)||_2^2 = E\int(f(y;x) - \hat{f}(y;x))^2 dy$$

denotes the **mean integrated square error** of the density estimator $\hat{f}(y;x)$.

In the same way we get

$$E(\frac{\partial P}{\partial x_k}(x) - \frac{\partial \hat{P}}{\partial x_k}(x))^2 \leq (y_u - y_{\ell})MISE_{\frac{\partial \hat{f}}{\partial x_k}} \qquad (27.1)$$

$$= (y_u - y_{\ell})E\int(\frac{\partial f}{\partial x_k}(y;x) - \frac{\partial \hat{f}}{\partial x_k}(y;x))^2 dy,$$

and

$$E(D_{\underline{\ell}}P(x) - \hat{D}_{\underline{\ell}}P(x))^2 \leq (y_u - y_{\ell})MISE_{D_{\underline{\ell}}\hat{f}} \qquad (27.2)$$

$$= (y_u - y_{\ell})E\int(D_{\underline{\ell}}f(y;x) - \hat{D}_{\underline{\ell}}f(y;x))^2 dy.$$

Thus, we still have to consider the MISE of the estimators $\hat{f}, \frac{\partial \hat{f}}{\partial x_k}, \hat{D}_{\underline{\ell}}f$ defined by (25)-(25.2). Because of

$$f(y;x) - \hat{f}(y;x) = \sum_{j=0}^{q(n)} (c_j(x) - \hat{c}_j(x))\varphi_j(y)$$

$$+ \sum_{j=q(n)+1}^{\infty} c_j(x)\varphi_j(y)$$

and (9.3.2) we get

$$||f(\cdot;x) - \hat{f}(\cdot;x)||_2^2 = \sum_{j=0}^{q(n)} (c_j(x) - \hat{c}_j(x))^2$$

$$+ \sum_{j=q(n)+1}^{\infty} c_j^2(x).$$

Hence, with (22) we have

$$MISE_{\hat{f}} = \sum_{j=0}^{q(n)} \frac{1}{n} \operatorname{var}(\varphi_j(y(a(\cdot),x))) + \sum_{j=q(n)+1}^{\infty} c_j^2(x)$$

$$- \frac{1}{n} \sum_{j=0}^{q(n)} E(\varphi_j(y(a(\omega),x)) - c_j(x))^2 \qquad (28)$$

$$+ \sum_{j=q(n)+1}^{\infty} c_j^2(x),$$

and in the same way with (22.1),(22.2), resp., we obtain

$$MISE_{\frac{\partial \hat{f}}{\partial x_k}} = \sum_{j=0}^{q(n)} \frac{1}{n} \operatorname{var}(\varphi_j'(y(a(\cdot),x))\frac{\partial y}{\partial x_k}(a(\cdot),x))$$

$$+ \sum_{j=q(n)+1}^{\infty} c_{kj}^2(x) \qquad (28.1)$$

$$- \frac{1}{n} \sum_{j=0}^{q(n)} E(\varphi_j'(y(a(\omega),x))\frac{\partial y}{\partial x_k}(a(\omega),x)$$

$$- c_{kj}(x))^2 + \sum_{j=q(n)+1}^{\infty} c_{kj}^2(x),$$

$$MISE_{D_{\underline{\ell}}\hat{f}} = \sum_{j=0}^{q(n)} \frac{1}{n} \operatorname{var}(D_{\underline{\ell}}\varphi_j(y(a(\cdot),x))) + \sum_{j=q(n)+1}^{\infty} c_{\underline{\ell}j}^2(x)$$

$$- \frac{1}{n} \sum_{j=0}^{q(n)} E(D_{\underline{\ell}}\varphi_j(y(a(\omega),x)) - c_{\underline{\ell}j}(x))^2 \qquad (28.2)$$

$$+ \sum_{j=q(n)+1}^{\infty} c_{\underline{\ell}j}^2(x).$$

Using (9.3.3) and (20),(20.1), for $MISE_{\hat{f}}$ and $MISE_{\frac{\partial \hat{f}}{\partial x_k}}$ we

find then the bounds

$$MISE_{\hat{f}} \le d_2^2 \frac{q(n)+1}{n} + \sum_{j=q(n)+1}^{\infty} c_j^2(x), \qquad (29)$$

$$MISE_{\frac{\partial \hat{f}}{\partial x_k}} \le 2d_2^2 \frac{(q(n)+1)(q(n)+2)}{n} E(\frac{\partial y}{\partial x_k}(a(\omega),x))^2$$

$$+ \sum_{j=q(n)+1}^{\infty} c_{kj}^2(x) \qquad (29.1)$$

and similar bounds can be derived for $MISE_{\hat{D_{\ell}f}}$. Conse-

quently, we have the following **consistency result**:

Theorem 3.2. For a given, fixed vector x suppose that
$f(\cdot;x) \in L^2(\Re)$, $\frac{\partial f}{\partial x_k}(\cdot;x) \in L^2(\Re)$, resp., and the second moments under consideration are finite. a) If $q=q(n)$ is
chosen such that $q(n) \longrightarrow \infty$ and $\frac{q(n)}{n} \longrightarrow 0$ as $n \longrightarrow \infty$,
then $MISE_{\hat{f}} \longrightarrow 0$ and $E(P(x)-\hat{P}(x))^2 \longrightarrow 0$ as $n \longrightarrow \infty$.
b) If $q=q(n)$ is selected such that $q(n) \longrightarrow \infty$ and
$\frac{q(n)^2}{n} \longrightarrow 0$ as $n \longrightarrow \infty$, then $MISE_{\frac{\partial \hat{f}}{\partial x_k}} \longrightarrow 0$ and

$E(\frac{\partial P}{\partial x_k}(x)-\frac{\partial \hat{P}}{\partial x_k}(x))^2 \longrightarrow 0$ as $n \longrightarrow \infty$.

Proof. The assertions follow from (27),(28),(29) and
(27.1),(28.1),(29.1), resp., and the fact that $\sum_{j=0}^{\infty} c_j^2(x)$
$= ||f(\cdot;x)||_2^2$, $\sum_{j=0}^{\infty} c_{kj}^2(x) = ||\frac{\partial f}{\partial x_k}(\cdot;x)||_2^2$, resp., is a
convergent series.

Remark 3.1. Similar consistency properties can be
shown - of course - also for the estimator $\hat{D_{\ell}f}$, cf.
(19)-(21).

For the consideration of the convergence rates of the estimators (25)-(25.2) and (26)-(26.2), depending essentially on the second term in the equation (28),(28.1), (28.2), resp., we need the following result [15]:

Lemma 3.2. Let p>1 be a given integer. For g=g(y;x) denoting $f(y;x), \frac{\partial f}{\partial x_k}(y;x), D_{\ell}f(y;x)$, resp., suppose that the functions

$$y \longrightarrow y^i \frac{\partial^{p-1}}{\partial y^{p-1}} g(y;x), \quad i=0,1,\ldots,p \tag{30}$$

are integrable. Then the function

$$\tilde{g}^{(p)}(y;x) := e^{\frac{y^2}{2}} \frac{\partial^p}{\partial y^p}(e^{-\frac{y^2}{2}} g(y;x)) \tag{30a}$$

exists and is integrable, and the Fourier coefficients $b_j(x):=c_j(x)$, $b_j(x):=c_{kj}(x)$, $b_j(x):=c_{\ell j}(x)$, resp., are bounded by

$$|b_j(x)| \le d_3(p,x) \prod_{\ell=1}^{p} (2(j+\ell))^{-\frac{1}{2}}, \quad j \ge 1, \tag{31}$$

where

$$d_3(p,x) := d_2 ||\tilde{g}^{(p)}(\cdot;x)||_1. \tag{31a}$$

Proof. (cf. [15]). Using the relation $H'_{j+1}(y) = 2(j+1)H_j(y)$, by iteration - starting from the definition (12)-(12.1.1) - we get

$$b_j(x) = (-1)^p \prod_{\ell=1}^{p} (2(j+\ell))^{-\frac{1}{2}} \int \varphi_{j+p}(y) \tilde{g}^{(p)}(y;x)dy$$

which yields the assertion, cf. (9.3.3).

If $g(y;x) = f(y;x)$, $g(y;x) = \frac{\partial f}{\partial x_k}(y;x)$, resp., fulfills - for an integer p>1 - the assumptions in the above lemma, then

$$\sum_{j=q(n)+1}^{\infty} b_j^2(x) \le d_3^2(p,x)2^{-p} \sum_{j=q(n)+1}^{\infty} j^{-p}$$

$$\le d_3^2(p,x)2^{-p} \int\limits_{q(n)}^{+\infty} t^{-p}dt$$

$$= d_3^2(p,x)\frac{2^{-p}}{p-1}q(n)^{-(p-1)} .$$

Consequently, according to (29),(29.1), resp., we have that

$$\text{MISE}_{\hat{f}} \le d_2^2 \frac{q(n)+1}{n} + d_3^2(p,x)\frac{2^{-p}}{p-1}q(n)^{-(p-1)} \qquad (32)$$

and

$$\text{MISE}_{\frac{\partial\hat{f}}{\partial x_k}} \le d_2^2 \frac{(q(n)+1)(q(n)+2)}{n} E(\frac{\partial y}{\partial x_k}(a(\omega),x))^2$$

$$+ d_3^2(p,x)\frac{2^{-p}}{p-1}q(n)^{-(p-1)} , \qquad (32a)$$

which yield the following **convergence rates**:

<u>Theorem 3.3.</u> If $f(\cdot;x)$, $\frac{\partial f}{\partial x_k}(\cdot;x)$, resp., satisfies the above assumptions with an integer $p\ge1$, then

$$E(P(x)-\hat{P}(x))^2 \le (y_u-y_e)(d_2^2 \frac{q(n)+1}{n} + \sum_{j=q(n)+1}^{\infty} c_j^2(x))$$

$$\le (y_u-y_e)(d_2^2 \frac{q(n)+1}{n}$$

$$+ d_3^2(p,x)\frac{2^{-p}}{p-1}q(n)^{-(p-1)}), \qquad (33)$$

and

$$E(\frac{\partial P}{\partial x_k}(x) - \frac{\partial\hat{P}}{\partial x_k}(x))^2 \le (y_u-y_e)(2d_2^2 \frac{(q(n)+1)(q(n)+2)}{n}$$

$$\times E(\frac{\partial y}{\partial x_k}(a(\omega),x))^2 + \sum_{j=q(n)+1}^{\infty} c_{kj}^2(x))$$

$$\le (y_u-y_e)(2d_2^2 \frac{(q(n)+1)(q(n)+2)}{n}$$

$$\times E(\frac{\partial y}{\partial x_k}(a(\omega),x))^2 + d_3^2(p,x)\frac{2^{-p}}{p-1}q(n)^{-(p-1)})$$

$$\qquad (33a)$$

Concerning the selection of $q=q(n)$ we have the following consequence:

Corollary 3.1. Let the assumptions of Theorem 3.3 be fulfilled.

a) If $q(n)=O(n^{1/p})$, then $E(P(x)-\hat{P}(x))^2=O(n^{-(p-1)/p})$.

b) If $q(n)=O(n^{1/2p})$, then $E(\frac{\partial P}{\partial x_k}(x)-\frac{\partial P}{\partial x_k}(x))^2=O(n^{-(p-1)/p})$.

Proof. a) The first part is an immediate consequence of (33). b) If $q(n)=O(n^{1/2p})$, then $\frac{(q(n)+1)(q(n)+2)}{n}=O(n^{1/p})$, and the rest follows as in (a).

3.2.1. The integrated square error (ISE) of deterministic approximations.

Corresponding to the estimators (25),(25.1) and (26), (26.1) of $f(y;x)$, $\frac{\partial f}{\partial x_k}(y,x)$ and $P(x)$, $\frac{\partial P}{\partial x_k}(x)$, resp., **deterministic** approximations of these functions can be defined by

$$\tilde{f}(y,x):= \sum_{j=0}^{q(n)} \tilde{c}_j(x)\varphi_j(y), \qquad (34)$$

$$\frac{\tilde{\partial f}}{\partial x_k}(y;x):= \sum_{j=0}^{q(n)} \tilde{c}_{kj}(x)\varphi_j(y) \qquad (34.1)$$

and

$$\tilde{P}(x):= \int_{y_e}^{y_u} \tilde{f}(y;x)dy - \sum_{j=0}^{q(n)} \tilde{c}_j(x)\int_{y_e}^{y_u} \varphi_j(y)dy, \qquad (35)$$

$$\frac{\tilde{\partial P}}{\partial x_k}(x):= \int_{y_e}^{y_u} \frac{\tilde{\partial f}}{\partial x_k}(y;x)dy - \sum_{j=0}^{q(n)} \tilde{c}_{kj}(x)\int_{y_e}^{y_u} \varphi_j(y)dy, \quad (35.1)$$

where $\tilde{c}_j(x)$, $\tilde{c}_{kj}(x)$ are the approximative Fourier coefficients defined by (24.2),(24.6) and (24.4), respectively. Thus, for the integrated square error $ISE_{\tilde{f}}$, $ISE_{\frac{\tilde{\partial f}}{\partial x_k}}$ of $\tilde{f}(y;x)$, $\frac{\tilde{\partial f}}{\partial x_k}(y;x)$, resp., we get

$$ISE_{\tilde{f}}:= ||f(\cdot;x)-\tilde{f}(\cdot;x)||_2^2 = \sum_{j=0}^{q(n)} (c_j(x)-\tilde{c}_j(x))^2$$

$$+ \sum_{j=q(n)+1}^{\infty} c_j^2(x), \qquad (36)$$

and

$$ISE_{\frac{\tilde{\partial f}}{\partial x_k}} := \left|\left| \frac{\partial f}{\partial x_k}(\cdot;x) - \frac{\tilde{\partial} f}{\partial x_k}(\cdot;x) \right|\right|_2^2 = \sum_{j=o}^{q(n)} c_{kj}(x) - \tilde{c}_{kj}(x))^2$$

$$+ \sum_{j=q(n)+1}^{\infty} c_{kj}^2(x). \tag{36.1}$$

Since

$$(P(x) - \tilde{P}(x))^2 \leq (y_u - y_e) ISE_{\tilde{f}}, \tag{37}$$

$$(\frac{\partial P}{\partial x_k}(x) - \frac{\tilde{\partial} P}{\partial x_k}(x))^2 \leq (y_u - y_e) ISE_{\frac{\tilde{\partial f}}{\partial x_k}}, \tag{37.1}$$

bounds for $ISE_{\tilde{f}}$, $ISE_{\frac{\tilde{\partial f}}{\partial x_k}}$ and therefore for the error

$|P(x) - \tilde{P}(x)|$, $|\frac{\partial P}{\partial x_k}(x) - \frac{\tilde{\partial} P}{\partial x_k}(x)|$ follow, cf. Theorem 3.3,

from the error bounds (24.3) and (24.7), (24.5), resp.,
and using again Lemma 3.2.

4. Expansions in Hermite functions in case m>1

The results known from the univariate case (m=1) can
be transfered easily to the multivariate case (m>1) by
considering instead of the univariate Hermite functions
φ_j the product functions

$$\varphi_\lambda(y) := \prod_{i=1}^{m} \varphi_{\lambda_i}(y_i), \quad y \in \mathfrak{R}^m, \tag{38}$$

where $\lambda = (\lambda_1, \ldots, \lambda_m)' \in Z_+^m$ denotes a multiple index and
φ_{λ_i} is one of the univariate Hermite functions given by
(9.2). Hence,

$$\varphi_\lambda \in L^1(\mathfrak{R}^m) \cap L^2(\mathfrak{R}^m) \text{ for all } \lambda \in Z_+^m, \tag{38.1.1}$$

cf. (9.3.1), and it is known [13],[14] that

(φ_λ) forms a complete, orthonormal set in $L^2(\mathfrak{R}^m)$,

$$\tag{38.1.2}$$

see (9.3.2).

Assuming again, cf. (10)-(10.1.1), that

$$f(\cdot;x) \in L^2(\mathfrak{R}^m) \tag{39}$$

$$\frac{\partial f}{\partial x_k}(\cdot;x) \in L^2(\mathfrak{R}^m) \tag{39.1}$$

$$D_{\underline{\ell}}f(\cdot;x) \in L^2(\mathfrak{R}^m), \tag{39.1.1}$$

resp., then we have the following expansions in series of orthogonal functions in $L^2(\mathfrak{R}^m)$:

$$f(y;x) = \sum_{||\lambda||^2=0}^{\infty} c_\lambda(x)\varphi_\lambda(y) \tag{40}$$

$$\frac{\partial f}{\partial x_k}(y;x) = \sum_{||\lambda||^2=0}^{\infty} c_{k\lambda}(x)\varphi_\lambda(y) \tag{40.1}$$

$$D_{\underline{\ell}}f(y;x) = \sum_{||\lambda||^2=0}^{\infty} c_{\underline{\ell}\lambda}(x)\varphi_\lambda(y), \tag{40.1.1}$$

cf. (11)-(11.1.1), where the corresponding Fourier coefficients are defined, see (12)-(12.1.1), by

$$c_\lambda(x) := \int f(y;x)\varphi_\lambda(y)\,dy \tag{41}$$

$$c_{k\lambda}(x) := \int \frac{\partial f}{\partial x_k}(y;x)\varphi_\lambda(y)\,dy \tag{41.1}$$

$$c_{\underline{\ell}\lambda}(x) := \int D_{\underline{\ell}}f(y;x)\varphi_\lambda(y)\,dy. \tag{41.1.1}$$

Obviously, cf. (13),

$$c_\lambda(x) = E\varphi_\lambda(y(a(\omega),x)), \tag{42}$$

and under assumptions corresponding to (14),(18), resp., we get, cf. (14.1) and (19),(14.2) and (21),

$$c_{k\lambda}(x) = \frac{\partial}{\partial x_k}c_\lambda(x) = E\frac{\partial}{\partial x_k}\varphi_\lambda(y(a(\omega),x))$$

$$= \sum_{i=1}^{m} E \prod_{j\neq i} \varphi_{\lambda_j}(y_j(a(\omega),x))\varphi_{\lambda_i}'(y_i(a(\omega),x))$$

$$\times \frac{\partial y_i}{\partial x_k}(a(\omega),x), \tag{42.1}$$

$$c_{\underline{\ell}\lambda}(x) = D_{\underline{\ell}}c_\lambda(x) = ED_{\underline{\ell}}\varphi_\lambda(y(a(\omega),x)). \tag{42.2}$$

The inequalities (15)-(15.2) can be generalized easily to the present case: Replacing the index "j" by the multiple index λ, we find

$$|P(x) - \sum_{||\lambda||^2=0}^{n} c_\lambda(x) \int_{y_\ell}^{y_u} \varphi_\lambda(y)dy| \leq \mu_0 ||f(\cdot;x)$$

$$- \sum_{||\lambda||^2=0}^{n} c_\lambda(x)\varphi_\lambda(\cdot)||_2, \tag{43}$$

$$|\frac{\partial P}{\partial x_k}(x) - \sum_{||\lambda||^2=0}^{n} c_{k\lambda}(x) \int_{y_\ell}^{y_u} \varphi_\lambda(y)dy| \leq \mu_0 ||\frac{\partial f}{\partial x_k}(\cdot;x)$$

$$- \sum_{||\lambda||^2=0}^{n} c_{k\lambda}(x)\varphi_\lambda(\cdot)||_2, \tag{43.1}$$

$$|D_{\underline{\ell}}P(x) - \sum_{||\lambda||^2=0}^{n} c_{\underline{\ell}\lambda}(x) \int_{y_\ell}^{y_u} \varphi_\lambda(y)dy| \leq \mu_0 ||D_{\underline{\ell}}f(\cdot;x)$$

$$- \sum_{||\lambda||^2=0}^{n} c_{\underline{\ell}\lambda}(x)\varphi_\lambda(\cdot)||_2, \tag{43.1.1}$$

where $\mu_0 := (\prod_{i=1}^{m} (y_{ui}-y_{\ell i}))^{1/2}$ with $y_\ell, y_u \in \mathbb{R}^m$. Hence, we have now the following generalization of Theorem 3.1:

Theorem 4.1. Under the above assumptions, the probability function $P(x)$ and its derivatives can be represented by the following convergent series:

$$P(x) = \sum_{||\lambda||^2=0}^{\infty} c_\lambda(x)\int_{y_\ell}^{y_u} \varphi_\lambda(y)dy, \tag{44}$$

$$\frac{\partial P}{\partial x_k}(x) = \sum_{||\lambda||^2=0}^{\infty} c_{k\lambda}(x)\int_{y_\ell}^{y_u} \varphi_\lambda(y)dy, \tag{44.1}$$

$$D_{\underline{\ell}}P(x) = \sum_{||\lambda||^2=0}^{\infty} c_{\underline{\ell}\lambda}(x)\int_{y_\ell}^{y_u} \varphi_\lambda(y)dy. \tag{44.1.1}$$

Since

$$\int_{y_\ell}^{y_u} \varphi_\lambda(y)dy = \int_{y_{\ell 1}}^{y_{u1}} \ldots \int_{y_{\ell m}}^{y_{um}} \prod_{i=1}^{m} \varphi_{\lambda_i}(y_i)dy_i$$

$$= \prod_{i=1}^{m} \int_{y_{\ell i}}^{y_{ui}} \varphi_{\lambda_i}(y_i)dy_i, \qquad (45)$$

the integrals in (44)-(44.1.1) can be obtained also by means of the simple recursion (17.1)-(17.3.1) derived for the case m=1.

Since the Fourier coefficients $c_\lambda(x)$, $c_{k\lambda}(x)$, $c_{\ell\lambda}(x)$, resp., are given by the expectations (42)-(42.2), consistent estimators $\hat{c}_\lambda(x)$, $\hat{c}_{k\lambda}(x)$, $\hat{c}_{\ell\lambda}(x)$, of $c_k(x)$, $c_{k\lambda}(x)$, $c_{\ell\lambda}(x)$ and therefore also consistent estimators $\hat{P}(x)$, $\frac{\partial \hat{P}}{\partial x_k}(x)$, $\hat{D}_\ell P(x)$ of $P(x)$ and its derivatives can be obtained similar to the case m=1. Further details are given in a subsequent paper.

5. Expansions in trigonometric, Legendre and Laguerre series

Since the generalization to the case m≥1 can be carried out relatively easy, see Sections 3 and 4, in the following we concentrate to the case m=1.

5.1. Expansions in trigonometric and Legendre series

Expansions in trigonometric and Legendre series are suitable if we know in advance that

$$y_0 < y(a(\omega),x) < y_1 \text{ a.s. (almost sure) for all } x \in \Delta, \qquad (46)$$

where $y_0 < y_1$ are given, fixed bounds and Δ is an open neighborhood of a point x_0 at which the probability function or one of its derivatives should be calculated.

Up to the rate of convergence for estimators based on Hermite expansions, given by Theorem 3.3, the formulas and results from Section 3 can be transfered directly to the present situation by replacing the Hermite functions (9.2) by the **trigonometric functions**

$$\varphi_{co}(y):= \frac{1}{\sqrt{y_1 \cdot y_0}}, \quad \varphi_{cj}(y):= (\frac{2}{y_1 \cdot y_0})^{1/2} \cos(j\pi \frac{y \cdot y_0}{y_1 \cdot y_0}),$$

$$\text{(46.1)}$$

$$\varphi_{sj}(y):= (\frac{2}{y_1 \cdot y_0})^{1/2} \sin(j\pi \frac{y \cdot y_0}{y_1 \cdot y_0}), \quad j=1,2,\ldots,$$

or by the **Legendre polynomials**

$$\varphi_j(y):= (\frac{2j+1}{y_1 \cdot y_0})^{1/2} \frac{1}{j!(y_1 \cdot y_0)} \frac{d^j}{dy^j}((y \cdot y_0)(y \cdot y_1)), \text{(46.2)}$$

where $y_0 \leq y \leq y_1$. Since the range of $y_x(\omega)$ is bounded here, the assumptions (14) and (18) can be weakened considerably!

5.2. Expansions in Laguerre series

Expansions in Laguerre series are of interest if it is known a priori that $y_x(\omega)$ is bounded from one side. In the case

$$y_x(\omega) = y(a(\omega),x) \geq 0 \text{ a.s.} \tag{47}$$

the corresponding sequence of **Laguerre functions** reads

$$\varphi_j(y) = C_j e^{\frac{y}{2}} y^{-\frac{\alpha}{2}} \frac{d^j}{dy^j}(y^{j+\alpha} e^{-y}), \quad y \geq 0, \quad j=0,1,\ldots, \tag{47.1}$$

where $\alpha > -1$ is a given, fixed parameter, and C_j, $j=0,1,\ldots$, are normalizing constants. As mentioned in Section 5.1, the formulas and results from Section 3, up to Theorem 3.3, can be transfered directly to this case by replacing the Hermite functions (9.2) by the Laguerre functions (47.1). Further details are shown in a subsequent paper.

References

[1] Barner, M., Flohr, F.: Analysis II. Walter de Gruyter, Berlin-New York 1983

[2] Breitung, K., Hohenbichler, M.: Asymptotic approximations for multivariate integrals with applications to multinormal probabilities. J. of Multivariate Analysis 30 (1989), 80-97

[3] Breitung, K.: Asymptotische Approximation für Wahrscheinlichkeitsintegrale. Habilitationsschrift, Fakultät für Philosophie, Wissenschaftstheorie und Statistik der Universität München 1990; to appear in the Springer Lecture Notes Series in Mathematics

[4] Gajewski, A., Zyczkowski, M.: Optimal Structural Design under Stability Constraints. Kluwer, Dordrecht-Boston-London 1988

[5] Gollwitzer, S., Rackwitz, R.: An efficient solution to the multinormal integral. Prob. Eng. Mech. 3(2) (1988), 98-101

[6] Hohenbichler, M., Rackwitz, R.: Non-normal dependent vectors in structural safety. J. Eng. Mech. Div. ASCE 107(6) (1981), 1227-1249

[7] Kall, P., Wallace, S.: Stochastic Programming. J. Wiley, Chichester (UK) 1994

[8] Marti, K.: Stochastic Optimization Methods in Structural Mechanics. ZAMM 70 (1990), T742-T745

[9] Marti, K.: Approximations and Derivatives of Probability Functions. In: G. Anastassiou, S.T. Rachev (eds.): Approximation, Probability and Related Fields. Plenum Press, New York 1994, 367-377

[10] Melchers, R.E.: Structural Reliability Analysis and Prediction. J. Wiley, New York 1987

[11] Richter, H.: Wahrscheinlichkeitstheorie. Springer-Verlag, Berlin 1966

[12] Sansone, G.: Orthogonal Functions. Interscience Publishers Inc., New York - London 1959

[13] Schüler, L.: Schätzungen von Dichten und Verteilungsfunktionen mehrdimensionaler Zufallsvariabler auf der Basius orthogonaler Reihen. Dissertation Naturwissenschaftliche Fakultät der TU Braunschweig, Braunschweig 1974

[14] Schüler, L., Wolff, H.: Schätzungen mehrdimensionaler Dichten mit Hermiteschen Polynomen und lineare Verfahren zur Trendverfolgung. Forschungsbericht BMVg-FBWT 76-23, TU Braunschweig, Institut für Angewandte Mathematik, Braunschweig 1976

[15] Schwartz, S.C.: Estimation of Probability Density by an Orthogonal Series. Ann. Math. Statist. 38 (1967), 1261-1265

[16] Streeter, V.L., Wylie, E.B.: Fluid Mechanics. McGraw Hill, New York 1951 (first edition)

[17] Szegö, G.: Orthogonal Polynoms. American Math. Soc., Providence (R.I., USA) 1939

[18] Thoft-Christensen, P., Murotsu, Y.: Application of Structural Systems Reliability Theory. Springer-Verlag, Berlin-Heidelberg-New York-Tokyo 1986

[19] Tricomi, F.G.: Vorlesungen über Orthogonalreihen. Springer-Verlag, Berlin-Heidelberg-New York 1970

[20] Uryas'ev, S.: A differentiation formula for integrals over sets given by inclusion. Numer. Funct. Anal. and Optimiz. 10 (1989), 827-841

[21] Uryas'ev, S.: Derivatives of Probability Functions and Integrals Over Sets Given by Inequalities. J. Comp. and Appl. Vol. 56 (1994), in press

Computer Support for Modeling
in Stochastic Linear Programming

P. Kall and J. Mayer
Institute for Operations Research, University of Zurich
Moussonstr. 15, CH-8044 Zurich

1. Introduction

The purpose of the paper is to discuss the modeling process in stochastic linear programming (SLP) and to point out the SLP-specific features of computer support to this process.

In the next section SLP models and solution algorithms will be considered from the modeling point of view. Section 3 is devoted to a discussion of computer support to modeling, by considering the life cycle of SLP models. Section 4 presents an overview on how modeling support has been implemented in the model management system SLP-IOR.

2. Stochastic linear programming models

In this section a brief summary of stochastic linear programming models will be given. The discussion will be restricted to two-stage models and chance-constrained models with separate and with joint chance constraints. For a detailed presentation and basic properties of the models see Kall [25] and Wets [52].

<u>Two-stage models</u> with fixed recourse can be formulated as follows:

$$
\begin{aligned}
\min \quad & \{c^T x + E_\omega Q(x,\omega)\} \\
\text{s.t.} \quad & Ax && \propto b \\
& x && \in [l, u],
\end{aligned}
\tag{1}
$$

where

$$
\begin{aligned}
Q(x,\omega) = \quad \min \quad & q^T(\omega)y \\
\text{s.t.} \quad & Wy && \propto h(\omega) - T(\omega)x \\
& y && \geq 0.
\end{aligned}
\tag{2}
$$

Any one of the relations $=,\leq,\geq$ is permitted row-wise; this fact is indicated by the symbol $\propto$.

The matrix W is called the recourse matrix of the model. In the special case when it has the form W=(I,-I) with I standing for an identity matrix of the appropriate size, (1) is called a simple recourse problem.

Chance constrained models with a joint chance constraint have the following form:

$$\min c\,x$$

$$P(\{\omega \mid Tx \geq h(\omega)\}) \geq \alpha$$

$$Ax \propto b \tag{3}$$

$$x \in [l,u],$$

with $0 \leq \alpha \leq 1$ being a prescribed (high) probability level.

Chance constrained models with separate chance constraints are models formulated as:

$$\min c\,x$$

$$P(\{\omega \mid t_i^T(\omega)x \geq h_i(\omega)\}) \geq \alpha_i,\ \forall i$$

$$Ax \propto b \tag{4}$$

$$x \in [l,u],$$

with $0 \leq \alpha_i \leq 1, \forall i$ being given (high) probability levels for the separate constraints. $t_i^T(\omega)$ denotes the i-th row of the matrix $T(\omega)$.

In the models above $\omega \in \Omega$, $(\Omega, \mathcal{F}, P)$ is a probability space; $q(\omega), h(\omega), t_i(\omega)$ are random vectors and $W(\omega), T(\omega)$ random matrices. The random entries in the various arrays are represented through affine sums as follows:

$$\begin{cases} q(\omega) & = q^0 & + \sum_{j=1}^k q^j \xi_j(\omega) \\ h(\omega) & = h^0 & + \sum_{j=1}^k h^j \xi_j(\omega) \\ T(\omega) & = T^0 & + \sum_{j=1}^k T^j \xi_j(\omega) \\ t_i(\omega) & = t_i^0 & + \sum_{j=1}^k t_i^j \xi_j(\omega) \end{cases} \tag{5}$$

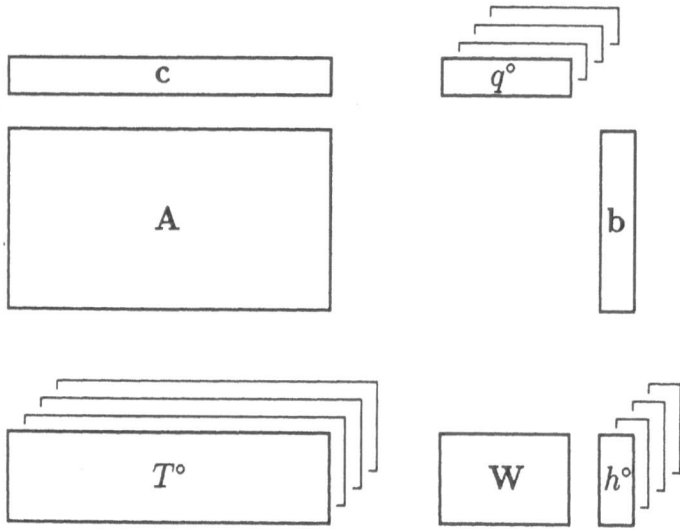

Figure 1: Two-stage model: Data arrays

with $\xi_j(\omega)$, j=1,..k being random variables.

Let us consider a two-stage model. The array-type data needed for model specification are shown on Figure 1. We assume to have three random variables; the terms corresponding to these in the affine sums (5) are shown behind the constant term of this sum.

The deterministic arrays c, A, b and W, as well as the constant terms in the affine sums q^0, h^0, T^0 together with the relations denoted by $\propto$ define the *underlying algebraic structure* of the model.

The affine sums (5) establish a mapping between the random variables and the random entries of the various arrays in the model. They determine which of these arrays are random and for the random arrays which of the components are random variables. Notice that equations (5) express relations between functions on Ω, i.e. between random variables. This is to be understood in a probabilistic sense: The relations imply a transformation of the probability distribution of the random vector $\xi(\omega) = (\xi_1(\omega), ...\xi_k(\omega))$ thus determining the probability distribution of the random entries in the model-arrays. Because of this feature, the structure expressed by (5) is inherently non-algebraic. Apart from the constant terms we do not consider (5) to be part of the underlying algebraic structure.

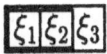

Figure 2: Two-stage model: Random variables

Notice that the linearity in (5) is immaterial; the relations can be replaced by measurable nonlinear mappings as long as the implied transformation of the probability distribution of $\xi(\omega)$ results in a distribution of the random entries which can be handled within SLP numerically.

The modeling idea behind (5) is the following: These relations provide means for expressing the random entries in terms of fewer "basic" random variables. This is a quite important feature when considering solvers for SLP models. To see this let us consider two stage models with a finite discrete distribution and solvers based on approximation methods working with subdivisions of the support of the random variables, see e.g. Kall and Stoyan [26]. These solvers require the solution of LP's for each vertex of k-dimensional intervals. Assume that we have 20 random entries in the model expressed in terms of 5 random variables (k=5 in (5)). This means working with intervals having $2^5 = 32$ vertices whereas in the space of random entries we would have intervals with $2^{20} = 1048576$ vertices which would clearly be beyond our numerical capabilities.

The relations (5) also provide means for representing a special kind of stochastic dependence implied by those relations. The most general way of representing stochastic dependence is by specifying the joint probability distribution of a random vector. This feature is accounted for as follows:

The components of the random vector $\xi(\omega) = (\xi_1(\omega), ...\xi_k(\omega))$ are assumed to be partitioned into mutually stochastically independent groups, see Figure 2 with two groups.

The probability distribution of $\xi(\omega)$ is determined by the probability distributions of the separate groups.

The dependency structure and probability distribution of the random entries of the models are clearly determined by the dependency structure and probability distribution of $\xi(\omega)$ and by the relations (5). We split up this structure into two structures: One corresponds to the the dependency structure and probability distribution of $\xi(\omega)$ and the other to the affine relations (5). The reasons for doing so can be summarized as follows:

- Besides of specifying a certain kind of stochastic dependence, relations (5) also serve for selecting those arrays which are considered to be stochastic, a feature which greatly influences the properties of the SLP models as well as solver selection. On the other hand, the dependency structure and probability distribution influences model properties and solver selection too. The point is that the way of influence is different for the two structures and it is desirable to provide separate manipulation facilities for the two structures.

- As mentioned above relations (5) are means for reducing the number of random variables. In approximation methods support-subdivision concerns $\xi(\omega)$ and not the random entries directly. These features again suggest that the two structures should be accessible as separate entities.

- Specifying stochastic dependence by affine relations or by a joint probability distribution are quite different approaches from the modeling point of view.

As a possible origin of relations (5) might be some kind of regression analysis, the nonconstant part of relations (5) will be called the *underlying regression structure*. The dependency structure expressed by the groups and the probability distributions of the groups of the random vector $\xi(\omega)$ will be called the *underlying random variable structure*.

Notice that the distribution and dependency structures of the random entries in $q(\omega), T(\omega), h(\omega)$ are determined by the interaction of the random variable- and regression structures, e.g. a $\xi(\omega)$ with independent components may lead to a model with highly dependent random entries.

Important structural properties of SLP-models, like convexity and separability, depend on the interaction of the three underlying structures discussed above, see e.g. Kall [25] and Prékopa [40].

SLP problems are in general numerically hard to solve. The source of difficulties is the expected value term in two-stage models and the probabilities appearing in the chance-constrained models. These lead to multivariate integrals in the case of continuous distributions and to sums usually having a large amount of terms in the discretely distributed case.

For overviews on SLP algorithms see Kall [24], Kall et al. [29], Mayer [38], Prékopa [41] and Wets [51]. The solution algorithms can roughly be subdivided into two classes:

- Algorithms in the first class are designed to solve an algebraic equivalent. By an algebraic equivalent we mean an equivalent reformulation of an SLP-model as a mathematical programming model completely specifyed in terms of explicit algebraic formulas. Depending on the three underlying structures there may, or may not exist such an equivalent. Two-stage models with a finite discrete distribution can be reformulated as (usually large-scale) LP models, see e.g. Kall [25], or chance-constrained models with separate constraints can be transformed into nonlinear programming models for certain types of continuous distributions, see Marti [36]. On the other hand, there exists no algebraic equivalent for a jointly chance-constrained model with a multivariate normal distribution. In the list below some examples for algorithms belonging to the first class are presented. For two-stage models with a finite discrete distribution:

- The L-shaped method of Van Slyke and Wets [48];
- the basis reduction method of Strazicky [45];
- the regularized decomposition method of Ruszczynski [42].

For chance constrained models with separate chance constraints transformations into an algebraic equivalent are e.g. possible:

- For the normally distributed case, Van de Panne and Popp [39];
- for the Cauchy-distribution, Marti [36].

• Algorithms of the second class work directly with the original formulation. Examples for algorithms belonging to this class are listed below. For two-stage models:

- Algorithms based on successively approximating the probability measure: The algorithm of Kall [23], Kall and Stoyan [26]; the methods of Frauendorfer and Kall [10] and Frauendorfer [11];
- the stochastic quasigradient method of Ermoliev [9] and Gaivoronski [12];
- the method of Dantzig and Glynn [7] based on importance sampling;
- the stochastic decomposition method of Higle and Sen [22].
- For simple recourse problems with only the RHS being stochastic the resulting separability property can be utilized, as e.g. in the approximation method of Kall and Stoyan [26].

Algorithms for jointly chance-constrained models:

- The method of Szántai [47] based on the supporting hyperplane method;
- the method of Mayer [37] based on the reduced gradient algorithm.

3. The modeling life-cycle in SLP

In this section a simplified version of the modeling life-cycle for SLP-models will be considered from the point of view of computer support. In a recent survey on model management systems Bharadwaj et al. [2] consider the modeling life cycle for mathematical programming. They discuss a basic life cycle consisting of four steps, the approach being mainly motivated by linear programming (LP) models. In this section we consider the same four steps and point out the additional features which enter the scheme in the SLP case. A thorough discussion of the modeling life cycle in operations research can be found in Geoffrion [14], [15].

Before discussing the separate steps in the modeling life cycle let us consider the main specific features of SLP modeling.

Assume that an instance of an SLP model has been formulated. To solve it we have to choose a solver, i.e. a computer implementation of a solution algorithm.

- For LP models there exist general solution algorithms, e.g. the simplex method. This is not the case in SLP: Appropriateness of SLP algorithms depends on the model type and on the interaction of the underlying structures as discussed in the previous section.

- For LP there exists a great variety of commercially available solvers. When considering solver selection, one of the implementations of the simplex method is usually readily available. It may happen that this is not the best choice, e.g. for an LP with a network structure, but anyhow the set of solvers is not empty. In SLP, solvers for the various model types are scattered around the world at various academic institutions. At the time of writing this paper none of them was commercially available. This fact implies that modeling possibilities are narrowed down by the availability of SLP-solvers in the given computing environment.

- When an LP problem has a solution then a good LP-solver typically finds it. For SLP solvers several runs might be necessary within a tuning cycle to produce a solution. For an analogous situation in nonlinear programming see Schittkowski [44].

From the model formulation point of view obviously for SLP models all of the three underlying structures are to be considered with the underlying algebraic structure corresponding to LP.

- In LP we are dealing with a single general modeling paradigm and various special cases of this general class. In SLP at least the three quite different basic model classes, specified in Section 2, are to be considered.

- Within LP modeling, any models fitting into the LP framework are without further consideration subjects of model formulation and solution. For SLP models the situation is different: Depending on the interaction of the underlying structures models may emerge for which no solution algorithm exists, not to speak of solvers. As an example consider a jointly chance constrained model with random entries having a non-quasiconcave probability distribution function. This leads to a nonconvex optimization problem with no existing solution algorithm. As mentioned above model formulation is also constrained by the availability of solvers.

Next the separate steps in the modeling life cycle will be considered and SLP-specific issues commented on, see Figure 3.

Step 1. Problem Identification/Revision

In this step the decision problem is to be clearly and concisely identified or an already formulated problem is to be revised within the decision making environment. Main actors at this step are the decision makers themselves. In addition to the discussion in Bharadwaj et al. [2] the following SLP-specific features appear. Quantities playing the role of variables later on in the mathematical model should clearly be separated from quantities considered as parameters, i.e. those playing the role of coefficients in the formal model. This latter class should further be subdivided into deterministic and random parameters. Relationships between random parameters should be explored thus providing the basis for the formulation of the regression structure in the next (mathematical modeling) step. The distribution and stochastic dependency of the random parameters should be investigated possibly by utilizing statistical techniques.

From the modeling support point of view this step seems to be the most difficult one as it is to be performed in the decision makers' natural environment. One of the research directions aiming this subject concerns decision support systems, see Sprague and Carlson [46]. Another research direction is via model management systems, for a survey see Bharadwaj et al. [2]. A recent approach is due to Geoffrion within the general concept of a modeling environment [15]. The idea is to consider Steps 1 and 2 together and to provide a model definition language which is easy-to-learn for decision makers and at the same time covers a sufficiently wide range of OR/MS models [16].

Step 2. Model Formulation

In this step the mathematical model is formulated. This is intimately related to the problem of model representation, i.e. to a computer-executable concise framework for representing models. The formulation of the mathematical model runs in parallel with setting up a mapping between model parameters and data sources. This mapping is quite important in the SLP context: As discussed in Section 2, for some solvers the model instant is to be transformed into an algebraic equivalent. Notice that model formulation may be constrained by the availability of solvers.

From the computer-support point of view this is the most thoroughly investigated step of the modeling life cycle of mathematical programming models. The different approaches can roughly be grouped around the following general frameworks:

- Algebraic modeling languages and systems; for a comparative survey see Greenberg and Murphy [20].

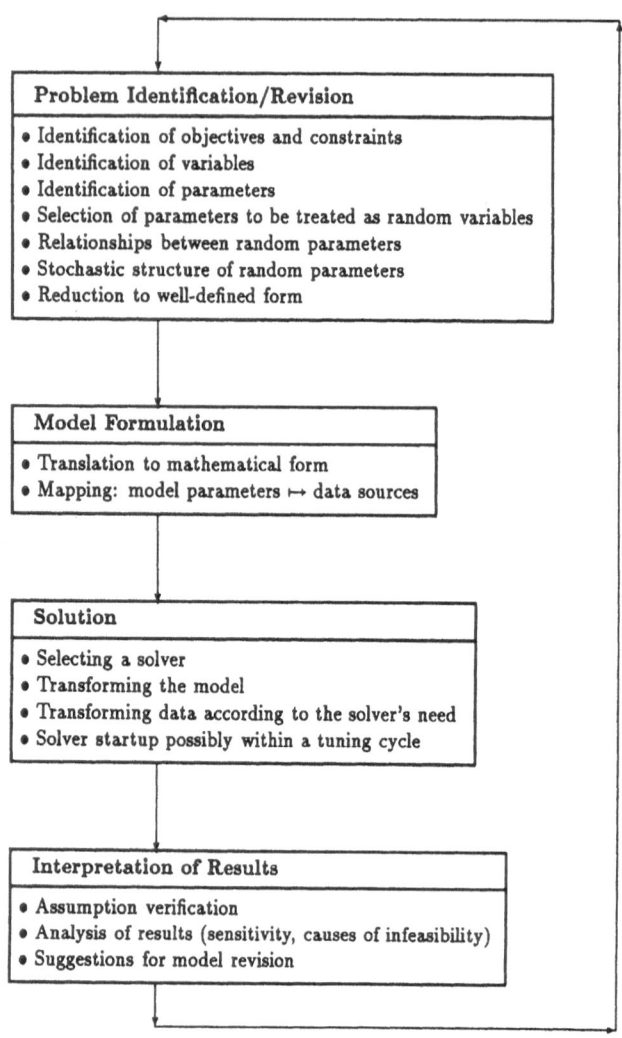

Problem Identification/Revision

- Identification of objectives and constraints
- Identification of variables
- Identification of parameters
- Selection of parameters to be treated as random variables
- Relationships between random parameters
- Stochastic structure of random parameters
- Reduction to well-defined form

Model Formulation

- Translation to mathematical form
- Mapping: model parameters $\mapsto$ data sources

Solution

- Selecting a solver
- Transforming the model
- Transforming data according to the solver's need
- Solver startup possibly within a tuning cycle

Interpretation of Results

- Assumption verification
- Analysis of results (sensitivity, causes of infeasibility)
- Suggestions for model revision

Figure 3: The modeling life cycle of SLP models

- Model management systems; a survey can be found in Bharadwaj [2].

- Modeling environments; see Geoffrion [14], [15], [16].

- Recent approaches based on the idea of integrating algebraic modeling languages and model management systems; see Greenberg [18] for LP, Gassmann and Ireland [13] for multistage scenario-based stochastic programming.

Step 3. Solution

This is a critical step in the SLP case.

- *Selecting a solver.* As mentioned in the previous section there exists no universal SLP solver; solver selection depends on the model type and on the interaction of the underlying structures. It is also influenced by dimensionalities of the model instance, e.g. considering two-stage models the algorithms based on the algebraic equivalent LP may be excluded if the number of joint realizations of the random variables is too high. The criteria in solver selection are of a highly technical nature requiring computer support. For nonlinear programming Schittkowski [43], [44] developed a methodology based on expert system ideas for supporting solver selection.

- *Transforming the model.* As already discussed above one class of solvers is designed to solve algebraic equivalents. This means that prior to starting them, the equivalent algebraic mathematical programming problems are to be generated. These transformations are far from being just changes in dataformat. The specific form of the transformation depends on the solver, on the model type and on the underlying structures. Considering computer support, the same comment applies as for solver selection.

- *Transforming data according to the solver's needs.* This means formatting model data (perhaps those of an algebraic equivalent) in accordance with the input dataformat of the solver. In LP modeling this usually means producing an MPS file. For SLP models an extension of the MPS dataformat exists, see Birge et al. [4], unfortunately not covering chance constrained models. Computer support here means an automatic dataformat transformation.

- *Solver startup.* As SLP problems are typically numerically difficult, several runs with changed solver parameters may be needed to get a solution (tuning cycle). Selecting appropriate solver parameters needs detailed knowledge on the solver and consequently requires computer support. A solver is to be understood in a broad sense: e.g. for large scale problems a solver run may just mean computing bounds on the optimal objective; for a survey on bounds see Kall [28].

Step 4. Interpretation

At this step the results of the previous step are analyzed. Ideally, the results of the analysis should be formulated as suggestions for model revision. Notice that a result can be that there exists no solution, or the objective is unbounded on the feasible domain. The cause of this may be modeling error (e.g. contradictory or missing constraints) or simply bugs in the numerical data. Similarly to computer programs, models must also be "debugged". For LP, Greenberg's ANALYZE [19] can be used to detect causes of infeasibility or unboundedness. In the SLP context an immediate application of ANALYZE can be to detect causes for incomplete recourse (for the notion see Kall [25]) for two-stage models with a discrete distribution. Some of the further SLP-specific issues are the following:

- Checking the complete recourse property; Kall [25], Wallace and Wets [49], [50].

- Computing general indicators for assessing the stochastic model; see Avriel and Williams [1], Birge [3], Hausch and Ziemba [21], Kall [27].

- In a two-stage setting the decision maker may be interested in the reliability of his first-stage decision, i.e. in the probability that second-stage actions will be necessary. This can be computed provided that in (2) all of the relations are inequalities.

4. SLP-IOR: A model management system for SLP

In the first part of this section a short summary of the basic features of model management systems in mathematical programming will be given. The second part is devoted to illustrate modeling support as implemented in SLP-IOR, a model management system for SLP, being under development by the authors. For SLP-IOR see Kall and Mayer [30], [31], [32], [33].

The objective of model management systems is to provide computer based support for the entire modeling cycle, see Bharadwaj et al. [2] and Dolk [8]. Model management systems originate in database management systems (DBMS). The original intention was to extend the capabilities of DBMS by handling models as a highly structured data type and extending the usual DBMS operations through model specific operations like "Solve". The basic design requirement for model management systems is that the model representation should be independent on data and on solvers. The major model manipulation operations a model management system is supposed to provide can be summarized as follows.

- Database operations: Save, retrieve, export/import, archivate, document.

- Main model-specific operations: Solve, perturb, analyze, randomly generate.

- Facilities for transforming models.

- Operations for building models by utilizing components of existing models.

- Facilities for enforcing model integrity: Consistency and completeness.

- Facilities for guiding user actions by utilizing knowledge concerning models and solvers ("intelligent" component).

SLP-IOR is designed to support Steps 2 through 4 of the modeling cycle, with an interactive menu driven interface for providing support to Step 1. The first step is only supported for users having a basic knowledge on SLP models, i.e. for modeling professionals. This is in accordance with the objective of SLP-IOR: It should support modeling professionals in using SLP models and it should serve as a workbench for algorithmic research in SLP.

The design efforts for SLP-IOR have been concentrated on the SLP-specific features. Deterministic LP is clearly a special case of SLP. For LP an enormous amount of excellent model management tools and ideas exist; instead of trying to reinvent or re-implement some of them, an open system architecture has been chosen which facilitates the integration with model management systems for LP.

The main vehicles for achieving openness are the object-oriented programming style and the idea on relying on an algebraic modeling language. The relative independence of model representation and data/solvers can easily be achieved this way too: For modeling languages this is one of their basic design principles; for the object-oriented approach the distinction between a class and its instances can be utilized along with handling solvers as separate classes.

The algebraic modeling language and system GAMS, Bisschop and Meeraus [5], Brooke et al. [6], has been chosen. It has the property of openness with respect to adding new solvers, an extremely important feature in the SLP case. GAMS provides a uniform interface facility to solvers which has been utilized for connecting the SLP-solvers to the system. For starting up a solver SLP-IOR first generates a GAMS input file, this is submitted to GAMS for execution and after solver termination the results are returned to SLP-IOR via GAMS. Being an algebraic modeling system, GAMS facilitates the handling of the underlying algebraic structure as well as the algebraic equivalent problems. Having a language component is also important from the model documentation point of view.

Next some of the major model manipulation operations of SLP-IOR will be discussed. For a detailed description see Kall and Mayer [30], [31], [32], [33]. The discussion will be restricted to model specific operations because the database operations are quite obvious. The only exception might be export/import meaning the communication of models with the outside world. In SLP-IOR this is either possible in the SLP standard format of Birge et al. [4] or in GAMS form.

Model construction is supported by a building block system. The three underlying structures can be used as building blocks for models. The main constituents of the underlying structures (arrays, probability distributions) serve as building blocks for the underlying structures themselves. These elementary entities can be changed by using editors.

A modeling toolbox serves for manipulating models and their underlying structures. It contains a facility for randomly generating complete recourse problems based on GENSLP [34] and further developed by E. Keller and the authors. For randomly generated problems in LP see Greenberg [17]. Another facility of the toolbox serves for sampling and for discretizing continuous distributions. The development of a facility for imposing random perturbations onto a given problem is in progress.

Model transformations are possible between the three basic model types. Missing data in the resulting problems are replaced by defaults.

Solvers. As discussed in the previous section, in SLP several different solvers are needed according to model type and the underlying structures. For a list of the presently connected solvers see Kall and Mayer [33]; linking further solvers to SLP-IOR is an ongoing activity.

Solver selection means that the system produces a list of solvers which are appropriate with respect to model type, underlying structures and dimensionalities of the current model instance. This list is presented to the user for selecting a single item. Supporting the user in this selection will be implemented by utilizing similar ideas as those developed by Schittkowski [43], [44] for nonlinear programming solvers.

Analyze facilities: In the present version of SLP-IOR the reliability of a two-stage solution can be computed. The implementation of the other ideas discussed above will be done in the near future.

Rules. As discussed above, building, solving and analyzing SLP models presupposes quite a lot of knowledge of a technical nature. This knowledge is implemented in SLP-IOR in the form of polymorphic Boolean functions within the class-hierarchies of models and solvers. The construction is analogous to the artificial intelligence concept of frames containing rules.

Some HW/SW characteristics of the present version are as follows.

Programming language:	Borland Pascal 7.0
Computer:	IBM PC/AT 386 (or higher)
Storage:	at least 4MB
Arithmetic coprocessor:	required
Operating system :	DOS 5.0

For portability reasons we plan to develop a C++ version too.

References

[1] M. Avriel and A. C. Williams. The value of information and stochastic programming. *Operations Research*, 18:947–954, 1970.

[2] A. Bharadwaj, A. Lo. Choobineh, and B. Shetty. Model management systems: A survey. *Annals of Operations Research*, 38:17–67, 1992.

[3] J. R. Birge. The value of stochastic solution in stochastic linear programs with fixed recourse. *Mathematical Programming*, 24:314–325, 1982.

[4] J. R. Birge, M. A. H. Dempster, H. Gassmann, E. Gunn, A. J. King, and S. W. Wallace. A standard input format for multiperiod stochastic linear programs. Working Paper WP-87-118, IIASA, 1987.

[5] J. Bisschop and A. Meeraus. On the development of a general algebraic modeling system in a strategic planning environment. *Mathematical Programming Study*, 20:1–29, 1982.

[6] A. Brooke, D. Kendrick, and A. Meeraus. *GAMS. A User's Guide*. The Scientific Press, 1988.

[7] G. B. Dantzig and P. W. Glynn. Parallel processors for planning under uncertainty. Technical Report SOL 88-8R, Department for Operations Research, Stanford University, 1989.

[8] D. R. Dolk. Model management systems for operations research: A prospectus. In G. Mitra, editor, *Mathematical Methods for Decision Support*, pages 347–373. Springer Verlag, 1988.

[9] Y. Ermoliev. Stochastic quasigradient methods and their application to systems optimization. *Stochastics*, 9:1–36, 1983.

[10] K. Frauendorfer and P. Kall. A solution method for SLP recourse problems with arbitrary multivariate distributions - the independent case. *Problems of Control and Information Theory*, 17:177–205, 1988.

[11] K. Frauendorfer. Solving SLP recourse problems with arbitrary multivariate distributions - the dependent case. *Mathematics of Operations Research*, 13:377–394, 1988.

[12] A. Gaivoronski. Stochastic quasigradient methods and their implementation. In Y. Ermoliev and R.J-B. Wets, editors, *Numerical Techniques for Stochastic Optimization*, pages 313–351. Springer Verlag, 1988.

[13] H. I. Gassmann and A. M. Ireland. Scenario formulation in an algebraic modelling language. Working Paper WP-92-7, School of Business Administration, Dalhousie University, Halifax, 1992.

[14] A. M. Geoffrion. An introduction to structured modeling. *Management Science*, 33:547–588, 1987.

[15] A. M. Geoffrion. Computer-based modeling environments. *European Journal on Operational Research*, 41:33–43, 1989.

[16] A. M. Geoffrion. FW/SM: A prototype structured modeling environment. *Management Science*, 37:1513–1538, 1991.

[17] H. J. Greenberg. RANDMOD: A system for randomizing modifications to an instance of a linear program. *ORSA Journal on Computing*, 3:173–175, 1991.

[18] H. J. Greenberg. *Modeling by object-driven linear elemental relations: A user's guide to MODLER*. Kluwer Academic Publishers, 1993.

[19] H. J. Greenberg. *A computer-assisted analysis system for mathematical programming models and solutions: A user's guide to ANALYZE*. Kluwer Academic Publishers, 1993.

[20] H. J. Greenberg and F. H. Murphy. A comparison of mathematical programming modeling systems. *Annals of Operations Research*, 38:177–238, 1992.

[21] D. B. Hausch and W. T. Ziemba. Bounds on the value of information in uncertain decision problems II. *Stochastics*, 10:181–217, 1983.

[22] J. L. Higle and S. Sen. Stochastic decomposition: An algorithm for two-stage linear programs with recourse. *Mathematics of Operations Research*, 16:650–669, 1991.

[23] P. Kall. Approximations to stochastic programs with complete fixed recourse. *Numerische Mathematik*, 22:333–339, 1974.

[24] P. Kall. Computational methods for solving two-stage stochastic linear programming problems. *Zeitschrift für angewandte Mathematik und Physik*, 30:261–271, 1979.

[25] P. Kall. *Stochastic linear programming*. Springer Verlag, 1976.

[26] P. Kall and D. Stoyan. Solving stochastic programming problems with recourse including error bounds. *Mathematische Operationsforschung und Statistik, Ser. Optimization*, 13:431–447, 1982.

[27] P. Kall. Towards computing the expected value of perfect information. In M. J. Beckmann, W. Eichhorn, and W. Krelle, editors, *Mathematische Systeme in der Ökonomie*, pages 277–287. Athenäum, 1983.

[28] P. Kall. Stochastic programming with recourse: Upper bounds and moment problems – a review. In J. Guddat, B. Bank, H. Hollatz, P. Kall, D. Klatte, B. Kummer, K. Lommatzsch, K. Tammer, M. Vlach, and K. Zimmermann, editors, *Advances in Mathematical Optimization (Dedicated to Prof. Dr.Dr.hc. F. Nožička)*, pages 86–103. Akademie-Verlag, Berlin, 1988.

[29] P. Kall, A. Ruszczynski, and K. Frauendorfer. Approximation techniques in stochastic programming. In Y. Ermoliev and R.J-B. Wets, editors, *Numerical Techniques for Stochastic Optimization*, pages 33–64. Springer Verlag, 1988.

[30] P. Kall and J. Mayer. SLP-IOR: A model management system for stochastic linear programming — system design —. In A.J.M. Beulens and H.-J. Sebastian, editors, *Optimization-Based Computer-Aided Modelling and Design*, pages 139–157. Springer Verlag, 1992.

[31] P. Kall and J. Mayer. A model management system for stochastic linear programming. In P. Kall, editor, *System Modelling and Optimization*, pages 580–587. Springer Verlag, 1992.

[32] P. Kall and J. Mayer. SLP-IOR: On the design of a workbench for testing SLP codes. Preprint, IOR, University of Zurich, 1992.

[33] P. Kall and J. Mayer. Model management for stochastic linear programming. *Mathematical Programming, Series B*, 1994. Submitted for publication.

[34] E. Keller. GENSLP: A program for generating input for stochastic linear programs with complete fixed recourse. Manuscript, IOR, University of Zurich, 1984.

[35] A. J. King. Stochastic programming problems: Examples from the literature. In Y. Ermoliev and R.J-B. Wets, editors, *Numerical Techniques for Stochastic Optimization*, pages 543–567. Springer Verlag, 1988.

[36] K. Marti. Konvexitätsaussagen zum linearen stochastischen optimierungsproblem. *Zeitschrift für Wahrscheinlichkeitstheorie und verw. Geb.*, 18:159–166, 1971.

[37] J. Mayer. Probabilistic constrained programming: A reduced gradient algorithm implemented on PC. Working Paper WP-88-39, IIASA, 1988.

[38] J. Mayer. Computational techniques for probabilistic constrained optimization problems. In K. Marti, editor, *Stochastic Optimization: Numerical Methods and Technical Applications*, pages 141–164. Springer Verlag, 1992.

[39] C. van de Panne and W. Popp. Minimum cost cattle feed under probabilistic problem constraint. *Management Science*, 9:405–430, 1963.

[40] A. Prékopa. Logarithmic concave measures and related topics. In M. A. H. Dempster, editor, *Stochastic Programming*, pages 63–82. Academic Press, 1980.

[41] A. Prékopa. Numerical solution of probabilistic constrained programming problems. In Y. Ermoliev and R.J-B. Wets, editors, *Numerical Techniques for Stochastic Optimization*, pages 123–139. Springer Verlag, 1988.

[42] A. Ruszczynski. A regularized decomposition method for minimizing a sum of polyhedral functions. *Mathematical Programming*, 35:309–333, 1986.

[43] K. Schittkowski. EMP: An expert system for mathematical programming. Technical report, Mathematisches Institut, Universitat Bayreuth, 1987.

[44] K. Schittkowski. Some experiments on heuristic code selection versus numerical performance in nonlinear programming. *European Journal on Operational Research*, 65:292–304, 1993.

[45] B. Strazicky. On an algorithm for solution of the two-stage stochastic programming problem. *Methods of Operations Research*, XIX:142–156, 1974.

[46] R. H. Sprague and E. D. Carlson. *Building effective decision support systems*. Prentice-Hall Publ. Co., 1982.

[47] T. Szántai. A computer code for solution of probabilistic-constrained stochastic programming problems. In Y. Ermoliev and R.J-B. Wets, editors, *Numerical Techniques for Stochastic Optimization*, pages 229–235. Springer Verlag, 1988.

[48] R Van Slyke and R. J-B. Wets. L-shaped linear program with applications to optimal control and stochastic linear programs. *SIAM J. Appl. Math.*, 17:638–663, 1969.

[49] S. W. Wallace and R. J-B. Wets. Preprocessing in stochastic programming: The case of uncapacitated networks. *ORSA Journal on Computing*, 1:252–270, 1989.

[50] S. W. Wallace and R. J-B. Wets. Preprocessing in stochastic programming: The case of linear programs. *ORSA Journal on Computing*, 4:45–59, 1992.

[51] R. J-B. Wets. Stochastic programming: Solution techniques and approximation schemes. In A. Bachem, M. Grötschel, and B. Korte, editors, *Mathematical programming: The state of the art*, pages 566–603. Springer Verlag, 1983.

[52] R. J-B. Wets. Stochastic programming. In G. L. et al. Nemhauser, editor, *Handbooks in OR and MS, Vol. 1*, pages 573–629. Elsevier, 1989.

Structural Design via Evolution Strategies

Bernd Kost[1]

Department of Bionics & Evolution Technique
Technical University of Berlin

Abstract. The structure of technical systems can have strong influences on their function and costs. However, till now there are no general rules for the structural layout of systems. It is desirable to have procedures which are as much as possible free from external operations and able to arrange the design elements appropriately to optimal structures.

The evolution strategy proves to be successfull to control the design process of several different technical systems, e.g. the structure of neural networks, digital filters and minimum weight trusses.

Keywords. structural optimization, topology optimization, evolution strategy

1 Introduction

In all engineering disciplines design processes can be interpreted as optimization procedures. Starting with an initial model of the problem to be treated the design engineer analyses a sequence of model modifications unless the result is optimal or acceptable.

The purpose of computational optimization methods is to assist the design engineer in his effort to find the optimal design. In the last decades a lot of work has been devoted to structural optimization to develop methods which are able to improve and to optimize technical structures, e.g. [8-10, 29, 32, 37, 39, 46, 48]. With the increase of computational power a growing interest in topology optimization, where the structural model itself is sought, can be noticed, e.g. [1-4, 23, 25-28, 47]. That is due to the fact that small modifications in the topology can lead to large economical advantages.

The biologically inspired evolutionary approaches for function optimization, like the genetic algorithms (see HOLLAND [22] and DE JONG[5]) and the

---

[1]Now with the Fachhochschule Hamburg, Fachbereich Maschinenbau und Chemieingenieurwesen, Berliner Tor 21, D - 20 099 Hamburg, Germany

evolution strategies (see RECHENBERG [40-44] and SCHWEFEL [50, 51]), have been shown to be an interesting alternative to conventional optimization methods and have been successfully used for a large number of applications. In the last years structural optimization as a field of application of such strategies has gained some interest [11-15, 24, 30, 31, 34, 35]. Besides the general capability to solve structural optimization problems it should be mentioned that the inherent parallel evolutionary approaches lead to algorithms which are well suited for parallel computers.

The present paper concerns the optimal design of network-like systems on the basis of evolution strategies and gives a short survey of some investigations which have been carried out at the Department of Bionics & Evolution Techniques of the Technical University of Berlin.

2 A Short Review of Evolution Strategies

Evolution strategies lean upon the Darwinian theory of the evolution of species, using the biological principles of inheritance, mutation, recombination and selection to obtain optimal solutions. Against the background of stochastic optimization an evolution strategy can be interpreted as a natural optimization method.

2.1 The Mutation-Selection Strategy

The evolution strategy has been established by RECHENBERG in the sixties nearly thirty years ago [40]. The very first problem he solved was the experimental drag minimization of a folded plate. Using only the basic mechanisms of biological evolution, mutation and selection, a flat configuration could be obtained after few more than 300 mutations. Because the solution of the plate problem was known beforehand this experiment could only demonstrate that his method really works. More convincing experiments with unknown solutions followed, e.g. the S-shaped plate or the optimal shape of a 90° - pipe coupling, which is not a quarter of a circle as one might believe [41, 42], and showed for the first time the capability of the evolution strategy to achieve unexpected solutions.

A further historical experiment was the evolution of a two-phase jet nozzle which was carried out by SCHWEFEL [49] in cooperation with industry. The experiment will be specified here some more in detail because it already had some features of the algorithm described below.

The problem which could not be solved analytically was to obtain the internal nozzle shape for a fluid-steam mixture with maximum thrust. Based on an idea of SCHWEFEL the experimental set-up of the nozzle was given by a series of ring segments with conical borings. As initial or parent configuration served a Laval-type nozzle. The diameters of the nozzle were mutated and from time to time the length of the nozzle could be changed by adding or dropping a ring segment. Within the scope of this text this could already be interpreted as a certain kind of discrete structural mutations.

Figure 1 shows the whole process of the evolution experiment which produced a completly unexpected optimal shape. As it was said above, this configuration could not be predicted by analytical methods, all one could do was to explain afterwards the strange internal shape of the nozzle and the fluidmechanical role of the single chambers. It is worth mentioning that the efficiency could be raised from 55% to nearly 80% (see also [43]).

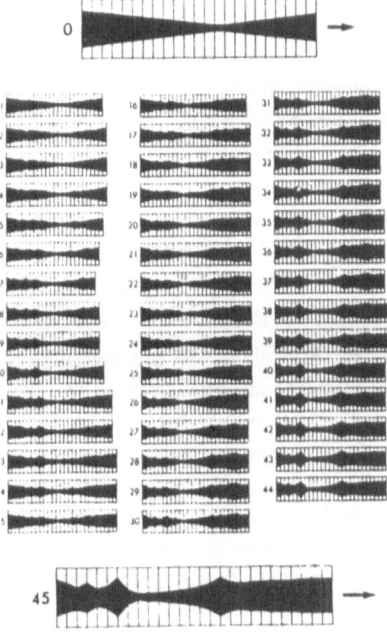

Figure 1. Development of a two-phase jet nozzle from the initial form 0 to the optimal form 45

Considering a minimization problem the formal desciption of the simplest evolution strategy is as follows:

☐ mutation

$$x_d^{(g)} = x_p^{(g)} + \delta z^{(g)} \tag{1}$$

☐ selection

$$\textbf{if} \quad Q(x_d^{(g)}) \leq Q(x_p^{(g)}) \quad \textbf{then}$$

$$x_p^{(g+1)} = x_d^{(g)}$$

else

$$x_p^{(g+1)} = x_p^{(g)}$$

end if

where

- $x_p \in \mathbf{R}^n$ is the parent vector[2]
- $x_d \in \mathbf{R}^n$ is the descendant vector
- $z \in \mathbf{R}^n$ is a normalized random vector, which represents the mutation
- $\delta \in \mathbf{R}$ is the stepsize
- $g \in \mathbf{N}$ is the number of generations
- $Q \in \mathbf{R}$ is a quality measurement, the objective function
- $n \in \mathbf{N}$ is the number of variables.

Against the biological background of the algorithm equ. (1) can be interpreted as the replication and mutation mechanisms. The replication is the information inheritance from parent to offspring. From nature one knows that this copying process does not work perfectly and so one has to add some random noise to model the mutation. From the biological point of view normally distributed random numbers[3] seemed reasonable to serve as mutations because the descendant vector should be with higher probability in the vicinity of the parent vector (i.e. the descendant has a certain similarity to the parent, like in nature).

The iteration process is being governed by an exogenously controlled stepsize. On the basis of two rather different model functions RECHENBERG

---

[2]It should be noticed that evolution strategies can deal with discrete variables too, see for example [17].

[3]In the discrete case appropiate mutation operators have to be formulated.

derived the socalled 1/5-success rule for the stepsize control, for details see [41].

2.2 The Population Model of the Evolution Strategy

Although the mutation-selection strategy works quite well for certain problems there are some points which are not very satisfying, e.g. a parent could survive forever or the exogenous stepsize control.

In order to include some more features of the biological evolution in the evolution strategy a population model was introduced [41, 50, 51].

After SCHWEFEL [50, 51] an evolution strategy can be described formally in the following notation:

$$(\mu / \rho +, \lambda) \text{ - evolution strategy}$$

where

- μ is the number of parents
- ρ is the recombination number
- λ is the number of descendants
- $+$ or $,$ denotes the type of selection.

Nowadays the general formulation of the population model describes a class of evolution strategies with mutative stepsize control and can be given in the following way:

❏ mutation

$$
\begin{aligned}
x_{d1}^{(g)} &= x_{pr}^{(g)} + \delta_{d1}^{(g)} z_1 \qquad \text{with} \qquad \delta_{d1}^{(g)} = \delta_{pr}^{(g)} \xi_1 \\
x_{d2}^{(g)} &= x_{pr}^{(g)} + \delta_{d2}^{(g)} z_2 \qquad\qquad\qquad \delta_{d2}^{(g)} = \delta_{pr}^{(g)} \xi_2 \\
&\ \ \vdots \qquad\qquad\qquad\qquad\qquad\qquad \vdots \\
x_{d\lambda}^{(g)} &= x_{pr}^{(g)} + \delta_{d\lambda}^{(g)} z_\lambda \qquad\qquad\quad \delta_{d\lambda}^{(g)} = \delta_{pr}^{(g)} \xi_\lambda
\end{aligned}
$$

where

$$r = \text{rand}\{ 1, \ 2, \ \cdots, \ \mu \}$$

denotes a uniformly distributed random number in the range from 1 to μ and the stepsize factors ξ_j $(j = 1, \cdots, \lambda)$ are multiplicative stepsize mutations, which make a stepsize adaptation possible.

That means the mutations of the individuals will be controlled by changeable stepsizes which will be passed to the offspring similar to the object variables. On the basis of two rather different model functions RECHENBERG [41] found analytically very similar functional relations between the rate of progress in optimization and the stepsize for both model functions. He concluded that an evolution window exists and during the optimization process an optimal stepsize will be found. That means, due to the stepsize inheritance, the stepsize mutations and together with the selection principle the evolution strategy is capable of a stepsize adaptation during the optimization itself. **Figure 2** shows the general result of the theory: the rate of progress versus the mutation stepsize. The figure indicates that evolution takes place only within a very narrow band of the mutation stepsize.

It is worth mentioning that the stepsize adaptation is not only limited to general stepsizes as in the description above. Individual stepsizes for each component of the vector z can be used to scale the problems [50, 51]. At present, this procedure is being investigated anew [38]. The results show that the concept is very promising, but this is beyond the scope of this paper.

❑ selection

"+" - strategies

$$S : \left\{x_{p1}^{(g)}, \ldots, x_{p\mu}^{(g)}, x_{d1}^{(g)}, \ldots, x_{d\lambda}^{(g)}\right\} \xrightarrow{\text{the } \mu \text{ best}} \left\{x_{p1}^{(g+1)}, \ldots, x_{p\mu}^{(g+1)}\right\}$$

"," - strategies

$$S : \left\{x_{d1}^{(g)}, \ldots, x_{d\lambda}^{(g)}\right\} \xrightarrow{\text{the } \mu \text{ best}} \left\{x_{p1}^{(g+1)}, \ldots, x_{p\mu}^{(g+1)}\right\}$$

Beside a lot of variants two main selection operators S are commonly used. In the socalled "+" - strategies as selection unit serves a set of $\mu + \lambda$ individuals, the population of the μ parents and the new created λ offsprings. The μ individuals with the best quality values (together with the associated stepsizes) will be selected to be parents of the next generation. In the "," - strategies the μ best individuals will be taken from the set of λ offsprings only. Using this type of strategies it is possible to leave local optima.

In the mathematical description above a further biological mechanism can be easily included for a better modelling of natural evolution, the recombination or crossover. The recombination operator simulates the sexual reproduction of individuals in evolution strategies. Assuming a set of μ vectors is given, with the intention of creating new offsprings a certain number ρ of vectors $(\rho \le \mu)$ will be chosen at random to interchange their information.

In the genetic algorithms [5, 22], the counterpart to the evolution strategies, the recombination or crossover is the main operator, whereas in the evolution strategies the attention has been concentrated on the mutations. Nevertheless, both operaters are useful mechanisms in the optimization process.

Using the notation mentioned above the simple mutation-selection strategy can be identified as a $(1/1+1)$-strategy.

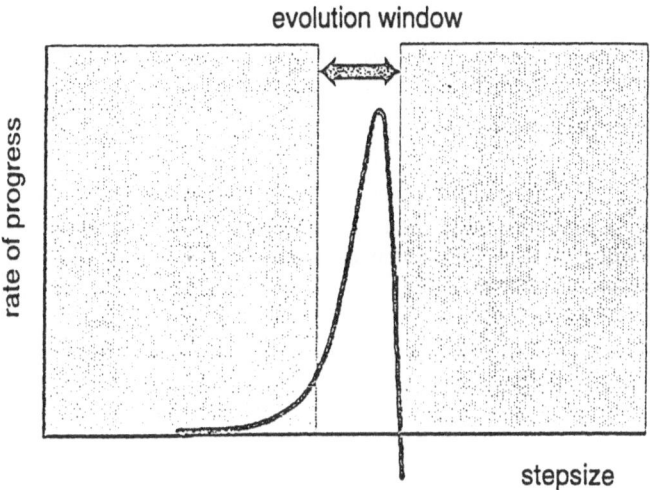

Figure 2. Evolution window

2.3 The Multi Population Model of the Evolution Strategy

The classical evolution strategies are based on the competition among the individuals within one population. Multi population evolution strategies imitate a higher level of biological evolution, assuming the competition takes place not only between individuals but also between parallel operating populations. The general concept was suggested for the first time by SC'IWEFEL [50] in the year 1975, specified by RECHENBERG [44] in 1978, but never used successfully till 1989. LOHMANN [34] introduced formally the isolation principle in the general notation which was already described in the early work of SCHWEFEL [50], and implemented the algorithm. By that a further biological mechanism, the isolation time, was added to the evolution strategy as a new strategy parameter.

Now the general formulation for a class of complex algorithms can be given as a formal extension of the classical notation:

$$\left[\; \mu' / \rho' \; +, \; \lambda' (\mu / \rho \; +, \; \lambda)^{\gamma} \; \right] \text{ - evolution strategy}$$

where

- μ' is the number of parental populations
- ρ' is the recombination number on population level
- λ' is the number of offspring populations
- γ is a period of isolation (time or generations).

How this approach can be extended to multi level strategies in order to adapt further strategy parameters was shown by HERDY [18, 19].

2.4 Structure Evolution

A structure optimization problem can often be formulated as a mixed discrete-continuous problem. Using evolution strategy this case can be described formally by a multi population strategy. LOHMANN [34] interpreted the set of populations as different structures and introduced beside the parameter variation of the continuous variables the discrete mutations on structural level. The whole procedure, called structure evolution, can be identified as a two-stage method, see **Figure 3**.

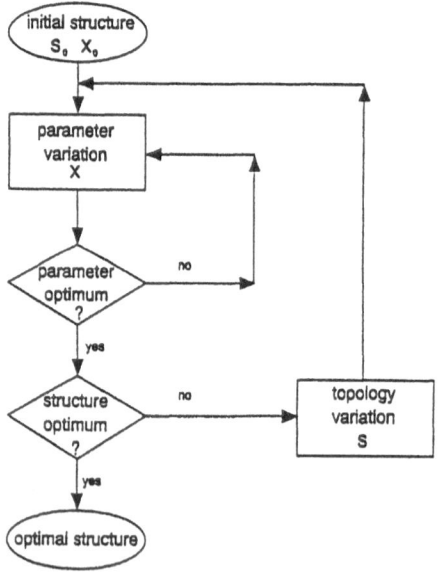

Figure 3. The two-stage method

The evolutionary procedure on the structural level is similar to the usual evolution strategy. The structure evolution starts with an initial set of μ' structure vectors (in the figure denoted by s_0) which components are binary. The quality of each structure depends on the appropriate adjustment of the associated parameter vector x with real components. Starting with an initial set of μ parameter vectors x_0 in the inner loop the parameter variations can be done as in the classical evolution strategy for a certain isolation time. After the parameter adjustment has been finished the different structures can be assigned quality values which are the basis of the following selection on structural level. Then in the outer loop topology mutations take place and the cycle continues until the optimal structure or a sufficient result is found.

It should be mentioned that besides the different mutation procedures on both levels, which are evident, the quality measurements and the selection operators may be different too.

3   Applications to Structural Design Problems

The two-stage structure evolution procedur was successfully used for the first time in the field of visual systems. LOHMANN [34] developed the structure of local feature detection filters in image processing. Since then the method has been employed in a wide range of applications. In [35] LOHMANN reported on the evolution of "framework" topologies. However, it should be noticed that because of the absence of forces the problem was not really a framework or truss optimization but a minimization of the total length of "bar elements". Nevertheless, his example is in addition to the local filter development a further good demonstration of the capability of structure evolution. In the following some other applications , as artificial neural networks, digital filters and trusses, will be described.

Using evolution techniques in structural optimization in any case some heuristical work has to be done because the discrete mutation operators depend on the problem to be solved and cannot be established in general. In **Figure 4** some simple mutation procedures are shown which can be used for the topology variation of network-like systems.

3.1 Neural Networks

The motivation for studying neural computation is due to the fact that the human brain is superior to computers at many tasks. The main inspirations for creating artificial neural networks come from neuroscience.

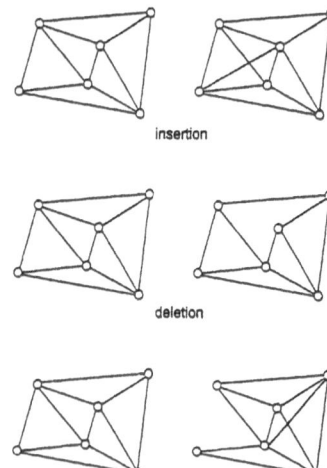

insertion

deletion

shifting

Figure 4. The mutation prodedures

Artificial neural networks can be considered as information processing systems. They have to be trained to respond on the input vectors with appropriate output vectors. In **Figure 5** an example of an artificial neural network is shown which consists of the input layer, the output layer and one hidden layer. The black dots represent the artificial neurons or processor elements, that means the nonlinear transfer functions. The connections between the neurons are the artificial synapses. Their weight coefficients have to be adjusted in the training phase of the network. This is a minimization problem, the minimization of the training error. With respect to the structure evolution the minimization of the training error has to be done in the inner loop and for the outer loop an appropriate design criterium, e.g. design of minimal networks, has to be chosen.

First tests in structuring neural networks with evolution strategies can be found in [45], e.g. TRINT developed the optimal network structure for logical problems (XOR, parity-problems) and OSTERMEIER used the Fourier transform as a test problem. Certainly, the Fourier transform is not a good task for a neural network because of the linear property but the structuring experiment has a known solution which can be used as a reference structure. **Figure 6** shows the fast Fourier transform structure as solution of the structural optimization which has been started with the structure of the discrete Fourier transform.

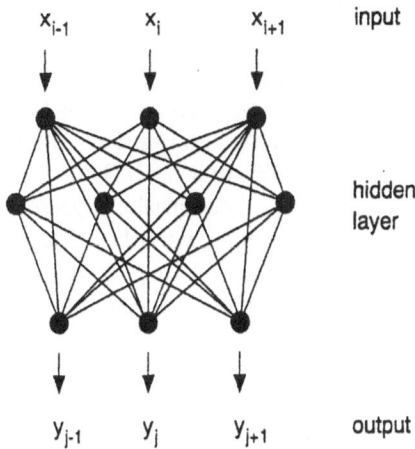

Figure 5. An artificial neural network

Initial Network

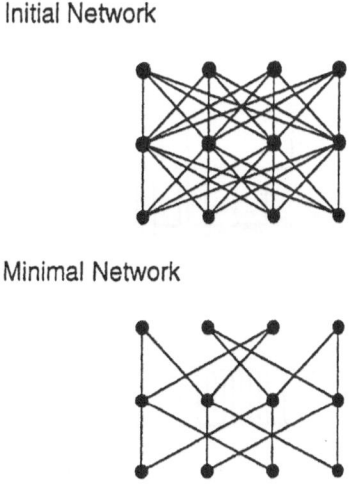

Minimal Network

Figure 6. The Fourier transform network development; initial network: discrete Fourier transform, minimal network: fast Fourier transform, reprinted from [45] by permission of A. OSTERMEIER

3.2 Digital Filters

The evolutionary design of non-trivial digital filters is a further structure optimization experiment which was carried out at the Department of Bionics & Evolution Techniques in collaboration with the Institute of Communication Science of the Technical University of Berlin [11].

A recursive digital filter with infinite impulse response can be described by a set of recursive and non-recursive real-valued filter coefficients and an associated filter structure.

The filter design based on structure evolution works in the usual way and appropriate structural mutation procedures like the insertion or deletion of coefficients have to be formulated similar to Figure 4. On the first level the filter coefficients have to be determined by minimizing the deviation from a given impulse response. Then the structure can be assessed and on the second level the filter length, that means the number of non-zero coefficients, can be minimized. That leads to filters with maximum speed of computation. In **Figure 7** an example is shown.

Random Initial Filter

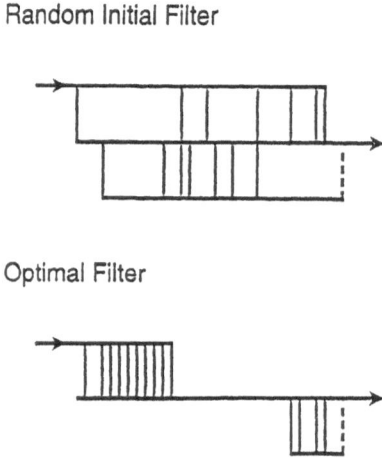

Optimal Filter

Figure 7. The digital filter design, arrows denote the signal flow, the upper part of the filter is the non-recursive branch and the lower part the recursive branch, reprinted from [45] by permission of T. GÖRNE

3.3 Trusses

Truss systems are the classical field of application in structural optimization. In the last decades most of the work done on structural optimization has been devoted to trusses, e.g. see the references in [29, 46]. The reason might be that a truss is a relatively simple structure but the related structural optimization problems are not trivial at all. **Figure 8** displays the three most important branches of optimum structural design with regard to trusses, the optimization of cross-sections, the shape optimization and the topology optimization. The major part of the work has been done on the cross-section optimization of the truss members and the minor part is related to topology optimization.

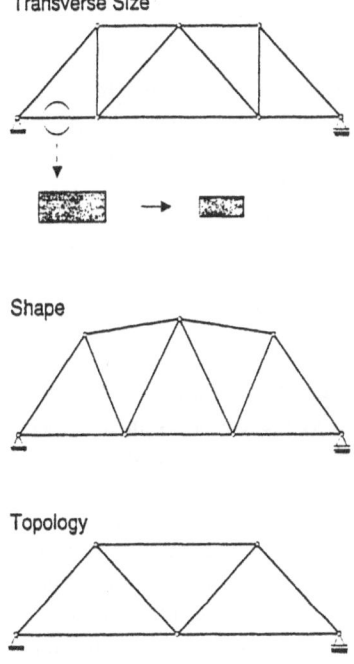

Figure 8. Hierarchy of structure optimization

The contribution of the evolution strategy to structural optimization was limited in the past to shape optimization problems [7, 20, 21, 33] where configurational variables like node coordinates have been sought.

First experiments indicate that the simultaneous optimization of geometrical and topological variables by means of the structure evolution is apparently a difficult task. At the beginning it seems to be more promising to separate the problem in two subproblems and to use each stage of the structure evolution alone.

As it has been mentioned above, the general capability of the evolution strategy to solve the subproblem shape optimization was shown earlier and shall not be discussed further here.

With respect to the topology optimization only the outer loop of the structure evolution was realized. For the test the simple sum of the bar weights, representing the structure weight W, served as the objective function:

$$W = \omega \sum_{i=1}^{m} A_i(s) \, l_i$$

where

- $W \in \mathbf{R}$ is the the structural weight
- $\omega \in \mathbf{R}$ is the weight per unit volume
- $A_i \in \mathbf{R}$ is the transverse size of member i
- $l_i \in \mathbf{R}$ is the length of member i
- $s \in \mathbf{S}^m$ is the structure vector
- $m \in \mathbf{N}$ is the maximum number of members
- $\mathbf{S}^m = \{0, 1\} \times \{0, 1\} \times \cdots \times \{0, 1\}$
  is the stock of structures.

The structural analysis to get the bar forces according to the external load was carried out with the finite element method based on unit cross-sections. Afterwards the new cross-sections A_i of the members were calculated using a prescribed maximal stress value and the weight W of each structural mutant could be determined.

The structural mutations have been carried out according to Figure 4, i.e. between parent and offspring structures Hamming distances one and two have been realized. During the optimization process it has to be assured that the structures remain non-kinematic. With a view to evolution strategies a kinematic structure can be interpreted as a lethal mutation.

In the following some results of the topological layout of trusses with the evolution strategy will be given. All examples are based on a grid with 30 nodes. In the case of complete connection of all nodes, that means each node is linked with all the others, the maximum number of members is m = 435. With only 30 nodes the number of possible structures 2^{435} is immense, approximately 10^{131}, and complete enumeration certainly would not be a good choice.

Figure 9 displays some approximate solutions of MICHELL structures. A MICHELL structure can be viewed as a borderline case of a truss, a quasi-continuous orthogonal network structure with infinitesimal bar lengths. It is known to be a least weight structure for a given external load [16, 36]. Starting the evolution strategy from two different initial points in the search space, i.e. from an extremly statically indeterminate or redundant structure and from a statically determinate truss, slightly different final structures have been found. In the figure a comparison with the solution presented by HÖFLER [21] is given. It should be noted that he used linear programming [6, 16] to get the set of maximal strained bars and selected the shown structure without doing a real topology optimization.

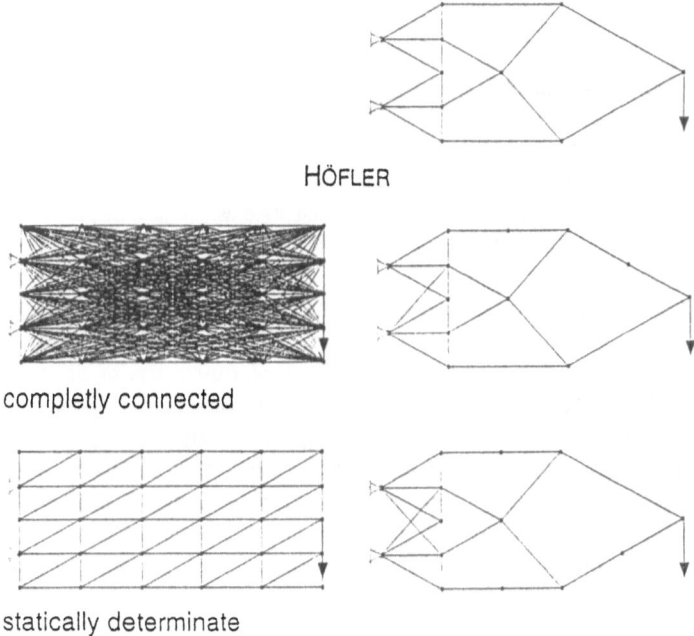

HÖFLER

completly connected

statically determinate

Figure 9. Comparison of approximated MICHELL structures, arrows denote external loads, left: initial structures, right: final structures, the completely connected initial structure has been analysed with a $(1/1 + 1(1/1,1))^1$-strategy, the statically determinate initial structure has been analysed with a $(10/1 + 20(1/1,1))^1$-strategy. With regard to the remaining single joints, it should be mentioned that there are a few zero-force bars, not shown in the figure

Although the three final structures are structurally different in the vicinity of the supports, all structures have the same weight, i.e. a problem with multiple optimal solutions. In **Figure 10** the evolution of the weight for this example is drawn.

Further results of topology optimization with evolution strategy are given in Figure 11 and 12. **Figure 11** shows a structure which looks like one half of a roof construction. Starting from the completly connected initial structure and from a statically determinate structure different solutions have been found. In both cases the final weight was the same. The optimal topology displayed in **Figure 12** is some kind of suspension bridge. The shown result is based on the completly connected initial topology. Starting from a statically determinate initial structure, in analogy to the examples before, basically the same structure, with a few single joints different, has been found.

4 Final Remarks

Evolution strategies have been applied to a diverse range of optimization problems and proved their universality. It could be shown that in optimum design problems they are applicable too. The price which must be paid for this universality can be seen in Figure 10. Normally, the evolution strategies need a great deal of computational effort compared to conventional optimization methods. In this context should not be forgotten that evolution strategies need only a short implementation time due to their simple algorithmic structure.

With respect to the topology optimization of trusses one of the today's favourite methods is the homogenization method [3, 47] based on a microstructural approach. It is a material distribution method using an artificial composite material made of substance and voids. Besides the evident difference in computation time between the homogenization method and the evolution strategy there is at least one important difference in favour of the latter. It is easy to include stability phenomena in the evolutionary topology design whereas with respect to the homogenization method it seems to be impossible.

A further common approach in structural optimization is to start from a highly connected initial structure and eliminate uneconomical links during the course of optimization. Little attention has been dedicated to methods which introduce new members into the structure. Concerning structural optimization problems evolution strategies, just the contrary, derive benefit from the interplay of adding and deleting structural elements, the structural mutations. Results achieved till now are promising and indicate that evolution strategies have the potential to be valuable tools for the structural design.

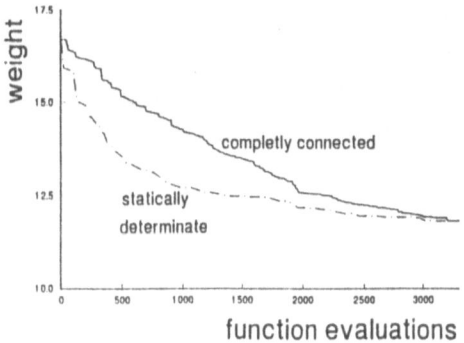

Figure 10. Weight development of the approximated MICHELL structures of Fig. 9

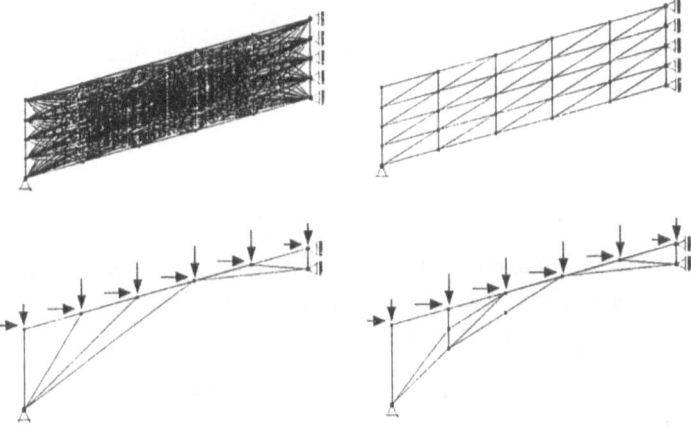

Figure 11. "Roof construction", top: initial structures, bottom: final structures, the completely connected initial structure has been analysed with a $(1/1+1(1/1,1)^1$-strategy, the statically determinate initial structure has been analysed with a $(6/1+12(1/1,1)^1$-strategy

88

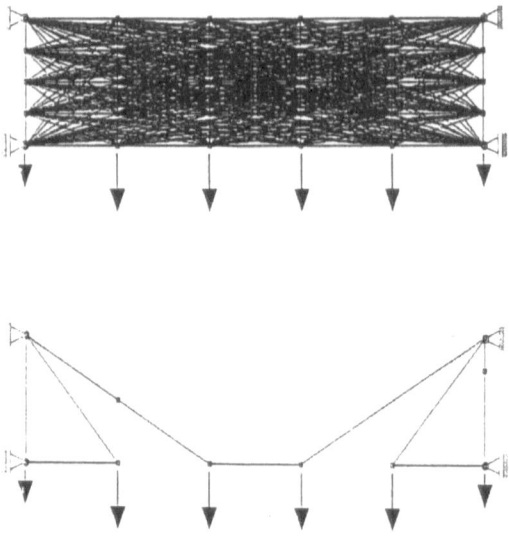

Figure 12. "Suspension bridge", top: initial structure, bottom: final structure, the structure has been analysed with a $(1/1 + 1(1/1,1)^1$-strategy

5 Acknowledgement

The author wishes to express his appreciation to Professor I. Rechenberg for giving him the chance to dedicate to topology optimization of trusses. He also gratefully acknowledges the support of his former colleagues.

6 References

[1] ACHTZIGER, W.; BENDSØE, M.; BEN-TAL, A.; ZOWE, J.: *Equivalent Displacement Based Formulations for Maximum Strength Truss Topology Design*, Report No. 338, University of Bayreuth, Mathematical Institute, 1991.

[2] ACHTZIGER, W.: *Truss Topology Design Under Multiple Loading*, Report No. 367, University of Bayreuth, Mathematical Institute, 1992.

[3] BENDSØE, M.P.; KIKUCHI, N.: *Generating optimal topologies in structural design using a homogenization method*, Computer Methods in Applied Mechanics and Engineering 71, 1988, 197 - 224.

[4] BEN-TAL, A.; KOČVARA, M.; ZOWE, J.: *Two Nonsmooth Approaches to Simultaneous Geometry and Topology Design of Trusses*, Report No. 383, University of Bayreuth, Mathematical Institute, 1992.

[5] DE JONG, K.A.: *An Analysis of the Behavior of a Class of Genetic Adaptive Systems*, Dissertation, University of Michigan, Ann Arbor, 1975.

[6] DORN, W.S.; GOMORY, R.E.; GREENBERG, H.J.: *Automatic design of optimal structures*, Journal de Mécanique 3, 1964, 25 - 52.

[7] EGGERT, H.: *Gewichts - und Durchbiegungsminimierung von Fachwerkträgern durch Anwendung der Evolutionsstrategie*, Diploma Thesis, Technical University of Berlin, 1990.

[8] ESCHENAUER, H.; POST, P.U.; BREMICKER, M.: *Einsatz der Optimierungsprozedur SAPOP zur Auslegung von Bauteilkomponenten*, Bauingenieur 63, 1988, 515 - 526.

[9] ESCHENAUER, H.: *Multidisciplinary Modeling and Optimization in Design Processes*, Zeitschrift für angewandte Mathematik und Mechanik ZAMM 72, 1992, T438 - 447.

[10] FLEURY, C.; BRAIBANT, V.: *Structural optimization: A new dual method using mixed variables*, International Journal for Numerical Methods in Engineering 23, 1986, 409 - 428.

[11] GÖRNE, T.; SCHNEIDER, M.: *Design of Digital Filters with Evolutionary Algorithms*, in: R.F. Albrecht, C.R. Reeves and N.C. Steele (Eds.): Proceedings of the International Conference on Artificial Neural Nets and Genetic Algorithms, Springer, Wien 1993, 368 - 374.

[12] GRIERSON, D.E.; PAK, W.H.: *Discrete optimal design using a genetic algorithm*, in: M.P. Bendsøe and C.A. Mota Soares (Eds.): Topology Design of Structures, Kluwer Academic Publishers, 1993, 89 - 102.

[13] HAJELA, P.: *Genetic Search - An Approach to the Nonconvex Optimization Problem*, AIAA Journal 28, 1990, 1205 - 1210.

[14] HAJELA, P.; LIN, C.-Y.: *Genetic search strategies in multicriterion optimal design*, Structural Optimization 4, 1992, 99 - 107.

[15] HAJELA, P.; LEE, E.; LIN, C.-Y.: *Genetic algorithms in structural topology optimization*, in: M.P. Bendsøe and C.A. Mota Soares (Eds.): Topology Design of Structures, Kluwer Academic Publishers, 1993, 117 - 133.

[16] HEMP, W.S.: *Studies in the theory of Michell structures*, Proc. Int. Congr. Appl. Mech., Munique, 1964, 621 - 628.

[17] HERDY, M.: *Application of the Evolutionsstrategie to Discrete Optimization Problems*, in: H.-P. Schwefel and R. Männer (Eds.): Lecture Notes in Computer Science, Vol. 496, Proceedings of the 1. Workshop Parallel Problem Solving from Nature, Springer, Berlin 1990, 188 - 192.

[18] HERDY, M.: *Reproductive Isolation as Strategy Parameter in Hierarchically Organized Evolution Strategies*, in: R. Männer and B. Manderick (Eds.): Proceedings of the Second Conference on Parallel Problem Solving from Nature, Elsevier Science Publishers B. V. (North-Holland), Amsterdam 1992, 207 - 217.

[19] HERDY, M.: *The Number of Offspring as Strategy Parameter in Hierarchically Organized Evolution Strategies*, sig bio newsletter 13, 1993, 2 - 7.

[20] HÖFLER, A.; LEYSSNER, U.; WIEDEMANN, J.: *Optimization of the layout of trusses combining strategies based on Michell's theorem and on the biological principles of evolution*, AGARD Conference Proceedings 123, Second Symposium on Structural Optimization, 1973, A1 - A8.

[21] HÖFLER, A.: *Formoptimierung von Leichtbaufachwerken durch Einsatz einer Evolutionsstrategie*, Dissertation, Technical University of Berlin, 1976.

[22] HOLLAND, J.H.: Adaptation in Natural and Artificial Systems, University of Michigan Press, Ann Arbor 1975.

[23] HÖRNLEIN, H.: Ein Algorithmus zur Strukturoptimierung von Fachwerkkonstruktionen, Diploma Thesis, University of Munique, 1979.

[24] JENKINS, W.M.: *Towards structural optimization via the genetic algorithm*, Computers & Structures 40, 1991, 1321 - 1327.

[25] KIRSCH, U.: *Optimal topologies of truss structures*, Computer Methods in Applied Mechanics and Engineering 72, 1989, 15 - 28.

[26] KIRSCH, U.: *Optimal topologies of structures*, Appl. Mech. Rev. 42, 1989, 223 - 239.

[27] KIRSCH, U.: *On the relationship between optimum structural topologies and geometries*, Structural Optimization 2, 1990, 39 - 45.

[28] KIRSCH, U.: *On singular topologies in optimum structural design*, Structural Optimization 2, 1990, 133 - 142.

[29] KIRSCH, U.: *Structural Optimization, Fundamentals and Applications*, Springer, Berlin 1993.

[30] KOST, B.: *Topology Optimization of Trusses by Evolution Techniques*, in: R.F. Sincovec et. al. (Eds.): Proceedings of the Sixth SIAM Conference on Parallel Processing for Scientific Computing, Norfolk/ Virginia 1993, 628 - 631.

[31] KOUMOUSIS, V.K.: *Lay-out and sizing design of civil engineering structures in accordance with the EuroCodes*, in: M.P. Bendsøe and C.A. Mota Soares (Eds.): Topology Design of Structures, Kluwer Academic Publishers, 1993, 103 - 116.

[32] LAWO, M.: Optimierung im Konstruktiven Ingenieurbau, Friedr. Vieweg & Sohn, Braunschweig 1987.

[33] LEYSSNER, U.: *Über den Einsatz Linearer Programmierung beim Entwurf optimaler Leichtbaustabwerke*, Dissertation, Technical University of Berlin, 1974.

[34] LOHMANN, R.: *Selforganization by Evolution Strategy in Visual Systems*, in: H.-M. Voigt, H. Mühlenbein and H.-P. Schwefel (Eds.): Evolution and Optimization '89, Akademie-Verlag, Berlin 1990, 61 - 68.

[35] LOHMANN, R.: *Structure Evolution and Incomplete Induction*, in: R. Männer and B. Manderick (Eds.): Proceedings of the Second Conference on Parallel Problem Solving from Nature, Elsevier Science Publishers B.V. (North-Holland), Amsterdam 1992, 175 - 185.

[36] MICHELL, A.G.M.: *The Limits of Economy of Material in Frame-structures*, Phil. Mag. S. 6, Vol. 8, No. 47, 1904, 589 - 597.

[37] MOE, J.: *Fundamentals of optimization*, Computers & Structures 4, 1974, 95 - 113.

[38] OSTERMEIER, A.; GAWELCZYK, A.; HANSEN, N.: *A Derandomized Approach to Self Adaptation of Evolution Strategies*, Technical Report TR-93-003, Technical University of Berlin, Department of Bionics & Evolution Techniques, July 1993.

[39] PEDERSEN, P.: *On the optimal layout of multi-purpose trusses*, Computers & Structures 2, 1972, 695 - 712.

[40] RECHENBERG, I.: *Cybernetic Solution Path of an Experimental Problem*, Royal Aircraft Establishment, Library Translation 1122, Farnborough 1965.

[41] RECHENBERG, I.: *Evolutionsstrategie - Optimierung technischer Systeme nach Prinzipien der biologischen Evolution*, Frommann Holzboog, Stuttgart 1973.

[42] RECHENBERG, I.: *Evolution Strategy and Human Decision Making*, in: H.P. Willumeit (Ed.): Human Decision Making and Manual Control, Elsevier Science Publishers B.V. (North-Holland), Amsterdam 1986, 349 - 359.

[43] RECHENBERG, I.: *Evolution Strategy: Nature's Way of Optimization*, in: H.W. Bergmann (Ed.): Lecture Notes in Engineering, Vol. 47, Optimization: Methods and Applications, Possibilities and Limitations, Springer, Berlin 1989.

[44] RECHENBERG, I.: *Evolutionsstrategien*, in: B. Schneider and U. Ranft (Eds.): Simulationsmethoden in der Medizin und Biologie, Springer, Berlin 1978, 84 - 114.

[45] RECHENBERG, I. et. al.: *Evolutionsstrategische Strukturbildung in neuronalen Systemen*, in: SALGON - Report 2.92-8.92, Technical University of Berlin, Department of Bionics & Evolution Techniques, 1992, to be published in Informatik - Fachberichte.

[46] ROZVANY, G.I.N.; MROZ, Z.: *Analytical Methods in Structural Optimization*, Applied Mechanics Reviews 30, 1977, 1461 - 1470.

[47] ROZVANY, G.I.N. et. al.: *Optimal topology of trusses or perforated deep beams with rotational restraints at both ends*, Structural Optimization 5, 1993, 268 - 270.

[48] SCHMIT, L.A.; FLEURY, C.: *Discrete-Continuous Variable Structural Synthesis Using Dual Methods*, AIAA Journal 18, 1980, 1515 - 1524.

[49] SCHWEFEL, H.-P.: *Experimentelle Optimierung einer Zweiphasendüse*, Ber. 35 AEG Forsch. Inst. Proj. MHD - Staustahlrohr, (Nr. 11034/68), Berlin 1968.

[50] SCHWEFEL, H.-P.: *Numerische Optimierung von Computer-Modellen mittels der Evolutionsstrategie*, Birkhäuser, Basel 1977, see also: *Evolutionsstrategie und numerische Optimierung*, Dissertation, Technical University of Berlin, 1975.

[51] SCHWEFEL, H.-P.: *Numerical Optimization of Computer Models*, Wiley & Sons, Chichester 1981.

On the Regularized Decomposition Method for Stochastic Programming Problems

Andrzej Ruszczyński

International Institute for Applied Systems Analysis, 2361 Laxenburg, Austria

Abstract. Application of the regularized decomposition method to large scale structured linear programming problems arising in stochastic programming is discussed. The method uses a quadratic regularizing term to stabilize the master but is still finitely convergent. Its practical performance is illustrated with numerical results for large real world problems.

Key words: Stochastic Programming, Decomposition.

1. Introduction

A large class of operations research problems lead to linear programming models of the form

$$\min \; c^T x$$
$$M x = r, \tag{1.1}$$
$$x^{\min} \leq x \leq x^{\max}.$$

In many applications, however, some coefficients of the resource/demand vector r or some entries of the matrix M are uncertain. They can be modeled as random variables $r(\omega)$ and $M(\omega)$ with $\omega \in \Omega$, where (Ω, B, P) is a probability space, but then the constraints in (1.1),

$$M(\omega)x = r(\omega), \quad \omega \in \Omega,$$

become prohibitively restrictive and usually impossible to satisfy for all realizations of the random entries.

One of modeling approaches to such a situation is the extension of (1.1) to a *stochastic programming problem with recourse*. We introduce recourse decisions (corrective activities) y_ω which can be taken after the realizations of the random entries are known, so as to fulfill the problem constraints.

To be more specific, let us split the constraints into the deterministic and the random parts:

$$M(\omega) = \left[\begin{array}{c} A \\ T_\omega \end{array} \right], \; r(\omega) = \left[\begin{array}{c} b \\ h_\omega \end{array} \right],$$

where A is an $m_1 \times n_1$ matrix and $b \in R^{m_1}$, T_ω is an $m_2 \times n_1$ random matrix and d_ω is an m_2-dimensional random vector over a probability space $(\Omega, \mathcal{B}, P)$. Finally, let $y_\omega \in R^{n_2}$ be the correction vector, $q \in R^{n_2}$ be the correction costs, and let W be an $m_2 \times n_2$ matrix describing our capabilities of corrections. The problem can be now reformulated as follows:

$$\min \left[c^T x + \int q^T y_\omega P(d\omega) \right]$$

subject to

$$Ax = b$$

$$T_\omega x + W y_\omega = d_\omega, \ \omega \in \Omega, \tag{1.2}$$

$$x^{\min} \leq x \leq x^{\max},$$

$$y^{\min} \leq y_\omega \leq y^{\max}, \ \omega \in \Omega.$$

In other words, we have to choose x so as to minimize the first stage cost $c^T x$ and the expected correction cost $\int q^T y_\omega P(d\omega)$, under the condition that the correction $W y_\omega$ compensates the shortage/surplus $d_\omega - T_\omega x$.

It should be stressed that the above reformulation is not just a formal trick to pose the problem correctly, but it reflects many real-life situations where shortage/surplus may occur, but they are connected with additional costs. There is also a broad class of application problems which already have the two-stage structure (1.2) with strategic decisions x and operating decisions y_ω - many examples can be found in [5] (see also section 5).

While there seems to be no doubt as to theoretical advantages of using models of form (1.2), their solution is much more difficult than for underlying deterministic models (1.1). For simplicity we shall focus our attention here on problems with discrete distributions; approximation of general distributions by discrete ones in stochastic programming is discussed in [2] and [6].

Let Ω be finite, $\Omega = \{1, 2, \ldots, L\}$, and let the realizations $(T_\omega, d_\omega), w \in \Omega$, be attained with probabilities $p_\omega > 0$, $(\sum_{\omega \in \Omega} p_\omega = 1)$. Then (1.2) can be rewritten as a large linear programming problem

$$\begin{array}{ccccccccl}
\min \ c^T x & + & p_1 q^T y_1 & + & p_2 q^T y_2 & + & \ldots & + & p_L q^T y_L \\
Ax & & & & & & & = & b, \\
T_1 x & + & W y_1 & & & & & = & d_1, \\
T_2 x & & & + & W y_2 & & & = & d_2, \\
\vdots & & & & & \ddots & & & \vdots \\
T_L x & & & & & & + \ W y_L & = & d_L,
\end{array} \tag{1.3}$$

$$x^{\min} \leq x \leq x^{\max},$$

$$y^{\min} \leq y_\omega \leq y^{\max}, \ \omega = 1, 2, \ldots, L.$$

There are several reasons for studying (1.3) thoroughly.

First of all, it is the remarkable size that makes this problem difficult from the practical point of view. Stochastic programs are usually extensions of deterministic linear models, so we should think of T having size of a constraint matrix in a typical linear program, and this size is multiplied in (1.3) by the number L of realizations of (T_ω, d_ω). For nontrivial problems with many independent random factors causing the stochasticity of the entries of T_ω and d_ω, L must be sufficiently large to reflect this randomness in our model. As a result, the dimension of (1.3) may go in hundreds of thousands.

Another difficulty is the possibility of ill-conditioning of (1.3). If the number of first stage activities x in the optimal basis exceeds $m_1 + m_2$, then similarity of the realizations T_ω, $\omega \in \Omega$, implies that the columns corresponding to these activities are close to being linearly dependent (for $T_\omega = T$ singularity would occur).

A very rich literature is devoted to solution methods for problems of form (1.3) or their duals. The first group of methods are variants of the simplex method which take advantage of the structure of the constraint matrix of (1.3) to construct compact representations of the basis inverse and to improve pivotal strategies (cf, e.g., [18]). The second group are linear decomposition methods coming down from the famous decomposition principle of Dantzig and Wolfe [4] (see [3, 16]). Finally, there is a possibility of reformulating (1.3) as a nonsmooth optimization problem and applying to it general non-differentiable optimization algorithms.

In the regularized decomposition (RD) method proposed for general large scale structured linear programming problems in [10] we combine the last two approaches: the problem is stated as a nonsmooth optimization problem, but for the purpose of solving it, we modify the general bundle method of [7] by taking full advantage of the problem's structure. As a result, a finitely convergent non-simplex method for large structured linear programs can be obtained. The method is described in sections 2,3 and 4 of the paper.

Finally, in section 5 we describe results of some computational tests showing that the method is capable of solving stochastic programs of considerable size.

2. The outline of the RD method

It can be readily seen that if x is fixed in (1.3) then minimization with respect to $y_1, y_2, .., y_L$ can be carried out separately. This leads to the following two-stage formulation

$$\min \left[F(x) = c^T x + \sum_{\omega \in \Omega} p_\omega f_\omega(x) \right] \qquad (2.1)$$

subject to

$$x \in X_0 = \{x : Ax = b, \ x^{\min} \le x \le x^{\max}\}, \qquad (2.2)$$

$$x \in X_\omega \ \text{for} \ \omega \in \Omega, \qquad (2.3)$$

with $f_\omega(x)$ defined as the optimal value of the second stage problem:

$$f_\omega(x) = \min \left\{ q^T y \mid Wy = d_\omega - T_\omega x, \ y^{\min} \le y \le y^{\max} \right\}, \qquad (2.4)$$

and with

$$X_\omega = \{x : f_\omega(x) < \infty\}.$$

We introduce condition (2.3) explicitly to the problem formulation, because we are going to use separate approximations of f_ω and of their domains X_ω. The functions f_ω are convex and polyhedral and the sets X_ω are convex closed polyhedra [17]. Thus (2.1)-(2.3) can, in principle, be solved by a method for piecewise linear problems or by a general algorithm for constrained nonsmooth optimization. Although the pieces of f_ω and the facets of X_ω are not given explicitly, it is possible to extract from the subproblems (2.4) at any x^k information about the piece $(\alpha_\omega^k, g_\omega^k)$ of f_ω active at x^k (an *objective cut*) or information about a constraint $(\bar{\alpha}_\omega^k, \bar{g}_\omega^k)$ defining X_ω and violated at x^k (a *feasibility cut*). The pieces (cuts) collected so far can be used to construct lower approximations of the functions f_ω,

$$f_\omega(x) \geq \bar{f}_\omega(x) = \max\{\alpha_\omega^j + (g_\omega^j)^T x, \ j \in J_\omega\},$$

and outer approximations of the sets X_ω

$$X_\omega \subset \bar{X}_\omega = \{x : \bar{\alpha}_\omega^j + (\bar{g}_\omega^j)^T x \leq 0, \ j \in \bar{J}_\omega\};$$

J_ω and $\bar{J}_\omega$ are some selected sets of cuts.

Crucial questions that arise in this respect are the following:

- how the successive points x^k are generated?

- how the cuts at x^k are constructed?

- how the approximations $\bar{f}_\omega$ and $\bar{X}_\omega$ are updated?

The most natural method for generating successive points x^k is to solve the linear approximation of (2.1)-(2.3) constructed on the basis of currently available information:

$$\min_{x \in \bar{X}} \left[\bar{F}(x) = c^T x + \sum_{\omega \in \Omega} p_\omega \bar{f}_\omega(x) \right], \tag{2.5}$$

where

$$\bar{X} = X_0 \cap \left(\bigcap_{\omega \in \Omega} \bar{X}_\omega \right).$$

After solving (2.5) we obtain cuts at the current solution, add them to the sets of cuts used previously, solve (2.5) again, etc..

Instead of constructing separate approximations for all f_ω in (2.5), we can also work with a piecewise linear approximation of their weighed sum $f(x) = \sum_{\omega \in \Omega} p_\omega f_\omega(x)$, as it was originally suggested in the L-shaped method of Van Slyke and Wets [16]. This would mean constructing objective cuts for f by averaging (with the weights p_ω) the objective cuts for f_ω.

The cutting-plane approach, however, has well-known drawbacks. Initial iterations are inefficient. The number of cuts increases after each iteration and there is no reliable rule for deleting them. The master problem (2.5) is sensitive with respect to changes in the set of cuts and its conditioning is getting worse when approaching the solution.

For these reasons in the RD method the linear master (2.5) is modified by adding to its objective a quadratic regularizing term:

$$\min_{x \in \hat{X}} \left[\eta(x) = \frac{1}{2\sigma} \|x - \xi^k\|^2 + c^T x + \sum_{\omega \in \Omega} p_\omega \bar{f}_\omega(x) \right]. \tag{2.6}$$

Here ξ^k is a certain regularizing point, and σ is a positive parameter. It turns out that this modification stabilizes the master and makes it possible to delete inactive cuts so that the size of (2.6) is limited. On the other hand, although we replace the linear master by a quadratic one, it is possible to arrange the algorithm for changing the regularizing points ξ^k in such a way that the whole method retains the finite convergence property of the linear approach. We shall present this algorithm in section 3.

Again, we could work here with a convex piecewise linear approximation of the expected second stage cost $f(x) = \sum_{\omega \in \Omega} p_\omega f_\omega(x)$, as in general algorithms of [7]. We use here more complicated separate approximations, because aggregation of cuts may slow down convergence of the method, as it was observed in [10] (this idea was also analyzed in [1]). We shall show in section 4 that it is possible to efficiently process separate approximations for each f_ω by exploiting the structural properties of (2.6).

Let us now pass to the question of obtaining objective and feasibility cuts. To discuss this matter in more detail we shall fix our attention on a specific method for solving the subproblems (2.4). Since for a given $x = x^k$ we have to solve (2.4) for all $\omega \in \Omega$ and only the right hand side in (2.4) varies, a reasonable choice is the dual simplex method. Upon termination, two cases may occur: optimality or infeasibility (dual unboundedness).

Objective cuts

Suppose that at $x = x^k$ problem (2.4) is solvable with an optimal basis B and simplex multipliers $\pi^T = q_B^T B^{-1}$. Then for any $x = x^k + \Delta x$, since B remains dual feasible, we get $f_\omega(x) \geq f_\omega(x^k) - \pi^T T_\omega \Delta x$. Hence, in the objective cut

$$\alpha_\omega^k + (g_\omega^k)^T x \leq f_\omega(x) \quad \text{for all } x \tag{2.7}$$

we have

$$g_\omega^k = -T_\omega^T \pi, \ \alpha_\omega^k = f_\omega(x^k) - (g_\omega^k)^T x^k. \tag{2.8}$$

Feasibility cuts

In case of dual unboundedness (for some $x = x^k$) we stop at a certain dual feasible basis B for which there is a basic variable y_{Br} whose value $\tilde{y}_{Br}$ is out of its bounds (e.g. $\tilde{y}_{Br} > y_{Br}^{\max}$) and which cannot be moved towards the feasibility interval by feasible changes of nonbasic variables. Clearly, $x^k \notin X_\omega$ in this case. On the other hand, for any $x \in X_\omega$ the value of y_{Br} with the basis B must not exceed $y_{Br}^{\max}$, because we shall not be able to decrease it by feasible changes of nonbasics. Denoting by π^T the r-th row of B we obtain the following estimate

$$X_\omega \subset \{x : \tilde{y}_{Br} - \pi^T T_\omega(x - x^k) \leq y_{Br}^{\max}\}.$$

If $\tilde{y}_{Br} < y_{Br}^{\min}$ we define π to be the negative of the r-th row of B and obtain

$$X_\omega \subset \{x : \tilde{y}_{Br} + \pi^T T_\omega(x - x^k) \geq y_{Br}^{\min}\}.$$

Consequently, at any $x = x^k$ for every $\omega \in \Omega$ for which (2.4) is not solvable, we get a feasibility cut

$$\bar{\alpha}_\omega^k + (\bar{g}_\omega^k)^T x \leq 0 \quad \text{for all} \quad x \in X, \tag{2.9}$$

where

$$\bar{g}_\omega^k = -T_\omega^T \pi, \quad \bar{\alpha}_\omega^k = \beta_\omega - (\bar{g}_\omega^k)^T x^k, \tag{2.10}$$

and β_ω is the distance to the violated bound.

It is not difficult to observe that there can be only finitely many objective and feasibility cuts, because the number of possible bases in (2.4) is finite.

3. The logic of the RD method

The method generates two sequences: a sequence ξ^k of regularizing points and a sequence x^k of trial points. Each iteration of the method consists in updating and solving the regularized master problem (2.6), which can be equivalently stated as follows. We introduce variables v_ω, $\omega \in \Omega$, to represent $\bar{f}_\omega(x)$ by inequalities involving objective cuts:

$$(g_\omega^j)^T x + \alpha_\omega^j \leq v_\omega, \ j \in J_\omega^k, \ \omega \in \Omega.$$

Using explicit formulations of feasibility cuts (2.9) and putting all the cuts together, we can rewrite the master (2.6) in a more compact form

$$\min \left[\frac{1}{2\sigma} \|x - \xi^k\|^2 + c^T x + p^T v \right] \tag{3.1}$$

subject to

$$(G^k)^T x + a^k \leq (E^k)^T v. \tag{3.2}$$

The constraints (3.2) (so called *committee*) comprise three groups of cuts:

(a) selected direct constraints from (2.2);

(b) selected feasibility cuts (2.9)-(2.10) collected at some previously generated trial points x^j, $j \in \bar{J}_\omega^k \subset \{0, 1, .., k\}$, $\omega \in \Omega$;

(c) selected objective cuts (2.7)-(2.8) collected at some previously generated trial points x^j, $j \in J_\omega^k \subset \{0, 1, .., k\}$, $\omega \in \Omega$.

Thus each column of the matrix E^k in (3.2) is either a null vector, if the cut is of class (a) or (b), or the l-th unit vector if the cut belongs to class (c) and approximates $f_l(x)$. There are never more than $n + 2L$ cuts in the committee.

There are two phases of the method. At Phase 1, we seek a point which satisfies (2.2)-(2.3). It serves then as a starting point for Phase 2, where we aim at solving (2.1)-(2.3). Since the Phase 1 algorithm is in fact a special case of the main Phase 2 method, we shall now describe in detail the latter.

Let ξ^0 be a starting point satisfying (2.2)-(2.3) and let the initial committee be given by

$$G^0 = [g_\omega]_{\omega \in \Omega}, \ a^0 = [\alpha_\omega]_{\omega \in \Omega}, \ E^0 = I,$$

with $(g_\omega, \alpha_\omega)$ describing objective cuts at ξ^0 for $\omega \in \Omega$. The committee may (but need not) also contain some constraints from (2.2) and some feasibility cuts of form (2.9) inherited from Phase 1.

The Regularized Master Algorithm

Step 1. Solve the master at ξ^k getting a trial point x^k and objective estimates v^k and calculate $\hat{F}^k = c^T x^k + p^T v^k$. If $\hat{F}^k = F(\xi^k)$, then stop (optimal solution found); otherwise continue.

Step 2. Delete from the committee some members inactive at (x^k, v^k) so that no more than $n_1 + L$ members remain.

Step 3. If x^k satisfies (2.2), then go to Step 4. Otherwise add to the committee no more than L violated constraints, set $\xi^{k+1} = \xi^k$, increase k by 1 and go to Step 1.

Step 4. For $\omega \in \Omega$ solve (2.4) at x^k and

    (a) if the constraints of (2.4) are inconsistent, append to the committee the feasibility cut (2.9)-(2.10);

    (b) else if $f_\omega(x^k) > v_\omega^k$ then append to the committee the objective cut (2.7)-(2.8).

Step 5. If all subproblems were solvable then go to Step 6; otherwise set $\xi^{k+1} = \xi^k$ and go to Step 7.

Step 6. If

    (a) $F(x^k) = \hat{F}^k$; or

    (b) $F(x^k) < F(\xi^k)$ and exactly $n + L$ members were active at (x^k, v^k),

then set $\xi^{k+1} = x^k$; otherwise set $\xi^{k+1} = \xi^k$.

Step 7. Increase k by 1 and go to Step 1.

If the starting point is not feasible, we put into the starting committee artificial cuts $v_\omega \geq -C$, where C is a very large constant, for all the functions $f_\omega(x)$, for which objective cuts are not yet available. We also set $F(\xi) = +\infty$.

It follows from the theory developed in [10] that after finitely many steps the method either discovers inconsistency in (2.1)-(2.3) or finds an optimal solution. Our proof for the case $p = [\ 1\ 1\ \ldots\ 1\ ]$, $\sigma = 1$ can be trivially extended to arbitrary $p > 0$, $\sigma > 0$. It is worth mentioning that the finite convergence property does not require any additional non-degeneracy assumptions typical for general cutting plane methods and bundle methods (see [15, 7]).

Few comments concerning implementation of the Regularized Master Algorithm are in order. The number of committee members may vary between L and $n_1 + 2L$, but in fact only cuts generated at Step 4 need to be stored (see section 4). If the number of linear constraints in (2.2) is large, various strategies can be used at Step 3, similarly to pricing strategies in linear programming. Finally, one can control the penalty coefficient σ on line, increasing it whenever steps are too short, and decreasing σ when $F(x^k) > F(\xi^k)$ (see section 5).

For the purpose of solving the regularized master problem, we suggested in [10] a special active set strategy, which we quickly summarize below. At each iteration we select a subset of constraints (3.2), defined by some submatrices G, a and E of G^k, a^k and E^k, such that E is of full row rank (at least one cut for each f_ω) and $\begin{bmatrix} G \\ E \end{bmatrix}$ is of full column rank. We treat these cuts as equalities and solve the resulting equality constrained subproblem by solving the system of its necessary and sufficient conditions of optimality

$$E\lambda = p, \tag{3.3}$$

$$E^T v + \sigma G^T G \lambda = G^T(\xi - \sigma c) + a \tag{3.4}$$

(for simplicity we drop the superscript k). The solution is given by

$$x = \xi - \sigma(c + G\lambda). \tag{3.5}$$

In the method, we alter the active set by adding or deleting cuts until the solution is optimal for (3.1)-(3.2).

4. Critical scenarios and reduced cuts

The system of necessary conditions of optimality (3.3)-(3.4) must be solved any time the active set is altered or ξ is changed by the master. The number of equations in (3.3) is equal to the number of blocks L, whereas the size of (3.4) is equal to the number of active cuts, which is $m + L$ with some $0 \leq m \leq n_1$. Thus the total size is $2L + m$: quite a large number when many realizations (scenarios) are taken into account. We need a special approach to this problem if we want to make our method competitive with standard techniques.

The key observation is that there must be at least one cut for each f_ω in the active set (E has full row rank). With the number of active cuts bounded by $L + n_1$, there may be at most m $(m \leq n_1)$ scenarios that are represented in the active set by more than one cut. We shall call them *critical scenarios*. The scenarios that are not critical have purely linear approximations. We shall exploit it in the numerical procedure to achieve substantial simplifications.

Formally, we select for each scenario one active cut and call it a *basic cut*. The basic cuts form the system

$$G_B^T x + a_B = v \qquad (4.1)$$

of dimension L. Other active cuts, which occur only for critical scenarios will be called *nonbasic*. Rearranging the order of cuts so that the basic cuts appear first, we shall have $E = \begin{bmatrix} I & N \end{bmatrix}$. The nonbasic cuts form the system

$$G_N^T x + a_N = N^T v \qquad (4.2)$$

of dimension m. Subtracting (4.1) multiplied by N^T from (4.2) yields *reduced cuts*:

$$\hat{G}^T x + \hat{a} = 0, \qquad (4.3)$$

where

$$\hat{G} = G_N - G_B N, \qquad (4.4)$$

$$\hat{a} = a_N - N^T a_B. \qquad (4.5)$$

In other words, each critical scenario is represented by the differences between its nonbasic cuts and its basic cut.

Next, partitioning λ into $\begin{bmatrix} \lambda_B \\ \lambda_N \end{bmatrix}$, we can use (4.1) to eliminate v and λ_B from (3.3)-(3.4), which yields

$$\sigma \hat{G}^T \hat{G} \lambda_N = \hat{G}^T x_B + \hat{a}, \qquad (4.6)$$

where x_B is the solution implied by basic cuts alone,

$$x_B = \xi - \sigma(c + G_B p). \qquad (4.7)$$

The other unknowns in (3.3)-(3.4) are defined by

$$\lambda_B = p - N \lambda_N,$$

$$x = x_B - \sigma \hat{G} \lambda_N. \qquad (4.8)$$

In this way the large system of necessary and sufficient conditions of optimality has been reduced to a relatively small system (4.6) whose order m never exceeds the number of first stage variables n_1, independently of the number of realizations taken into account. This is a substantial improvement over the LP formulation (1.3).

However, the other difficulty typical for stochastic programs, ill-conditioning, still remains. Its effect on our approach is that the reduced cuts (4.3) can be almost linearly dependent. Indeed, from (2.8) we see that a reduced objective cut is of the form

$$\hat{g}_\omega = T_\omega^T(\pi_{B\omega} - \pi_\omega), \qquad (4.9)$$

where $\pi_{B\omega}$ and π_ω are the vectors of simplex multipliers in block ω that generated the two cuts forming $\hat{g}_\omega$. The sets of possible multiplier vectors (basic solutions of duals to (2.4) are the same for each ω, so $\pi_{B\omega} - \pi_\omega$ may be identical for many reduced cuts. Then, by (4.9), the similarity of T_ω will cause similarity of the corresponding reduced cuts.

An established approach to such difficulties is the use of orthogonal factorization

$$\hat{G} = QR, \qquad (4.10)$$

where $Q^T Q = I$ and R is upper triangular. Then, defining an auxiliary vector w by

$$R^T w = \hat{G}^T x_B + \hat{a} \qquad (4.11)$$

we can reduce (4.6) to a triangular system

$$\sigma R \lambda_N = w. \qquad (4.12)$$

We can also use (4.10)-(4.12) to update the solution of (4.6) each time the active set is revised or ξ is changed by the master. This can be carried out by appropriate modifications of R, w, x_B and x, without storing Q explicitly. The details of the transformations used here can be found in [13].

5. Numerical results

A specialized code has been developed on the basis of the techniques described in this paper. For the purpose of solving lower level subproblems (2.4) we used subroutines from the XMP package of [8]. All computations were carried out on a Sun Sparcstation 2. The time reported is always in seconds.

The first series of experiments were carried out on a network planning problem with a random demand arising in telecommunication (see [14]). The first stage variables x are investment decisions that increase capacities of the arcs in the network. They are subject to a budget constraint. The second stage variables y_ω are on-line routing decisions dependent on the random demand for telecommunication connections. The routing decisions are subject to network constraints involving capacities defined by first stage decisions. The demand is modeled by 82 independent random variables, each taking 5 to 10 possible values, which results in an astronomical number of all possible scenarios. Therefore, we sampled from this distribution a smaller number of scenarios. The sizes of the original problem and the resulting LP approximations are given in Tables 1 and 2.

Table 1. Dimensions of Example 1.

	Rows	Columns
1st stage	1	89
2nd stage	175	706

Table 2. Size of LP formulations for Example 1.

Scenarios	Rows	Columns
1	176	795
10	1751	7149
20	3501	14209
50	8751	35389
100	17501	70689
200	35001	141289

Four versions of the method were run on these problems. The first one was our basic RD method described in this paper, with the penalty parameter σ controlled on-line as follows ($0.5 < \gamma < 1.0$):

- if $F(x^k) > \gamma F(\xi^k) + (1-\gamma)\hat{F}^k$ ("null step") then decrease σ;
- if $F(x^k) < (1-\gamma)F(\xi^k) + \gamma\hat{F}^k$ ("exact step") then increase σ;
- otherwise ("approximate step") keep σ unchanged.

The second version had a very large parameter σ in (2.6), which practically disabled the effect of the regularizing term (we used $\sigma = 10^6$). In this way, we aimed at assessing the effect of regularization on performance of the method. We still kept, however, separate cuts for each subproblem, as in the *multicut method* of [1].

The third version was the regularized method again, but with the cuts generated for the expected second stage cost

$$f(x) = \sum_{\omega \in \Omega} f_\omega(x)$$

instead for each f_ω separately. We used the same code again, but the objective cuts from subproblems were averaged before being passed to the master:

$$g^j = \sum_{\omega \in \Omega} g_\omega^j,$$

$$\alpha^j = \sum_{\omega \in \Omega} \alpha_\omega^j.$$

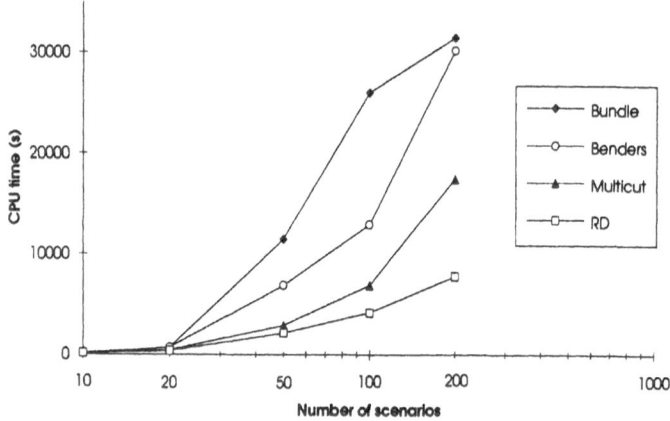

Fig. 1. Performance of the methods for Example 1.

As a result, we obtained the *bundle method* of [7].

The fourth version also accumulated aggregate cuts, as the previous one, but had the regularizing term disabled by $\sigma = 10^6$. So, the method was practically equivalent to the *L-shaped method* of [16], a specialized version of Benders decomposition.

The results are summarized in Figure 1.

Our RD method clearly outperforms all the other versions for larger problems. Comparing the performance of the regularized and the linear version, we see that the use of regularization substantially decreases the number of iterations of the RD method and, consequently, the number of dual simplex steps in subproblems. However, regularization alone, without the decomposition of cuts, does not help much, as the results for the bundle-type method show. It turns out that it is much better to work with our large master having separate cuts for each f_ω than with a bundle-type master of [7]. Both methods with aggregate cuts - the bundle method and the *L*-shaped method - were very slow near the solution: they had increasing difficulties in identifying attractive sets of cuts and good dirctions. Presumably, directional minimization for the bundle method might help a little, but in our case it would be prohibitively expensive. It is much better to have a more accurate approximation of the recourse function, as in the RD method, even if it requires solving a "large" master.

Apart from a much slower convergence, the methods with averaged cuts encountered numerical difficulties in the neigborhood of the solution: the cuts did not support the graph of the recourse function with the required accuracy. So, also from the numerical point of view, it is better to keep cuts for subproblems separately and to use probabilities explicitly in the master (see (4.7)), than to have a smaller number of averaged cuts.

Table 3. Dimensions of Example 2.

	Rows	Columns
1st stage	467	121
2nd stage	118	1259

Table 4. Size of LP formulations for Example 2.

Scenarios	Rows	Columns
1	585	1380
10	1647	12711
50	6376	63071
100	12267	126021
200	24067	251921
1000	118467	1259121

Since the size of the master could be the only possible argument against using separate cuts for the subproblems, we can safely discard methods with averaged cuts as clearly inferior (for two-stage stochastic programs).

The next example is a stochastic aircraft scheduling and transportation problem described in detail in [9]. There is a network of bases and a number of routes along which planes can move cargo between the bases. The first stage decisions x represent the number of flights of different aircraft types on the routes. They are subject to a number of constraints of military and technical nature, such as, e.g., airport capacities, numbers of flying hours, frequencies of connections, etc. The most important second stage variables represent the amounts of cargo carried by the planes; they have to be defined in such a way that a random demand for moving cargo in the network is satisfied in the cheapest possible way. The objective is to schedule the flights in such a way that the expected cost of moving cargo is minimum.

In Tables 3 and 4 we present problem statistics for various numbers of scenarios and in Figure 2 we summarize the performance of the RD method and of its linear counterpart (but still with decomposed cuts). Again, for these large problems our RD method is a clear winner, mainly because of a smaller number of null steps - long trial steps to points which are much worse than the current one. Additionally, in the 1000-scenario problem, the multicut method was not able to achieve the required accuracy. The methods with aggregate cuts failed on these problems.

There seems to be no doubt as to the efficiency of our approach and the usefulness of regularization. We were able to successfully solve problems that have deterministic equivalents of enormous sizes. This is mainly due to the use of regularization and cut decomposition in the master, and to our specialized

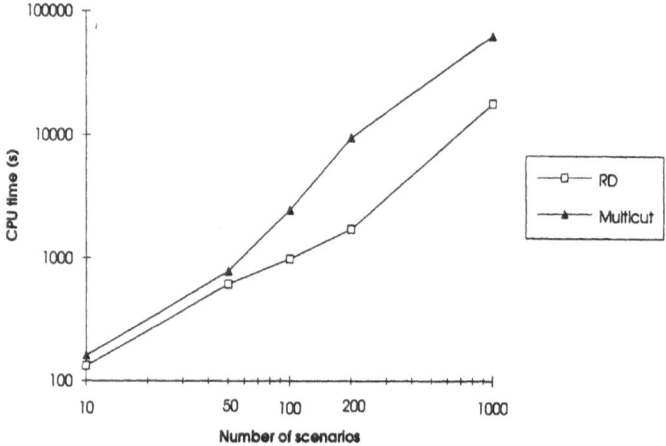

Fig. 2. Performance of the methods for Example 2.

technique for solving the master problem, which boils it down to a size that is practically independent on the number of scenarios.

6. Conclusions

The regularized decomposition method appears to be a rather efficient tool for solving large scale two-stage stochastic programming problems. Its efficiency is due to the following features:

1. The quadratic regularizing term stabilizes the master problem and helps to avoid aimless steps.

2. The use of separate approximations for scenario subproblems instead of aggregate (averaged) cuts speeds up convergence owing to a better description of the recourse function.

3. The special algorithm for solving the master problem based on dynamic selection of critical scenarios reduces it to a small numerical core whose size does not depend on the number of scenarios.

4. The use of the dual simplex method allows for rapid re-optimization of the subproblems.

There is also a disadvantage associated with our approach: the work with the reduced master requires a number of involved operations, so implementation of the method is not easy. But this is one of the keys to the good performance, because we use many closed-form formulae.

The method allows for a number of generalizations and extensions: its dual form turns out to be a decomposition method for augmented Lagrangians (see [11]); it can also be generalized to an asynchronous multistage version (see [12]). We hope to be able to discuss performance of these versions results for equally difficult examples as the ones discussed here.

References

[1] Birge J.R. and F.V. Louveaux, "A multicut algorithm for two-stage stochastic linear programs," *European Journal of Operations Research* 34(1988) 384-392.

[2] Birge J.R. and R.J.-B. Wets, "Designing approximation schemes for stochastic approximation problems, in particular for stochastic programs with recourse," in: *Stochastic Programming 1984*, A. Prekopa and R.J.-B. Wets (eds.), *Mathematical Programming Study* 27(1986) 54-102.

[3] Dantzig G. and A. Madansky, "On the solution of two-stage linear programs under uncertainty," in *Proceedings of the 4th Berkeley Symposium on Mathematical Statistics and Probability*, vol. 1, University of California Press, Berkeley 1961, pp. 165-176.

[4] Dantzig G. and P. Wolfe, "Decomposition principle for linear programs," *Operations Research* 8(1960) 101-111.

[5] Ermoliev Yu. and R.J.-B. Wets (eds.), *Numerical Techniques for Stochastic Optimization*, Springer Verlag, Berlin 1988.

[6] Kall P., A. Ruszczyński and K. Frauendorfer, "Approximation techniques in stochastic programming," in: *Numerical Techniques for Stochastic Optimization* (Yu. Ermoliev and R. Wets, eds.), Springer-Verlag, Berlin 1988, pp. 33-64.

[7] Kiwiel K.C., *Methods of Descent for Nondifferentiable Optimization*, Springer-Verlag, Berlin, 1985.

[8] Marsten R., "The design of the XMP linear programming library," *ACM Transactions of Mathematical Software* 7(1981) 481-497.

[9] Mulvey J.M. and A. Ruszczyński, "A new scenario decomposition method for large-scale stochastic optimization," technical report SOR-91-19, Deapartment of Civil Engineering and Operations Research, Princeton University, Princeton 1991 (to appear in *Operations Research*).

[10] Ruszczyński A., "A regularized decomposition method for minimizing a sum of polyhedral functions," *Mathematical Programming* 35(1986) 309-333.

[11] Ruszczyński A., "An augmented Lagrangian decomposition method for block diagonal linear programming problems," *Operations Research Letters* 8(1989) 287-294.

[12] Ruszczyński A., "Parallel decomposition of multistage stochastic programming problems," *Mathematical Programming* 58(1993) 201-228.

[13] Ruszczyński A., "Regularized decomposition of stochastic programs: algorithmic techniques and numerical results," working paper WP-93-21, International Institute for Applied Systems Analysis, Laxenburg 1993.

[14] Sen S., R.D. Doverspike and S. Cosares, "Network planning with random demand," technical report, Department of Systems and Industrial Engineering, University of Arizona, Tucson, 1992.

[15] Topkis J.M., "A cutting plane algorithm with linear and geometric rates of convergence," *Journal of Optimization Theory and Applications* 36(1982) 1-22.

[16] Van Slyke R. and R.J.-B. Wets, "*L*-shaped linear programs with applications to optimal control and stochastic programming," *SIAM Journal on Applied Mathematics* 17(1969) 638-663.

[17] Wets R.J.-B., "Stochastic programs with fixed recourse: the equivalent deterministic program," *SIAM Review* 16(1974) 309-339.

[18] Wets R.J.-B., "Large scale linear programming techniques," in: *Numerical Techniques for Stochastic Optimization* (Yu. Ermoliev and R. Wets, eds.), Springer-Verlag, Berlin 1988, pp. 65-94.

Multipoint Approximation Method for Structural Optimization Problems with Noisy Function Values

Vassili V. Toropov

Department of Civil Engineering, University of Bradford, Bradford, West Yorkshire, BD7 1DP, UK

Abstract. The multipoint approximation method is considered as a general iterative technique, which uses in each iteration simplified approximations of the original stochastic objective/constraint functions. They are obtained by the multiple regression analysis methods which can use information being more or less inaccurate. The technique allows to use in each iteration the information gained in several previous design points which are considered as a current design of numerical experiments (*multipoint approximations*). It allows to consider instead of the initial stochastic optimization problem a sequence of simpler mathematical programming problems and to reduce the total number of time-consuming structural analyses. The obtained approximations are assumed to be valid within a current subregion of the space of design variables, defined by move limits.

Several particular forms of approximations are considered including explicit expressions (linear, intrinsically linear, general nonlinear) and simplified implicit expressions. The proposed approach provides flexibility in choosing design variables and objective/constraint functions and allows the designer to use his/her experience and judgement in directing the optimization process.

Keywords. Stochastic optimization, noisy functions, multipoint approximations, response surface models.

1 Introduction

Many real-life structural optimization problems have the following characteristic features:

- the objective function and constraints are implicit functions of design variables,
- to calculate values of these functions means to use some numerical response analysis technique which usually involves a large amount of computer time,
- the function values and (or) their derivatives contain noise, i.e. can only be estimated with a finite accuracy.

The prohibitive computational cost of the direct combination of the numerical structural response analysis (mostly, the finite element method) with methods of mathematical programming stimulated the idea of approximation concepts based on the information from the first order design sensitivity analysis (Schmit and Farshi 1974). Nowadays this concept of sequential approximation of the initial optimization problem by explicit subproblems has proven to be very efficient and it is widely used in structural optimization (see a recent review by Barthelemy and Haftka 1993). Traditionally, approximation techniques use the information obtained by response analysis and first order design sensitivity analysis (i.e. values of functions and their derivatives) at a current point of the design variable space. They can be classified as *single point approximation* methods. Note that all the information from previous design points is discarded. At present, several first order approximation techniques are known, which are based upon the function value and its derivatives at the current and the previous design points (*two-point approximations*): Haftka et al. 1987, Fadel et al. 1990. The main purpose is to improve the quality of approximations and thus to reduce the number of iterations needed to solve the structural optimization problem and the total optimization time. The further generalization of this idea has been reported by Rasmussen 1990. The Accumulated Approximation technique (ACAP) uses function values and derivatives at a current design point and the function values obtained at all previous points as well. An example has been offered where the technique was able to cope with the noise level of 40% in the values of sensitivities of the objective and constraint functions. The drawback of these approaches is that the convergence of optimization algorithms can be slow if the objective/constraint function values present a considerable level of noise.

A different approach to structural optimization (Vanderplaats 1989) is to create approximate explicit expressions by analysing a chosen set of design points and using curve fitting or identification techniques. This approach is based on the multiple regression analysis methods, which can use information being more or less inaccurate. They are global in nature and allow to construct the explicit approximations valid in the entire design space but are restricted by relatively small optimization problems (up to ten design variables, Vanderplaats 1989). This approach was used for solving various structural optimization problems by Vanderplaats 1979, 1989, Brown and Nachlas 1981, White et al. 1985, Schoofs 1987, 1993, Schaller 1988, Müller and Tiefenthaler 1990, Rikards 1993. As stated by Vanderplaats 1989, the approach has some advantages. It provides a great deal of flexibility in choosing design variables and objective and constraint functions. It allows the designer to use his experience and judgement in directing the optimization process. It is important to note that the approach is based on an inherent parallel strategy. While simulation results are needed sequentially in applications of conventional nonlinear programming methodology, they are needed all at one time when the design of experiments methodology is used, thereby enabling better usage of parallel processors.

The remainder of this paper presents a general iterative technique which has been used for various complex engineering systems optimization problems. The simplified approximations of the original objective/constraint functions are obtained by the weighted least-squares method. It uses in each iteration the information about function values (and their derivatives, if available) gained in several previous design points (*multipoint approximations*, Toropov 1992, Toropov et al. 1993) which are considered as a current design of numerical experiments defined on a given set of parameters. It allows to consider instead of the initial stochastic optimization problem a sequence of simpler mathematical programming problems and to reduce the total number of time-consuming numerical analyses. We believe that it combines the advantages of two above mentioned basic approaches. A similar basic idea was used by Free et al. 1987 to construct a sequence of linear and quadratic response surfaces using the function values only. The stable performance was reported for the problems with the function values noise level of 1%. As the accurate calculation of gradients seriously influences the convergence of many optimization algorithms, Marti 1990 suggested to replace the gradients of original stochastic functions by the gradients of the polynomial response surface models. Linear models are used for the points located far away from the optimum, as the algorithm reaches the vicinity of the optimum point, it switches to quadratic models.

2 Formulation of a stochastic optimization problem

Consider a general stochastic structural optimization problem:

Minimize

$$F_0(x) \tag{1}$$

subject to

$$F_j(x) \leq C_j , \ (j = 1,..., M) \tag{2}$$

and

$$A_i \leq x_i \leq B_i , \ (i = 1,..., N), \tag{3}$$

where

$x \in R^N$ is a vector of design variables which is varied in the course of the design procedure,

$F_0(x)$ is an objective function which provides a basis for choice between alternative acceptable designs;

$F_j(x), \ (j = 1,..., M)$ are the constraint functions which impose limitations on various behavioural characteristics of the structure;

A_i and B_i are side constraints, describing physical upper and lower limitations on the design variables.

Depending on the specific problem under consideration, functions $F_j(x), \ (j = 0,..., M)$ can describe various structural response quantities, such as weight, stress, displacements, frequency, buckling load factor, dynamic and aeroelastic response etc. Often the function values and (or) their derivatives contain some level of noise due to approximation errors, round-off errors, etc., especially when some iterative technique is implemented (e.g. nonlinear structural behaviour) or semi-analytical sensitivity analysis is carried out (accuracy problems reported by Barthelemy and Haftka 1988). Besides, it is not uncommon to have non-smooth or "staircase" functions in the engineering design, Free et al. 1987.

3 Multipoint approximation technique

3.1 Basic approximation concept

The approximation concept leads to the iterative approximation of the original functions $F_j(x)$, $(j = 1,..., M)$ by the simplified functions $\tilde{F}_j(x)$. The initial stochastic optimization problem (1) - (3) is replaced with the succession of simpler mathematical programming subproblems as follows:

Find the vector x_*^k that maximizes the objective function

$$\tilde{F}_0^k(x) \tag{4}$$

subject to

$$\tilde{F}_j^k(x) \leq C_j \, , \, (j = 1,..., M) \tag{5}$$

and

$$A_i^k \leq x_i \leq B_i^k, \quad A_i^k \geq A_i, \quad B_i^k \leq B_i, \, (i = 1,..., N), \tag{6}$$

where k is the current iteration number. Note that the current move limits A_i^k and B_i^k define a subregion of the design space where the simplified functions $\tilde{F}_j^k(x)$ can be considered as adequate approximations of the initial functions $F_j(x)$. To estimate the order of the adequacy of simplified functions in comparison with the initial functions, the error parameters

$$r_j^k = |[\tilde{F}_j^k(x_*^k) - F_j^k(x_*^k)]/F_j^k(x_*^k)|, \, (j = 0, 1,..., M) \tag{7}$$

can be evaluated.

Various mathematical programming techniques can be used to solve the subproblem (4) - (6) because these functions are simple.

The next, $(k+1)$-th iteration is started from the obtained point x_*^k. The size and location of the next search subregion (i.e. the choice of move limits A_i^k and B_i^k depend on values of the error parameters (7). It will be discussed in Section 3.6.

3.3 Multipoint approximations

To construct the simplified expression $\tilde{F}_j^k(x)$ in the subproblem (4) - (6), we shall implement the methods of regression analysis, Draper and Smith 1981. These methods are intended for obtaining an expression that reflects the behaviour of an object considered as a function of its parameters, based on a discrete set of experimental results. Here *an experiment* means a numerical experiment using some numerical response analysis. It is essential to note that we do not intend to construct simplified expressions that are adequate in the whole of the search region determined by side constraints A_i and B_i in (3) because it takes too large number of numerical experiments in the case of a multiparameter problem. Therefore, we construct such expressions iteratively only for separate search subregions which are determined by move limits A_i^k and B_i^k at each k-th step of the problem in (4) - (6). Thus, the functions $\tilde{F}_j^k(x)$, $(j = 0,..., M)$ give piece-wise approximation of the initial functions $F_j(x)$.

To simplify notation, we will suppress the indices k and j on the functions $\tilde{F}_j^k(x)$. Assume that the functions (4), (5) are expressed in the following general form:

$$\tilde{F} = \tilde{F}(x, a) \tag{8}$$

The vector $a = (a_0, a_1,..., a_L)^T$ in expression (8) consists of *tuning parameters*, that is, free parameters, the value of which is determined on the basis of numerical experiments at points located in the design variable space in accordance with some design (plan) of experiments. Then the weighted least-squares method leads to the following problem:

Find the vector a that minimizes the function

$$G(a) = \sum_{p=1}^{P} w_p [F(x_p) - \tilde{F}(x_p, a)]^2, \tag{9}$$

where p is the number of a current point in the plan of experiments, P is the total number of such points, w_p is the weight coefficient that characterizes the relative contribution of the p-th experiment information. The solution of the optimization problem (9) is the vector a which makes up the simplified function (8).

Let us formulate the problem of vector a evaluation if in addition to values of function $F(x_p)$ their first order derivatives $F(x)_{,i} = \partial F(x) / \partial x_i$, $(i = 1, ..., N)$ at points x_p, $(p = 1,..., P)$ are known. In this case we shall minimize the function

$$G(a) = \sum_{p=1}^{P} \{w_p^{(0)} [F(x_p) - \tilde{F}(x_p, a)]^2 + \sum_{i=1}^{N} w_p^{(i)} [F(x_p)_{,i} - \tilde{F}(x_p, a)_{,i}]^2\}, \tag{10}$$

where $w_p^{(0)}$ and $w_p^{(i)}$ are the weight coefficients which characterize the relative contribution of information about $F(x_p)$ and $F(x_p)_{,i}$ correspondingly in the p-th experiment. *An experiment* means in this case a numerical analysis including the design sensitivity analysis. Note that the formula (10) could easily be extended if higher order sensitivities would become available.

3.4 Choice of the simplified expression structure

To construct the simplified expressions $\tilde{F}$, it is necessary to define them as a function of tuning parameters a. Apparently, the efficiency of the optimization technique depends greatly on the accuracy of such expressions. The simplest case is a linear function of parameters a :

$$\tilde{F}(a) = a_0 + \sum_{l=1}^{L} a_l \varphi_l .$$ (11)

The optimization problem (10) can be rewritten using matrix notation as follows:

Minimize

$$G(a) = (f - \tilde{f})^{\mathsf{T}} W (f - \tilde{f}),$$ (12)

where

$$f = \left\{ \begin{array}{c} F(x_1) \\ F(x_1)_{,1} \\ ... \\ F(x_1)_{,N} \\ ... \\ F(x_P) \\ F(x_P)_{,1} \\ ... \\ F(x_P)_{,N} \end{array} \right\}, \qquad \tilde{f} = \left\{ \begin{array}{c} \tilde{F}(x_1) \\ \tilde{F}(x_1)_{,1} \\ ... \\ \tilde{F}(x_1)_{,N} \\ ... \\ \tilde{F}(x_P) \\ \tilde{F}(x_P)_{,1} \\ ... \\ \tilde{F}(x_P)_{,N} \end{array} \right\},$$

$$W = diag \, [w_1^{(0)}, w_1^{(1)}, ..., w_1^{(N)}, ..., w_P^{(0)}, w_P^{(1)}, ..., w_P^{(N)}].$$

The vector f is the vector of $P(N+1)$ elements containing information from experiments obtained at P plan points, i.e. values of implicit functions $F(x_p)$ and their derivatives $F(x_p)_{,i}$ $(p = 1, ..., P$ and $i = 1, ..., N)$; the vector $\tilde{f}$ presents simplified function values and their derivatives with respect to design variables at the same plan points; W is the diagonal matrix $P(N+1) \times P(N+1)$ consisting of weight coefficients $w_p^{(0)}$ and $w_p^{(i)}$.

If we assume that the simplified function $\tilde{F}$ is a linear function (11) of tuning parameters $a = (a_0, a_1, ..., a_L)^{\mathsf{T}}$, then the vector $\tilde{f}$ can be expressed in the following form:

$$\tilde{f} = \Phi a ,$$ (13)

where

$$
\Phi = \begin{bmatrix}
1 & \varphi_{11} & \cdots & \varphi_{L1} \\
0 & \varphi_{11'1} & \cdots & \varphi_{L1'1} \\
\cdots & \cdots & \cdots & \cdots \\
0 & \varphi_{11'N} & \cdots & \varphi_{L1'N} \\
\cdots & \cdots & \cdots & \cdots \\
1 & \varphi_{1P} & \cdots & \varphi_{1P} \\
0 & \varphi_{1P'1} & \cdots & \varphi_{LP'1} \\
\cdots & \cdots & \cdots & \cdots \\
0 & \varphi_{1P'N} & \cdots & \varphi_{LP'N}
\end{bmatrix}, \qquad
a = \begin{Bmatrix}
a_0 \\
a_1 \\
\cdots \\
\cdots \\
\cdots \\
\cdots \\
\cdots \\
\cdots \\
a_L
\end{Bmatrix},
$$

$$
\varphi_{lp} = \varphi_l(x_p), \quad (l = 0, \dots, L \text{ and } p = 1, \dots, P).
$$

The necessary condition for the minimum of the quadratic form (10) with respect to the tuning parameters a leads to the linear system of $L+1$ normal equations with $L+1$ unknowns $a_l, (l=0,1,\dots,L)$:

$$
\Phi^T W \Phi a = \Phi^T W f, \tag{14}
$$

Note that the structure of the simplified expression (11) is rather general because the individual regressors φ_l can be arbitrary functions of design variables.

The procedure described above can be further generalized by the application of *intrinsically linear* functions, Draper and Smith 1981. Such functions are nonlinear, but they can be led to linear ones by simple transformations. The most useful functions among them are as follows:

- The multiplicative function

$$
\tilde{F}(a) = a_0 \varphi_1{}^{a_1} \dots \varphi_L{}^{a_L} \tag{15}
$$

with the logarithmic transformation

$$
\ln \tilde{F}(a) = \ln a_0 + \sum_{l=1}^{L} a_l \ln \varphi_l.
$$

In this case, the system of linear equations (14) contains the following matrix and vectors:

$$\Phi = \begin{bmatrix} 1 & \ln\varphi_{11} & \cdots & \ln\varphi_{L1} \\ 0 & \dfrac{F(x_1)}{\varphi_{11}}\varphi_{11,1} & \cdots & \dfrac{F(x_1)}{\varphi_{L1}}\varphi_{L1,1} \\ \cdots & \cdots & \cdots & \cdots \\ 0 & \dfrac{F(x_1)}{\varphi_{11}}\varphi_{11,N} & \cdots & \dfrac{F(x_1)}{\varphi_{L1}}\varphi_{L1,N} \\ \cdots & \cdots & \cdots & \cdots \\ 1 & \varphi_{1P} & \cdots & \varphi_{LP} \\ 0 & \dfrac{F(x_1)}{\varphi_{1P}}\varphi_{1P,1} & \cdots & \dfrac{F(x_1)}{\varphi_{LP}}\varphi_{LP,1} \\ \cdots & \cdots & \cdots & \cdots \\ 0 & \dfrac{F(x_1)}{\varphi_{1P}}\varphi_{1P,N} & \cdots & \dfrac{F(x_1)}{\varphi_{LP}}\varphi_{LP,N} \end{bmatrix}, \; f = \begin{Bmatrix} \ln F(x_1) \\ F(x_1),_1 \\ \cdots \\ F(x_1),_N \\ \cdots \\ \ln F(x_P) \\ F(x_P),_1 \\ \cdots \\ \cdots \\ \ln F(x_P),_N \end{Bmatrix}, \; a = \begin{Bmatrix} \ln a_0 \\ a_1 \\ \cdots \\ \cdots \\ \cdots \\ \cdots \\ \cdots \\ \cdots \\ \cdots \\ a_L \end{Bmatrix}.$$

- The power function

$$\tilde{F}(a) = \left(a_0 + \sum_{l=1}^{L} a_l\varphi_l\right)^\alpha \qquad (16)$$

with the transformation

$$[\ln \tilde{F}(a)]^{1/\alpha} = a_0 + \sum_{l=1}^{L} a_l \ln\varphi_l.$$

In this case

$$\Phi = \begin{bmatrix} 1 & \varphi_{11} & \cdots & \varphi_{L1} \\ 0 & \alpha F(x_1)^{(1-1/\alpha)}\varphi_{11,1} & \cdots & \alpha F(x_1)^{(1-1/\alpha)}\varphi_{L1,1} \\ \cdots & \cdots & \cdots & \cdots \\ 0 & \alpha F(x_1)^{(1-1/\alpha)}\varphi_{11,N} & \cdots & \alpha F(x_1)^{(1-1/\alpha)}\varphi_{L1,N} \\ \cdots & \cdots & \cdots & \cdots \\ 1 & \varphi_{1P} & \cdots & \varphi_{LP} \\ 0 & \alpha F(x_P)^{(1-1/\alpha)}\varphi_{1P,1} & \cdots & \alpha F(x_P)^{(1-1/\alpha)}\varphi_{LP,1} \\ \cdots & \cdots & \cdots & \cdots \\ 0 & \alpha F(x_P)^{(1-1/\alpha)}\varphi_{1P,N} & \cdots & \alpha F(x_P)^{(1-1/\alpha)}\varphi_{LP,N} \end{bmatrix}, \; f = \begin{Bmatrix} F(x_1)^{1/a} \\ F(x_1),_1 \\ \cdots \\ F(x_1),_N \\ \cdots \\ F(x_P)^{1/a} \\ F(x_P),_1 \\ \cdots \\ F(x_P),_N \end{Bmatrix}, \; a = \begin{Bmatrix} a_0 \\ a_1 \\ \cdots \\ \cdots \\ \cdots \\ \cdots \\ \cdots \\ \cdots \\ a_L \end{Bmatrix}$$

Several particular forms of intrinsically linear functions, including functions (15) and (16), were implemented for the solution of various complex engineering systems optimization problems, Toropov 1989. The obtained results indicate that the multiplicative function (15) is the most universal one.

There is an alternative approach based on so-called mechanistic models, which finds increasing application in empirical model-building, Box and Draper 1987. The parameter estimation of such models requires the implementation of the most general (nonlinear) form of the least-squares method, it becomes a standard problem of nonlinear programming. These models are constructed on the basis of physical considerations, which can sometimes provide clues to the nature of a phenomena under consideration. The designer of such a mechanistic model can typically use *a priori* information, such as analytical solutions for simplified geometrical shape, loading or boundary conditions of the structure. Clearly, in this approach, the researcher's experience and engineering judgement is essential to create high-quality approximations. An example of such a model application to the problem of material parameter identification can be found in Toropov and van der Giessen 1993.

It should be noted that the simplified functions $\tilde{F}$ *need not necessarily be explicit*. There can be numerical procedures involved in their formulation, such as numerical integration or even finite element modelling of the simplified object. But, the basic requirements to such simplified models are:

- its description of the simplified process must depend on the same parameters x as the initial numerical model, presented by functions $F(x)$;
- they have to contain some tuning parameters a, which are obtained by solving the optimization problem (9) or (10);
- they have to be simple enough to be used in numerous repeating calculations;
- in order to achieve fast convergence of the algorithm, they have to be accurate enough in comparison with the original functions (1), (2);
- they have to be noiseless or, at least, the level of noise must not cause problems with convergence of an algorithm used to solve the optimization problem (4) - (6).

3.5 Choice of weight coefficients

The quality of approximations depends strongly on values of the weight coefficients $w_p^{(i)}$ ($p = 1,...,P$ and $i = 0,...,N$) in (10), they reflect the inequality of data obtained in different design points. The correct choice of weights can substantially improve the properties of the functions $\tilde{F}(x)$. Usually, the point that corresponds to the solution of a structural optimization problem lies on a boundary of the feasible region, where at least one of constraints is active. Therefore, the quality of approximation of functions $F(x)$ by functions $\tilde{F}(x)$ should not be the same within the current search subregion. It is reasonable to set weights $w_p^{(i)}$ so that the quality of approximation will be the best in a domain of the current search subregion which is located along the boundary of the feasible region. This is possible if the weights depend on the distance of a current point from the boundary of the feasible region. Therefore, the weights can be defined, for example, by the following expression:

$$w'_p = \begin{cases} [F(x_p) / C]^{\alpha} & \text{if} \quad F(x_p) \leq C \\ [F(x_p) / C]^{-\alpha} & \text{otherwise,} \end{cases} \tag{17}$$

where $\alpha > 1$ is the parameter that defines the degree of inequality of data contributed by different experiments, C is the upper bound for the considered constraints (corresponding to the parameter C_j in (2)). Similarly, weights can reflect the difference in the contribution of the information given by different experiments depending on the objective function value at individual design points:

$$w''_p = [F_0(x_1) / F_0(x_p)]^{\beta} , \; (p = 2, ...,P) \tag{18}$$

where $\beta > 0$ is similar to the parameter α in (17), $F_0(x_1)$ is the objective function value at starting point. This formula gives the maximum value of the weight coefficient to the experiment at the design point with the minimum value of the objective function. Taking into account both (17) and (18), we can finally define weights $w_p^{(0)}$ in (10) by

$$w_p^{(0)} = w'_p w''_p , \; (p = 1, ...,P). \tag{19}$$

Next we have to determine the weight coefficients $w_p^{(i)}$ $(i = 1,...,N)$ which reflect in (10) the relative importance of inequality of data contributed by the response analysis (i.e. the value of $F(x_p)$) and the design sensitivity analysis (i.e. values of $F(x_p)_{,i}$) at the same p-th design point. It can be done by the expression as follows:

$$w_p^{(i)} = w_p^{(0)} \gamma \, \delta_p \quad (i = 1, ...,N \text{ and } p = 1, ...,P) \tag{20}$$

where $0 < \gamma < 1$ if the response analysis information is assumed to be more important than the design sensitivity information, δ_p is the coefficient intended to normalize the design sensitivity values:

$$\delta_p = 1 / \sum_{p=1}^{N} [F(x_p)_{,i}]^2 \tag{21}$$

3.6 Choice of a search subregion

After formulation of the simplified functions (8), the current mathematical programming problem in (4) - (6) is solved and the error parameters r_j^k in (7) for the point x_*^k are estimated.

Then the task is to determine the move limits A_i^k and B_i^k for the next iteration. First, the condition

$$r_j^k \leq e_j^k, \quad (j = 0,...,M) \tag{22}$$

is checked, where e_j^k are small positive values which define the feasible accuracy of approximation of functions $F(x)$ by functions $F_j^k(\tilde{x})$. If this condition is not satisfied even for one of the active constraints or objective function, (i.e. approximations are inaccurate), then the size of the search subregion of the $(k+1)$-th step must be reduced. When the conditions (22) are satisfied for all active constraints and the objective function, we must decide upon the movement of the search subregion. If the point obtained, x_*^k, is located inside the k-th search subregion (none of the move limits are active), then this point can be considered as the current approximation of the solution x_*. In that case, the next search subregion should be reduced and the other conditions of the search termination should be checked. Otherwise, the search must be continued. This means that the search subregion must be moved in the direction $x_*^k - x_*^{k-1}$. The search process is terminated when (i) the conditions (22) are satisfied, (ii) none of move limits are active and (iii) the subregion has reached a required small size.

4 Example

The above described optimization technique was used for the performance optimization of small Stirling engines developed by Dr Henrik Carlsen of the Technical University of Denmark (Bartczak and Carlsen 1991). They are designed to be used as individual combined heat and power plants (co-generation plants). Such engines driven from natural gas or biomass can supply a single family house with heat and power with a high level of energy utilization. It is considered a one of possible ways of decentalization of energy supply in rural areas.

The basic Stirling engine working cycle is a closed cycle, where the working gas is kept inside the cylinders, and heat is added and removed from the working space through heat exchangers. The Stirling engine differs therefore from conventional internal combustion engines by having external combustion, like in a boiler. All sorts of gas, liquid and solid fuels can be used depending on the proper external burner system only.

The ideal Stirling working cycle has the maximum obtainable efficiency defined by Carnot efficiency, and highly efficient Stirling engines can therefore be built, if designed properly. Heat exchangers, regenerators and cylinder volumes have to be optimized carefully for heat transfer, flow losses, dead volume, etc. To analyse the power output and the efficiency of a Stirling engine under consideration, numerical simulation programs (NSP) have been developed, which calculate power output, efficiency and heat flow from the thermodynamic equations. These differential equations are solved by means of the fourth order Runge-Kutta method. The initial pressure and temperatures for steady state operation are found by iterative procedure. To verify NSP, the numerical results have been compared to the results of laboratory experiments with a 10 kW natural gas driven Stirling engine. The comparison showed that NSP allows to predict the Stirling engine performance quite accurately.

To describe the engine design realistically for the optimization purposes, it is necessary to consider several tens of design variables. The objective and constraint functions of the optimization problem present a considerable level of noise, i.e. can only be estimated with a finite accuracy due to the iterative nature of NSP. For that reason, it is impossible to estimate the gradients by finite differences accurately enough.

The efficiency has been maximized, while the power output has been controlled by two constraints. The first constraint forces the power output not to be less than 10 kW, and the second constraint keeps the specific power above a certain minimum. Other constraints ensure the strength characteristics of the engine components.

The objective and constraint functions have been approximated by multiplicative expressions (15) in the simplest form:

$$\tilde{F}(a) \quad = \quad a_0 x_1{}^{a_1} \dots x_{18}{}^{a_{18}}.$$

The optimization was carried out using 18 design parameters, the results showed that the efficiency increased from 30.1% to 41.2% with the same power output without changing the important design parameters like bore and stroke by more than 20%. Solutions were found which had never occurred by the traditional "trial and error" approach. Also the optimization helped the designer to understand the fundamentals of Stirling engine performance in a completely new way, and the possibilities for improvements were found to be numerous.

The number of calls for the numerical simulation program in all runs has not been greater than 200.

5 Conclusion

The multipoint approximation method has been developed for the optimization problems with computationally expensive and noisy function values. It uses in each iteration simplified approximations of the original objective and constraint functions, obtained by the multiple regression analysis methods. The technique allows to use the information gained in several previous design points, which are considered as a current design (plan) of numerical experiments.

The method can be efficiently parallelized, because in each iteration the simulation results are needed all at one time; therefore, the numerical simulator can be run independently on individual processors corresponding to individual points of the current plan of experiments.

It should be also noted that, although the global convergence of the method cannot always be guaranteed, it has certain non-local convergence features. The successive approximation functions, obtained in each iteration, tend to neglect insignificant local minima and, therefore, allow to reach a non-local solution of the optimization problem without an excessive number of iterations.

References

1. Bartczak, L.; Carlsen, H.: An optimization study of Stirling engines based on advanced simulation. Proc. 5th Int. Stirling Engine Conference, Dubrovnik, 1991, 161-166

2. Barthelemy, B.; Haftka, R.T.: Accuracy analysis of the semi-analytical method for shape sensitivity calculation. In: AIAA Paper 88-2284, Proc. AIAA/ASME/ ASCE/AHS/ASC 29th Structures, Structural Dynamics and Materials Conf., Williamsburg, VA, April 18-20, 1988. Part 1, 572-581

3. Barthelemy, J.-F.M, Haftka, R.T.: Approximation concepts for optimum structural design - A review. *Structural Optimization* 5 (1993), 129-144

4. Box, G.E.P., Draper, N.R.: Empirical model-building and response surfaces. Wiley, 1987

5. Brown, R.T.; Nachlas, J.A.: Structural optimization of laminated conical shells. *AIAA J.* 23 (1985), 781-787

6. Draper, N.R.; Smith, H. : Applied regression analysis. 2nd ed. N.Y.: Wiley, 1981

7. Fadel, G.M.; Riley, M.F.; Barthelemy, J.M.: Two point exponential approximation method for structural optimization. *Structural Optimization* 2 (1990), 117-124

8. Free, J.W.; Parkinson, A.R.; Bryce, G.R.; Balling, R.J.: Approximation of computationally expensive and noisy functions for constrained nonlinear optimization. Transactions of the ASME 109 (1987), 528-532

9. Haftka, R.T.; Nachlas, J.A.; Watson, L.T.; Rizzo,T.; Desai, R.: Two - point constraint approximation in structural optimization. *Comp. Meth. Appl. Mech. Eng.* 60 (1987), 289-301

10. Marti, K.: Stochastic programming: Numerical solution techniques by semi-stochastic approximation methods. In: Slowinski, R.; Teghem, J. (eds.), *Stochastic Versus Fuzzy Approaches to Multiobjective Mathematical Programming under Uncertainty*, 1991, 23-43, Kluwer

11. Müller, G.; Tiefenthaler, P.: Design optimization with a commercial finite element program. In: Eschenauer, H.A.; Mattheck, C.; Olhoff, N. (eds.), *Engineering Optimization in Design Processes*. Proc. Int. Conf., Karlsruhe Nuclear Research Center, Germany, Sept. 3-4, 1990, 201-210, Springer

12. Rasmussen, J.: Structural optimization by accumulated function approximation. Institute of Mechanical Engineering, Aalborg University, Denmark. Report No.20, 1990.

13. Rikards, R.: Elaboration of optimal design models for objects from data of experiments. In: Pedersen, P. (ed.), *Optimal Design with Advanced Materials, The Frithiof Niordson Volume*, 1993, 149-162, Elsevier

14. Schaller, A.: Design optimization with the ANSYS program: An engineer's approach. Proc. Int. Conf. StruCoMe 88, Paris, 2-4 November 1988, Vol. 1, 381-395

15. Schmit, L.A.; Farshi, B.: Some approximation concepts for structural synthesis. *AIAA J.* **12** (1974), 692-699

16. Schoofs, A.J.G.: Experimental design and structural optimization. Ph.D. Thesis, Eindhoven University of Technology, The Netherlands, 1987

17. Schoofs, A.J.G.; Klink, M.B.M.; van Campen, D.H.: Approximation of structural optimization problems by means of designed numerical experiments. *Structural Optimization* **4** (1993), 206-212

18. Toropov, V.V.: Simulation approach to structural optimization. *Structural Optimization* **1** (1989), 37-46

19. Toropov, V.V.: Multipoint approximation method in optimization problems with expensive function values. In: A. Sydow (ed.), *Computational Systems Analysis 1992*, 207-212, 1992, Elsevier

20. Toropov, V.V.; van der Giessen, E. : Parameter identification for nonlinear constitutive models: Finite Element simulation - Optimization - Nontrivial experiments. In: Pedersen, P. (ed.), *Optimal Design with Advanced Materials, The Frithiof Niordson Volume*, 1993, 113-130, Elsevier

21. Toropov, V.V.; Filatov, A.A.; Polynkin, A.A: Multiparameter structural optimization using FEM and multipoint explicit approximations. *Structural Optimization* **6** (1993), 7-14

22. Svanberg, K.: The method of moving asymptotes - a new method for structural optimization. *Int. J. Num. Meth. Eng.* **24** (1987), 359-373

23. Vanderplaats, G.N.: Efficient algorithm for numerical airfoil optimization. *AIAA J., Aircraft* **16** (1979), 842-847

24. Vanderplaats, G.N.: Effective use of numerical optimization in structural design. In: *Finite Elements in Analysis and Design* **6** (1989), 97-112, Elsevier

25. White, K.P., Jr.; Hollowell, W.T.; Gabler, H.C. III; Pilkey, W.D.: Simulation optimization of the crashworthiness of a passenger vehicle in the frontal collisions using response surface methodology. *SAE Transactions* **94** (1985), 3.798-3.811

Sequential Convex Programming Methods

K. Schittkowski[1], C. Zillober[2]

Abstract. Sequential convex programming methods became very popular in the past for special domains of application, e.g. the optimal structural design in mechanical engineering. The algorithm uses an inverse approximation of certain variables so that a convex, separable nonlinear programming problem must be solved in each iteration. In this paper the method is outlined and it is shown, how the iteration process can be stabilized by a line search. The convergence results are presented for a special variant called *method of moving asymptotes*. The algorithm was implemented in FORTRAN and the numerical performance is evaluated by a comparative study, where the test problems are formulated through a finite element analysis.

Keywords. Nonlinear programming, structural optimization, convex approximation

1 Introduction

In this paper we proceed from the general formulation of a nonlinear programming problem

$$x \in \mathbb{R}^n : \quad \begin{aligned} &\min h_0(x) \\ &h_j(x) \leq 0 \quad , j = 1, ..., m \\ &x_l \leq x \leq x_u \end{aligned} \quad (1)$$

where all problem functions are continuously differentiable. We may imagine, for example, that the objective function describes the weight of a structure that is to be minimized subject to sizing variables, and that the constraints impose limitations on structural response quantities, e.g. upper bounds for stresses or displacements under static loads. Many other objectives or constraints can be modelled in a way so that they fit into the above general frame.

Especially in design optimization for mechanical engineering, sequential convex programming (SCP) algorithms as developed by Fleury (1989), Svanberg (1987) and others became very popular for several reasons:

---

[1]Mathematisches Institut, Universität Bayreuth, D–95440 Bayreuth, Germany
[2]IWR, Universität Heidelberg, D–69120 Heidelberg, Germany

1. The mathematical formulation of certain types of constraints, in particular of stress constraints, contains inverse optimization variables e.g. in case of cross-sections of bars. Thus an inverse approximation of these variables *linearizes* the problem functions defining the restrictions.

2. SCP methods are first order methods, which are attractive in situations, where round-off errors prevent the precise evaluation of gradients, and are therefore unable to update any second order information in a sufficiently accurate manner.

3. SCP methods are able to solve large optimization problems with hundreds or even thousands of variables, since the convex and separable subproblem to be solved in each iteration, can be adapted to large scale optimization and is solved easily also in these situations.

As a result of these observations, the finite element analysis of very many structural design software systems was extended by optimization modules based on convex approximation methods, see e.g. Hörnlein and Schittkowski (1992) for a review. Practical experience shows that SCP methods are often much more efficient than other methods, e.g. sequential quadratic programming, feasible direction or generalized reduced gradient methods.

However the *typical* implemetation of an SCP methods has a severe drawback: The methods are not stabilized in the sense that convergence towards a solution from an arbitray initial design, is not guaranteed. In Zillober (1993a), a stabilization by a line search procedure was proposed which allows to prove global convergence theorems. The main results are repeated in this paper.

Moreover we summarize the results of an extensive comparative study of structural optimization codes, where the analysis is based on a finite element formulation and performed by the software system MBB-LAGRANGE, see Kneppe, Krammer, and Winkler (1987) or Zotemantel (1993). Besides of the nine optimization algorithms included in the official version, two additional methods are added to the system, i.e. certain variants of convex approximation methods. The codes represent all major classes of algorithms that are in practical use at present.

To conduct the numerical tests, 79 design problems have been collected. Most of them are *academic* ones, i.e. are more or less simple design problems found in the literature. The remaining ones possess some practical *real life* background from project work or are suitable modifications to act as benchmark test problems for the development of the software system. In all situations, we minimize the weight of a structure subject to displacement, stress, strain, buckling, dynamic and other constraints. Design variables are sizing variables, e.g. cross sectional areas and thicknesses of skeletal as well as membrane or shell structures.

In the following section we will describe some convex approximations used for SCP methods. The optimization algorithm is outlined in Section 3 together with some convergence results. Section 4 contains a summary of computational results which allow a direct comparison with other methods.

2 Convex Approximations

By using reciprocal variables, Fleury and Braibant (1986) developed an optimization method called CONLIN (convex linearization). An approximation of a function is defined by separate linearization for each variable depending on the sign of the partial derivative at the expansion point. If the sign is positive then the linearization is performed with respect to the original variable. Otherwise the approximation is obtained with respect to the inverse variable, leading to a convex approximation of the original function.

In a more formal notation, we replace the original problem functions $h_j(x)$, $j = 0, \ldots, m$, with respect to a given iterate $x_k \in \mathbb{R}^n$ by

$$\overline{h}_j(x) := h_j(x_k) + \sum_{i \in I_{j,k}^+} \frac{\partial}{\partial x_i} h_j(x_k)(x_i - x_i^k) - \sum_{i \in I_{j,k}^-} \frac{\partial}{\partial x_i} h_j(x_k)(x_i^{k^2}/x_i - x_i^k) \quad (2)$$

where $x = (x_1, ..., x_n)^T$ and $x_k = (x_1^k, ..., x_n^k)^T$ and where

$$I_{j,k}^- := \{i : 1 \leq i \leq n, \frac{\partial}{\partial x_i} h_j(x_k) \leq 0\}$$

$$I_{j,k}^+ := \{i : 1 \leq i \leq n, \frac{\partial}{\partial x_i} h_j(x_k) > 0\}$$

for $j = 0, \ldots, m$, and where $\overline{h}_j(x)$ is defined for all $x > 0$.

The reason for inverting design variables in the above way is, that e.g. in design optimization stresses and displacements are exact linear functions of the reciprocal linear homogeneous sizing variables in case of a statically determinated structure. Moreover the numerical experience shows that also in other cases, convex linearization is applied quite successfully in practice, in particular in shape optimization, although a mathematical motivation cannot be given in this case.

After some reorganization of constant data, we get a convex and separable subproblem of the following form:

$$x \in \mathbb{R}^n : \quad \begin{array}{l} \min \sum_{i \in I_{0,k}^+} c_{0,k}^i x_i - \sum_{i \in I_{0,k}^-} c_{0,k}^i/x_i \\ \sum_{i \in I_{j,k}^+} c_{j,k}^i x_i - \sum_{i \in I_{j,k}^-} c_{j,k}^i/x_i + \overline{c}_{j,k} \leq 0, \quad j = 1, ..., m \\ x_l \leq x \leq x_u \end{array} \quad (3)$$

where $c_{j,k}^i$ and $\overline{c}_{j,k}^i$ are the constant parameters of the convex approximation with respect to objective function and constraints, i.e. are defined by

$$\overline{h}_j(x) := \overline{c}_{j,k}^i + \sum_{i \in I_{j,k}^+} c_{j,k}^i x_i - \sum_{i \in I_{j,k}^-} c_{j,k}^i/x_i$$

for $j = 0, \ldots, m$. Without loss of generality we assume that $x_l > 0$.

The solution of the above problem determines then the next iterate x_{k+1}. We do not investigate here the question how the mathematical structure of the subproblem can be exploited to get an efficient solution algorithm. As long as the problem is not too big, we may assume without loss of generality, that is solved by any standard nonlinear programming technique.

To control the degree of convexification and to adjust it with respect to the problem to be solved, Svanberg (1987) introduced so-called moving asymptotes U_i and L_i to replace x_i and $1/x_i$ by

$$\frac{1}{x_i - L_i} \ , \quad \frac{1}{U_i - x_i}$$

where L_i and U_i are given parameters, which can also be adjusted from one iteration to the next. The algorithm is called *method of moving asymptotes*. The larger flexibility allows a better convex approximation of the problem and thus a more efficient and robust solution.

In this case, the approximation is of the form

$$\bar{h}_j(x) \ := \ h_j(x_k) + \sum_{i \in I_{j,k}^+} \frac{\partial}{\partial x_i} h_j(x_k) \left(\frac{(U_i - x_i^k)^2}{U_i - x_i} - (U_i - x_i^k) \right) \tag{4}$$

$$- \sum_{i \in I_{j,k}^-} \frac{\partial}{\partial x_i} h_j(x_k) \left(\frac{(x_i^k - L_i)^2}{x_i - L_i} - (x_i^k - L_i) \right)$$

which is defined for all $x \in \mathbb{R}^n$ with $L_i < x_i < U_i$, $i = 1, \ldots, n$, and for all j with $0 \le j \le m$.

It is easy to verify that $\bar{h}_j(x)$ is a first order approximation of $h_j(x)$ at x_k, i.e. that

$$\bar{h}_j(x_k) = h_j(x_k) \quad \text{and} \quad \nabla \bar{h}_j(x_k) = \nabla h_j(x_k) \,,$$

and that $\bar{h}_j(x)$ is a convex and separable function, $j = 0, \ldots, m$. The first convex approximation (2) can be considered as a limit case of the approximation by moving asymptotes, since we get it back through $L_i = 0$ and $U_i \to \infty$.

Practical experience shows, that in some cases the approximation of the objective function is almost linear because of small constants $c_{0,k}^i$ e.g. in the neighbourhood of a solution. To avoid instabilities when solving the subproblem, Svanberg (1993) suggested to append a quadratic term to the approximation of the objective function which guarantees strict convexity of the objective function. Proceeding from the notation

$$\bar{h}_j(x) := \bar{c}_{j,k}^i + \sum_{i \in I_{j,k}^+} \frac{c_{j,k}^i}{U_i - x_i} - \sum_{i \in I_{j,k}^-} \frac{c_{j,k}^i}{x_i - L_i}$$

with suitable constants $c_{j,k}^i$ and $\bar{c}_{j,k}^i$, we get the modified objective function of the subproblem by

$$\bar{h}_0(x) := \bar{c}_{0,k}^i + \sum_{i \in I_{0,k}^+} \frac{c_{0,k}^i + \epsilon(x_i - x_i^k)^2}{U_i - x_i} - \sum_{i \in I_{0,k}^-} \frac{c_{0,k}^i + \epsilon(x_i - x_i^k)^2}{x_i - L_i} \tag{5}$$

with a suitable constant $\epsilon > 0$.

Moreover we have to avoid that a solution of the subproblem approaches one of the asymptotes. Thus we introduce a sufficiently small constant $\omega > 0$ and define new bounds by

$$
\begin{aligned}
\overline{x}_i^l &:= \max\{x_i^l, L_i + \omega(x_i^k - L_i)\} \\
\overline{x}_i^u &:= \min\{x_i^u, U_i - \omega(U_i - x_i^k)\}
\end{aligned}
$$

where $x_l = (x_1^l, \ldots, x_n^l)^T$, $x_u = (x_1^u, \ldots, x_n^u)^T$. With $\overline{x}_l = (\overline{x}_1^l, \ldots, \overline{x}_n^l)^T$ and $\overline{x}_u = (\overline{x}_1^u, \ldots, \overline{x}_n^u)^T$ we finally get the subproblem

$$
\begin{aligned}
& \min \overline{h}_0(x) \\
x \in \mathbb{R}^n : \quad & \overline{h}_j(x) \leq 0 \quad , j = 1, \ldots, m \\
& \overline{x}_l \leq x \leq \overline{x}_u
\end{aligned}
\tag{6}
$$

This approximation remains separable and is strictly convex with two major advantages:

- The subproblem has a unique solution, if it is at all solvable.

- Very efficient dual methods for solving the subproblem are applicable, cf. Fleury (1989), Svanberg (1987) or Zillober (1992).

Next we state a result which is important to identify a solution.

Lemma 2.1: *Let $x_k \in \mathbb{R}^n$ be given. x_k is a stationary point of (1) if and only if x_k is a stationary point of (6).*

The proof for this lemma as well as the proofs for all following statements can be found in Zillober (1993a) and Zillober (1992).

The asymptotes L_i and U_i, respectively, can be adapted during the iteration process to keep them as tight as possible on the one hand, and to extend them on the other hand whenever it turns out that the initial choice was too narrow. To be able to prove the desired convergence results, we consider only special asymptotes.

Definition 2.2: *A strategy for the choice of asymptotes is called continuous, if for any sequence $x_k \to x$, $x \in \mathbb{R}^n$, we have $L_i(x_k) \to L_i(x)$ and $U_i(x_k) \to U_i(x)$ for $i = 1 \ldots n$.*

By the notation $L_i(x_k), U_i(x_k)$ we identify the asymptotes resulting from the evaluation of the chosen strategy at the point $x_k \in \mathbb{R}^n$. These asymptotes may depend on the current iteration point, additionally on previous iteration points, or may be independent of the iteration point. In the theorems of the subsequent section we will always assume that the strategy for the choice of the asymptotes is continuous.

3 The Sequential Convex Programming Method

A sequential convex programming method with moving asymptotes consists basically of the following steps:

1. Starting from an initial iterate x_0 and initial asymptotes L_i and U_i, $i = 1, \ldots, n$, create a convex, separable subproblem of the form (6) and solve it by any optimization technique.

2. Update the asymptotes and repeat.

For the original SCP method based on subproblem (3), Nguyen et al. (1987) presented a convergence proof under very restrictive assumptions. They showed furthermore by some examples that a generalization of the result to arbitrary constraints is not possible.

It is well known that a line search with respect to a suitable merit function stabilizes an optimization algorithm in particular by preventing too large steps outside of the feasible region. In general a line search requires additional function evaluations, i.e. extra costs.

First we reformulate (1) to get a simplified notation for the theoretical analysis of this section, by assuming that upper and lower bounds for the variables are treated as part of the general inequality constraints. Without loss of generality, we proceed now from the problem

$$\min h_0(x)$$
$$x \in \mathbb{R}^n : \quad h_j(x) \leq 0 \quad , j = 1, ..., m \tag{7}$$

In this case the lower and upper bounds in subproblem (6) are defined by

$$\overline{x}_i^l \ := \ L_i + \omega(x_i^k - L_i)$$
$$\overline{x}_i^u \ := \ U_i - \omega(U_i - x_i^k)$$

Next we introduce an augmented Lagrangian function by

$$\Phi_r(x, u) = h_0(x) + \sum_{j=1}^{m} \left\{ \begin{array}{ll} u_j h_j(x) + \dfrac{r}{2} h_j^2(x) \, , & \text{if } h_j(x) \geq -\dfrac{u_j}{r} \\[2mm] \dfrac{u_j^2}{2r} \, , & \text{otherwise} \end{array} \right. \tag{8}$$

which is defined for all $x \in \mathbb{R}^n$ and $u \in \mathbb{R}^m$. The so-called penalty parameter r must be sufficiently large and controls the degree of *penalization* when leaving the feasible region. This function is also used in a sequential quadratic programming method for solving general nonlinear optimization problems, see Schittkowski (1981).

Now we formulate the algorithm, which we call SCP or sequential convex programming method to indicate the similarity to the SQP-method.

Algorithm 3.1:

Step 0 : Choose $x_0 \in \mathbb{R}^n$, $u_0 \geq 0$, $u_0 \in \mathbb{R}^m$, $0 < c < 1$, $0 < \psi < 1$, $r > 0$, $\epsilon > 0$, and let $k := 0$.

Step 1 : Compute $h_j(x_k)$, $\nabla h_j(x_k)$, $j = 0, \ldots, m$.

Step 2 : Compute L_i^k and U_i^k, $i = 1, \ldots, n$, by a suitable strategy, and define $\overline{h}_j(x_k)$ for $j = 0, \ldots, m$, by (4) and (5), respectively.

Step 3 : Solve (6) by any internal method and let y_k, v_k be the solution, where v_k denotes the corresponding vector of Lagrange multipliers.

Step 4 : If $y_k = x_k$, then stop, x_k is a stationary point.

Step 5 : Let $\delta_k := \|y_k - x_k\|$,
$$\eta_k := \tfrac{1}{2} \min \left\{ \min_{i=1..n} \left\{ 2\epsilon \frac{(U_i^k - x_i^k)^2}{(U_i^k - L_i^k)^3} \right\} , \min_{i=1..n} \left\{ 2\epsilon \frac{(x_i^k - L_i^k)^2}{(U_i^k - L_i^k)^3} \right\} \right\}$$

Step 6 : Compute $\Phi_r(x_k, u_k)$, $\nabla \Phi_r(x_k, u_k)$, and $\gamma_k := \nabla \Phi_r(x_k, u_k)^T \binom{x_k - y_k}{u_k - v_k}$.

Step 7 : If $\gamma_k < \tfrac{1}{4} \eta_k \delta_k^2$, let $r := 10r$ and goto step 6. Otherwise compute the smallest $j = 0, 1, 2 \ldots$, such that
$$\Phi_r(x_k - \psi^j(x_k - x_k), u_k - \psi^j(u_k - v_k)) \leq \Phi_r(x_k, u_k) - c\psi^j \gamma_k.$$
Set $\alpha_k := \psi^j$.

Step 8 : Let $x_{k+1} := x_k - \alpha_k(x_k - y_k)$, $u_{k+1} := u_k - \alpha_k(u_k - v_k)$, and $k := k + 1$. Then repeat with step 1.

Suitable constants for initializing the algorithm, are $c = 0.001$, $\psi = 0.5$, $\epsilon = 0.001$ and $r = 1$.

The difficulty in proving any global convergence result for the sequential convex programming method is to show, that the search direction defined by Step 3 of the algorithm, is a descent direction for Φ_r and that the resulting sequence of penalty parameters is bounded.

The next lemma shows that the sequence $(x_k, u_k)_{k=0,1,2,\ldots}$ is bounded under some reasonable assumptions.

Lemma 3.2: *Let the sequence $\{(x_k, u_k)\}_{k=0,1,2,\ldots}$ be produced by Algorithm (3.1), all subproblems be solvable and gradients of active constraints at the intermediate iterates x_k be linear independent as well as those at any possible accumulation point of $\{x_k\}_{k=1,2,3,\ldots}$. If the sequence $\{x_k\}$ is bounded, then also the sequence $\{u_k\}$ for $k = 1, 2, 3, \ldots$.*

Note that the boundedness of $\{x_k\}$ is guaranteed as soon as we include again our original bounds x_l and x_u for the variables. Now we are able to state the first convergence result.

Theorem 3.3: *Let the assumptions of Lemma (3.2) be valid for some iterates x_k and $u_k \geq 0$ of Algorithm (3.1), $k = 1, 2, 3, \ldots$, where none of the x_k is a stationary point for (6). Moreover let η_k, δ_k be defined as in Algorithm (3.1) and let the choice of asymptotes be continuous. Then*

1) there is a penalty parameter $r_k > 0$ such that (y_k, v_k) is a direction of descent for all $r \geq r_k$ with respect to the augmented Lagrange function Φ_r, in other words

$$\nabla \Phi_r(x_k, u_k)^T \begin{pmatrix} x_k - y_k \\ u_k - v_k \end{pmatrix} \geq \frac{\eta_k \delta_k^2}{4}$$

for all $r \geq r_k$,

2) for each $\delta > 0$ there is a finite r_δ such that for all x_k, u_k and $\delta_k \geq \delta$ we have

$$\nabla \Phi_r(x_k, u_k)^T \begin{pmatrix} x_k - y_k \\ u_k - v_k \end{pmatrix} \geq \frac{\eta_k \delta_k^2}{4} \geq \frac{\eta \delta^2}{4}$$

for all $r \geq r_\delta$.

The next theorem shows some conditions which guarantee the boundedness of the penalty parameter also in case of $\delta_k \to 0$.

Theorem 3.4: Let the asssumptions of Lemma (3.2) be valid and assume a continuous choice of the asymptotes in Step 2 of Algorithm (3.1). For $\delta_k \neq 0$ we define $\gamma_k := (\|u_k - v\|/\delta_k)^2$ with respect to a series of iterates x_k, y_k, u_k, and v_k of Algorithm (3.1), $k = 1, 2, 3, \ldots$, where none of the iterates x_k is a stationary point. Moreover assume that

a) there is a $\gamma \in \mathbb{R}$, such that $\gamma_k \leq \gamma < \infty$.

b) $h_j(x_k) \geq -u_j^k/r$ if and only if $\overline{h}_j(y_k) = 0$ for $j = 1, \ldots, m$, where $u_k = (u_1^k, \ldots, u_m^k)^T$,

Then there is a $\delta_r > 0$, such that

$$\nabla \Phi_r(x_k, u_k)^T \begin{pmatrix} x_k - y_k \\ u_k - v_k \end{pmatrix} \geq \frac{\eta_k \delta_k^2}{4}$$

with $\delta_k \leq \delta_r$, where $r := \min\{10^j : j = 0, 1, 2, \ldots \text{ with } 10^j \geq 2\gamma/\eta\}$.

As a special consequence of the above theorem, it can be shown that the penalty parameters of the augmented Lagrangian remain bounded, see Zillober (1993a). Assumption a) of Theorem (3.4) guarantees a uniform convergence of both sequences $\{x_k\}$ and $\{u_k\}$, where b) requires that the same active constraints are identified by the convex subproblem (6) and the augmented Lagrangian (8).

From the above descent properties of the search directions generated by Algorithm (3.1), the following convergence theorem can be proved.

Theorem 3.5: Let x_k and u_k be computed by Algorithm (3.1) satisfying the assumptions of Lemma (3.2) and Theorem (3.4). Then the algorithm either terminates at a stationary point, or every accumulation point of the iteration sequence is a stationary point for (6).

It is possible to omit the additional assumptions of Theorem (3.5). In this case it is only possible to prove that at least one accumulation point is stationary, see Zillober (1993a).

An important assumption for the convergence analysis outlined above, is the solvability of the convex, separable subproblems, i.e. that these problems possess non-empty feasible regions. This cannot be ensured in advance and is sometimes not fulfilled, especially in the first iterations, when we are still far away from a solution. However there are various techniques to overcome this situation, see e.g. Fleury, Braibant (1986), Svanberg (1987) or Schittkowski (1983).

4 Numerical Results

The FE-analysis of the comparative study is performed by the software system MBB-LAGRANGE, see Kneppe, Krammer, and Winkler (1987) or Zotemantel (1993). MBB-LAGRANGE is a computer aided structural design system based on the finite element technique and mathematical programming. The optimization model is characterized by the design variables and different types of restrictions.

Design variables are element thicknesses, cross sections, concentrated masses, and fiber angles. Besides of isotropic, orthotropic, and anisotropic applications, the analysis and optimization of composite structures is among the most important features. The design can be restricted with respect to a static, dynamic or aeroelastic analysis. The general aim is to minimize the structural weight subject to some of the following constraints:

- displacements

- stresses

- strains

- buckling

- local compressive stresses

- aeroelastic efficiencies

- flutter speed

- natural frequencies

- dynamic responses

- eigenmodes

- weight

- bounds for the design variables (gages)

Gradients are evaluated either analytically or semi-analytically by a special sensitivity analysis.

Besides of the nine optimization algorithms included in the official version, two additional methods are added to the system, i.e. certain variants of convex approximation methods. The codes represent all major classes of algorithms that are in practical use at present. Most of the methods have been developed, implemented, and tested outside of the MBB-LAGRANGE environment, and are taken over from external authors.

By the subsequent comments, some additional features of the algorithms and special implementation details are to be outlined. To identify the optimization codes we use the notation of the MBB-LAGRANGE documentation.

SRM: The stress ratio code belongs to the class of optimality criteria methods and is motivated by statically determinated structures. The algorithm is applicable to problems with stress constraints only, consists of a simple update formula for the design variables, and does not need any gradient information.

IBF: The inverse barrier function method is an implementation of a penalty method which needs an feasible design to start the algorithm. The unconstrained minimization is performed with respect to a quasi-Newton update (BFGS) and an Hermite interpolation procedure for the line search. It is recommended to perform only a relatively small number of iterations, e.g. 5 or 10, and to start another cycle by increasing the penalty parameter through a constant factor.

MOM: Proceeding from the same unconstrained optimization routine as IBF, a sequential unconstrained minimization technique is applied. The method of multipliers uses an augmented Lagrangian function similar to (8) for the subproblem and the usual update rules for the multipliers. Both methods, i.e. IBF and MOM, have a special advantage when evaluating gradients of the objective function in the subproblem. The inverse of the stiffness matrix obtained by a decomposition technique, is multiplied only once with the remaining part of the gradient, not for each restriction as required for most of the subsequent methods.

SLP: The sequential linear programming method was implemented by Kneppe (1985). The linear subproblem is solved by a simplex method. So-called move limits are introduced to prevent cycling and iterates too far away from the feasible area. They are reduced in each iteration by the formula $\delta_{k+1} = \delta_k/(1 + \delta_k)$ and an additional cubic line search is performed as soon as cycling is observed.

RQP1: The first recursive or sequential quadratic programming code is subroutine NLPQL of Schittkowski (1985/86). Subproblems are solved by a dual algorithm based on a routine written by Powell (1983). The augmented Lagrangian function (8) serves as a merit function and BFGS-updates are used for the quasi-Newton formula. The special implementation of NLPQL is capable to solve also problems with very many constraints, see Schittkowski (1992), and is implemented in MBB-LAGRANGE in reverse communication. The idea is to save as much working memory as possible by writing optimization data on a file during the analysis, and by saving analysis data during an optimization cycle.

RQP2: This is the original sequential quadratic programming code VMCWD of Powell (1978) with an L_1-merit function. Also in this case, the BFGS-update is used internally together with a suitable modification of the penalty parameter.

GRG: The generalized reduced gradient code was implemented by Bremicker (1986). During the line search an extrapolation is performed to follow the boundary of active constraints closer. The Newton-algorithm for projecting non-feasible iterates during the line search onto the feasible domain, uses the derivative matrix for the very first step. Subsequently a rank-1-quasi-Newton formula of Broyden is updated.

QPRLT: To exploit the advantages of SQP and GRG methods, a hybrid method was implemented by Sömer (1987). Starting from a feasible design, a search direction is evaluated by the SQP-approach, i.e. by solving a quadratic programming subproblem. This direction is then divided in basic and non-basic variables, and a line search is performed very similar to the generalized reduced gradient method GRG.

CONLIN: This is the original implementation of Fleury (1989), where a convex and separable subproblem is generated as outlined in Section 2. In particular only variables belonging to negative partial derivatives, are inverted. The nonlinear subproblem is solved by a special dual method.

SCP: The sequential convex programming method was implemented by Zillober (1993b) and added to the MBB-LAGRANGE-system for the purpose of this comparative study. The algorithm uses moving asymptotes and a line search procedure for stabilization with respect to the merit function (8).

MMA: The code is a reimplementation of the original convex approximation method of Svanberg (1987) with moving asymptotes. As for CONLIN and SCP, the subproblems are solved by a special dual approach. The adaption of moving asymptotes is described in Zillober (1993b).

To conduct the numerical tests, 79 design problems have been collected. Most of them are *academic* ones, i.e. are more or less simple design problems found in the literature. The remaining ones possess some practical *real life* background from project work or are suitable modifications to act as benchmark test problems for the development of the software system. In all situations, we minimize the weight of a structure subject to displacement, stress, strain, buckling, dynamic and other constraints. Design variables are sizing variables, e.g. cross sectional areas and thicknesses of skeletal as well as membrane or shell structures.

Since moreover all test examples are to be solvable by all available optimization algorithms, the dimension of the structures, i.e. number of elements and degrees of freedom, is relatively small compared to real life applications. More details about the test cases are found in Schittkowski, Zillober and Zotemantel (1993).

We believe that the present set of test cases is representative at least for small or medium size structural designs. It is also important to note that we do not want to test the analysis part of an FE-system. Instead the response of optimization routines when applied to solve structural optimization problems, is to be investigated.

All tests have been performed on a VAX 6000-510 running under VMS, at the Computing Center of the University of Bayreuth. The numerical codes are implemented in double precision FORTRAN. The intention behind our tests is to apply all optimization routines to all test examples that are available. To evaluate the results achieved, we need some information about the optimal solution, since the difference from the minimal weight of a test structure and the corresponding constraint violation serves as a measure for the accuracy of an actual iterate.

Thus we have to compute an optimal solution for each test case as accurate as possible. The most reliable codes were executed with a very small termination tolerance and a large number of iterations, until we got a stable and reliable solution. Test examples that did not lead to a clear solution point e.g. because of too many different local minimizers, have not been included in our set of test problems.

Having now an accepted reference value, it is possible to define whether an actual iterate x_k is sufficiently close to the optimal solution x^* subject to a given tolerance $\epsilon > 0$ or not. For each function or gradient evaluation during a test run, we store the corresponding objective function value $h_0(x_k)$ and the maximum constraint violation

$$r(x_k) := \max\{\max(0, h_j(x_k)) : j = 1, \ldots, m\}$$

together with some further data for analysis number and calculation time.

Now we are able to evaluate the performance of an algorithm subject to a given accuracy level ϵ. We sum up the performance criterion, e.g. calculation time or number of function and gradient evaluations, until for the first time the conditions

$$h_0(x_k) \leq h_0(x^*)(1 + \epsilon) \ , \quad r(x_k) \leq \epsilon \tag{9}$$

are satisfied. We should note here that the constraint functions are scaled internally by the analysis procedure of MBB-LAGRANGE.

Moreover there are some reasonable upper bounds for the number of iterations, and we must be aware of the fact that there are situations where a code is unable to find at least one solution in a test problem class within the given accuracy level and the maximum number of iterations.

For the purpose of our comparative study, we evaluate the performance criteria

- calculation time in seconds
- number of function evaluations, where an evaluation of objective and all constraints is counted as one function call
- number of gradient evaluations, i.e. evaluation of gradient of objective function and of all active constraints, where active constraints are determined by the internal active set strategy of MBB-LAGRANGE

A complete description of the test procedure and the numerical results is found in Schittkowski, Zillober and Zotemantel (1993). For the purpose of this review paper, we present some numerical comparative results in form of quotients of mean values of two optimization algorithms, where the mean value is taken over the common subset of successfully solved problems. Since the numerical figures for the performance criterion calculation time differ drastically, we use the geometric mean in this case.

Since we present only results for the complete test set without further restrictions, the optimization codes IBF and SRM are omitted in the subsequent tables. The first one requires a feasible initial design, and the second one is applicable only to problems with stress constraints.

For the accuracy levels $\epsilon = 0.01$ and $\epsilon = 0.0001$, the subsequent tables present the corresponding results. Table 1 and Table 5 show the number of test problems that could be solved successfully by two optimization algorithms, and the other ones show the corresponding mean values for the performance criteria calculation time, number of function evaluations and number of gradient evaluations, respectively.

To understand the meaning of these tables, let us denote the performance index, e.g. number of function evaluations, of optimization algorithm i and test example k by p_i^k. If I_{ij} is the index set of all test problems solved successfully by algorithm i and algorithm j with respect to a given termination tolerance ϵ, then each coefficient of the matrix contains the values $\sum_{k \in I_{ij}} p_i^k$ over $\sum_{k \in I_{ij}} p_j^k$. The normalized row sums of the corresponding quotients define the weights of the last column, that are considered as estimates for the performance criterium under consideration.

	MOM	SLP	RQP1	RQP2	GRG	QPRLT	CONLIN	SCP	MMA
MOM	49	44	47	45	44	43	35	44	41
SLP	44	59	58	56	46	53	41	54	54
RQP1	47	58	66	60	50	60	44	57	57
RQP2	45	56	60	61	49	57	43	57	56
GRG	44	46	50	49	53	49	39	48	47
QPRLT	43	53	60	57	49	64	43	56	55
CONLIN	35	41	44	43	39	43	45	42	43
SCP	44	54	57	57	48	56	42	60	56
MMA	41	54	57	56	47	55	43	56	58

Table 1: Number of problems solved successfully by two optimization codes w.r.t. $\epsilon = 0.01$

	MOM	SLP	RQP1	RQP2	GRG	QPRLT	CONLIN	SCP	MMA	Weight
MOM	21.96 21.96	19.17 5.08	20.56 10.81	16.43 6.04	18.55 7.34	17.87 5.30	15.84 3.23	16.77 6.85	16.05 4.03	27.16
SLP	5.08 19.17	6.62 6.62	6.52 12.87	6.20 7.70	4.66 7.07	6.51 7.68	4.14 3.44	6.31 8.26	6.04 5.59	7.28
RQP1	10.81 20.56	12.87 6.52	17.38 17.38	13.25 8.68	9.92 7.57	16.08 9.51	10.63 4.12	12.72 9.07	12.16 6.53	14.15
RQP2	6.04 16.43	7.70 6.20	8.68 13.25	8.69 8.69	6.37 6.99	9.22 8.59	6.87 3.88	8.29 8.99	8.07 6.36	9.39
GRG	7.34 18.55	7.07 4.66	7.57 9.92	6.99 6.37	8.51 8.51	8.64 6.16	7.09 3.69	8.38 6.97	7.24 4.32	11.20
QPRLT	5.30 17.87	7.68 6.51	9.51 16.08	8.59 9.22	6.16 8.64	10.46 10.46	6.79 4.42	9.74 10.75	8.93 6.93	8.61
CONLIN	3.23 15.84	3.44 4.14	4.12 10.63	3.88 7.09	3.69 6.79	4.42 6.79	4.29 4.29	4.11 6.52	4.11 4.47	5.82
SCP	6.85 16.77	8.26 6.31	9.07 12.72	8.99 8.29	6.97 8.38	10.75 9.74	6.52 4.11	10.22 10.22	8.93 6.80	9.54
MMA	4.03 16.05	5.59 6.04	6.53 12.16	6.36 8.07	4.32 7.24	6.93 8.93	4.47 4.11	6.80 8.93	6.63 6.63	6.86

Table 2: Geometric mean values for calculation time over common sets of successfully solved problems w.r.t. $\epsilon = 0.01$

	MOM	SLP	RQP1	RQP2	GRG	QPRLT	CONLIN	SCP	MMA	Weight
MOM	98.06 / 98.06	103.02 / 9.59	101.04 / 17.64	100.98 / 12.56	99.82 / 29.41	103.30 / 16.74	108.46 / 5.09	95.05 / 22.55	100.78 / 6.54	43.72
SLP	9.59 / 103.02	10.29 / 10.29	10.16 / 15.95	10.41 / 12.29	9.46 / 46.43	10.53 / 37.70	7.66 / 6.24	9.96 / 15.91	9.74 / 6.70	3.66
RQP1	17.64 / 101.04	15.95 / 10.16	18.65 / 18.65	15.72 / 13.13	16.18 / 45.02	17.88 / 48.53	15.09 / 6.45	14.68 / 16.95	13.35 / 7.25	5.59
RQP2	12.56 / 100.98	12.29 / 10.41	13.13 / 15.72	13.08 / 13.08	12.78 / 45.00	13.46 / 39.39	11.81 / 6.37	11.54 / 17.18	10.91 / 7.16	4.49
GRG	29.41 / 99.82	46.43 / 9.46	45.02 / 16.18	45.00 / 12.78	45.21 / 45.21	47.45 / 24.94	41.85 / 6.13	48.29 / 21.40	46.15 / 6.94	17.34
QPRLT	16.74 / 103.30	37.70 / 10.53	48.53 / 17.88	39.39 / 13.46	24.94 / 47.45	46.94 / 46.94	32.47 / 6.81	41.18 / 20.07	41.27 / 7.29	13.45
CONLIN	5.09 / 108.46	6.24 / 7.66	6.45 / 15.09	6.37 / 11.81	6.13 / 41.85	6.81 / 32.47	6.69 / 6.69	6.74 / 17.50	6.60 / 6.58	2.63
SCP	22.55 / 95.05	15.91 / 9.96	16.95 / 15.09	17.18 / 11.54	21.40 / 48.29	20.07 / 41.18	17.50 / 6.74	19.60 / 19.60	15.55 / 7.39	6.39
MMA	6.54 / 100.78	6.70 / 9.74	7.25 / 13.35	7.16 / 10.91	8.94 / 46.15	7.29 / 41.27	6.58 / 6.60	7.39 / 15.55	7.29 / 7.29	2.73

Table 3: Arithmetic mean values for number of function evaluations over common sets of successfully solved problems w.r.t. $\epsilon = 0.01$

	MOM	SLP	RQP1	RQP2	GRG	QPRLT	CONLIN	SCP	MMA	Weight
MOM	121.71 / 121.71	120.59 / 17.23	124.26 / 22.64	118.00 / 18.40	112.27 / 6.73	117.63 / 5.30	125.49 / 8.23	116.05 / 20.59	122.15 / 11.12	55.67
SLP	17.23 / 120.59	18.61 / 18.61	18.34 / 22.10	18.86 / 18.96	16.96 / 9.04	19.09 / 8.98	13.37 / 10.54	17.96 / 17.70	17.52 / 11.44	6.62
RQP1	22.64 / 124.26	22.10 / 18.34	24.45 / 22.10	20.90 / 19.53	20.88 / 8.88	23.33 / 10.13	20.32 / 10.95	19.79 / 18.91	18.84 / 12.53	7.69
RQP2	18.40 / 118.00	18.96 / 18.86	19.53 / 20.90	19.51 / 19.51	18.98 / 8.86	19.96 / 9.05	18.14 / 10.79	17.75 / 18.95	17.43 / 12.36	7.05
GRG	6.73 / 112.27	9.04 / 16.96	8.88 / 20.88	8.86 / 18.98	8.77 / 8.77	9.12 / 6.47	8.26 / 10.31	9.25 / 20.50	8.81 / 11.91	3.62
QPRLT	5.30 / 117.63	8.98 / 19.09	10.13 / 23.33	9.05 / 19.96	6.47 / 9.12	9.80 / 9.80	7.26 / 11.67	9.32 / 20.25	9.20 / 12.62	3.02
CONLIN	8.23 / 125.49	10.54 / 13.37	10.95 / 20.32	10.79 / 18.14	10.31 / 8.26	11.67 / 7.26	11.42 / 11.42	11.52 / 18.67	11.26 / 11.21	4.59
SCP	20.59 / 116.05	17.70 / 17.96	18.91 / 19.79	18.95 / 17.75	20.50 / 9.25	20.25 / 9.32	18.67 / 11.52	20.00 / 20.00	17.71 / 12.82	7.11
MMA	11.12 / 122.15	11.44 / 17.52	12.53 / 18.84	12.36 / 17.43	11.91 / 8.81	12.62 / 9.20	11.21 / 11.26	12.82 / 17.71	12.62 / 12.62	4.64

Table 4: Arithmetic mean values for number of gradient evaluations over common sets of successfully solved problems w.r.t. $\epsilon = 0.01$

	MOM	SLP	RQP1	RQP2	GRG	QPRLT	CONLIN	SCP	MMA
MOM	19	16	17	15	15	10	13	15	14
SLP	16	48	46	46	36	37	32	43	44
RQP1	17	46	60	54	43	48	34	51	52
RQP2	15	46	54	57	43	46	35	52	53
GRG	15	36	43	43	48	42	33	40	41
QPRLT	10	37	48	46	42	55	31	43	44
CONLIN	13	32	34	35	33	31	37	35	35
SCP	15	43	51	52	40	43	35	54	51
MMA	14	44	52	53	41	44	35	51	54

Table 5: Number of problems solved successfully by two optimization codes w.r.t. $\epsilon = 0.0001$

	MOM	SLP	RQP1	RQP2	GRG	QPRLT	CONLIN	SCP	MMA	Weight
MOM	42.20 / 42.20	29.70 / 5.61	28.32 / 12.64	19.00 / 6.88	24.93 / 8.58	16.32 / 5.23	24.33 / 4.24	24.42 / 5.73	25.51 / 4.76	31.58
SLP	5.61 / 29.70	5.48 / 5.48	5.25 / 9.80	5.08 / 5.67	4.82 / 6.49	5.69 / 5.74	4.64 / 3.54	4.99 / 6.32	4.84 / 4.40	7.30
RQP1	12.64 / 28.32	9.80 / 5.25	15.35 / 15.35	12.17 / 7.43	8.99 / 6.89	13.83 / 8.76	8.04 / 3.37	11.77 / 8.19	11.67 / 6.23	13.07
RQP2	6.88 / 19.00	5.67 / 5.08	7.43 / 12.17	8.42 / 8.42	5.94 / 7.35	7.50 / 7.74	5.69 / 3.44	7.46 / 8.33	7.41 / 6.25	8.31
GRG	8.58 / 24.93	6.49 / 4.82	6.89 / 8.99	7.35 / 5.94	9.50 / 9.50	9.98 / 7.01	8.23 / 3.89	7.93 / 6.37	6.90 / 4.36	10.68
QPRLT	5.23 / 16.32	5.74 / 5.69	8.76 / 13.83	7.74 / 7.50	7.01 / 9.98	11.33 / 11.33	6.23 / 3.97	9.09 / 9.14	7.76 / 6.48	8.17
CONLIN	4.24 / 24.33	3.54 / 4.64	3.37 / 8.04	3.44 / 5.69	3.89 / 8.23	3.97 / 6.23	4.19 / 4.19	4.24 / 7.76	3.84 / 4.28	5.33
SCP	5.73 / 24.42	6.32 / 4.99	8.19 / 11.77	8.33 / 7.46	6.37 / 7.93	9.14 / 9.09	7.76 / 4.24	9.79 / 9.79	8.97 / 6.40	9.03
MMA	4.76 / 25.51	4.40 / 4.84	6.23 / 11.67	6.25 / 7.41	4.36 / 6.90	6.48 / 7.76	4.28 / 3.84	6.40 / 8.97	6.69 / 6.69	6.54

Table 6: Geometric mean values for calculation time over common sets of successfully solved problems w.r.t. $\epsilon = 0.0001$

	MOM	SLP	RQP1	RQP2	GRG	QPRLT	CONLIN	SCP	MMA	Weight
MOM	131.63	139.00	137.35	145.00	133.27	99.00	157.77	146.07	148.57	47.76
	131.63	9.69	20.29	17.40	39.73	12.00	8.54	15.13	8.36	
SLP	9.69	13.42	13.33	13.74	13.11	14.62	11.78	14.05	13.70	4.22
	139.00	13.42	16.00	12.22	41.33	29.86	7.41	18.88	8.52	
RQP1	20.29	16.00	19.52	15.98	17.56	18.90	14.85	14.73	14.56	4.59
	137.35	13.33	19.52	13.57	49.00	56.65	7.35	19.80	9.88	
RQP2	17.40	12.22	13.57	13.88	12.65	12.54	12.74	13.52	12.94	3.88
	145.00	13.74	15.98	13.88	50.49	42.80	7.29	19.92	9.81	
GRG	39.73	41.33	49.00	50.49	55.44	56.76	55.91	59.65	47.93	15.71
	133.27	13.11	17.56	12.65	55.44	37.07	7.06	20.70	8.93	
QPRLT	12.00	29.86	56.65	42.80	37.07	60.31	35.65	55.37	43.84	12.72
	99.00	14.62	18.90	12.54	56.76	60.31	6.19	20.09	9.34	
CONLIN	8.54	7.41	7.35	7.29	7.06	6.19	7.35	7.49	7.17	2.35
	157.77	11.78	14.85	12.74	55.91	35.65	7.35	22.97	7.57	
SCP	15.13	18.88	19.80	19.92	20.70	20.09	22.97	20.76	19.27	6.01
	146.07	14.05	14.73	13.52	59.65	55.37	7.49	20.76	9.49	
MMA	8.36	8.52	9.88	9.81	8.93	9.34	7.57	9.49	9.81	2.75
	148.57	13.70	14.56	12.94	47.93	43.84	7.17	19.27	9.81	

Table 7: Arithmetic mean values for number of function evaluations over common sets of successfully solved problems w.r.t. $\epsilon = 0.0001$

	MOM	SLP	RQP1	RQP2	GRG	QPRLT	CONLIN	SCP	MMA	Weight
MOM	237.21	240.94	242.29	246.60	219.40	91.70	277.62	249.00	247.93	63.20
	237.21	17.44	29.12	24.87	8.60	4.50	15.15	20.47	14.79	
SLP	17.44	24.85	24.67	25.50	24.25	27.27	21.59	26.12	25.43	7.25
	240.94	24.85	23.15	19.76	8.14	7.81	12.84	21.74	15.07	
RQP1	29.12	23.15	27.42	22.91	23.51	25.73	20.50	21.78	21.67	5.75
	242.29	24.67	27.42	21.50	9.37	12.27	12.74	23.39	17.79	
RQP2	24.87	19.76	21.50	22.30	19.84	20.07	19.91	21.52	21.26	5.23
	246.60	25.50	22.91	22.30	9.74	10.63	12.60	23.44	17.64	
GRG	8.60	8.14	9.37	9.74	10.25	10.33	9.91	10.68	9.24	2.71
	219.40	24.25	23.51	19.84	10.25	8.10	12.15	22.77	15.88	
QPRLT	4.50	7.81	12.27	10.63	8.10	12.29	7.65	12.26	10.80	2.53
	91.70	27.27	25.73	20.07	10.33	12.29	10.42	22.81	16.70	
CONLIN	15.15	12.84	12.74	12.60	12.15	10.42	12.73	13.00	12.37	3.49
	277.62	21.59	20.50	19.91	9.91	7.65	12.73	24.14	13.17	
SCP	20.47	21.74	23.39	23.44	22.77	22.81	24.14	24.39	22.88	5.65
	249.00	26.12	21.78	21.52	10.68	12.26	13.00	24.39	17.00	
MMA	14.79	15.07	17.79	17.64	15.88	16.70	13.17	17.00	17.65	4.19
	247.93	25.43	21.67	21.26	9.24	10.80	12.37	22.88	17.65	

Table 8: Arithmetic mean values for number of gradient evaluations over common sets of successfully solved problems w.r.t. $\epsilon = 0.0001$

Conclusions

The paper describes a class of methods for nonlinear programming, which is applicable only in certain situations, where inverse variables in constraints dominate. The algorithm generates a sequence of convex, separable nonlinear subproblems, which must be solved in each iteration. It is shown how the algorithm can be stabilized by a line search with an augmented Lagrangian merit function. Some theoretical convergence results are mentioned.

Also the numerical results of a comparative study of optimization routines are included. The 79 test problems are taken from structural design optimization, where the FE-analysis is performed by the software system MBB-LAGRANGE. The results indicate, that the sequential convex programming method belongs not only to the most reliable approach, but is also more efficient than *classical* nonlinear programming algorithms, e.g. sequential quadratic programming.

References

Bremicker, M. (1986): *Entwicklung eines Optimierungsalgorithmus der generalisierten reduzierten Gradienten*. Report, Forschungslaboratorium für angewandte Strukturoptimierung, Universität-GH Siegen

Fleury, C. (1989): *An efficient dual optimizer based on convex approximation concepts*. Structural Optimization, Vol. 1, 81 - 89

Fleury, C.; Braibant, V. (1986): *Structural Optimization – a new dual method using mixed variables*. International Journal for Numerical Methods in Engineering, Vol. 23, 409 - 428

Hörnlein, H.; Schittkowski, K. (eds.) (1992): *Software systems for structural optimization*. Birkhäuser Verlag, Basel, ISNM 110

Kneppe, G. (1985): *Direkte Lösungsstrategien zur Gestaltsoptimierung von Flächentragwerken*. Reihe 1, No. 135, VDI-Verlag, Düsseldorf

Kneppe, G.; Krammer, J.; Winkler, F. (1987): *Structural Optimization of large scale problems using MBB LAGRANGE*. 5th World Congress and Exhibition on FEM, Salzburg, Austria

Nguyen, V.H.; Strodiot, J.J.; Fleury, C. (1987): *A mathematical convergence analysis for the convex linearization method for engineering design optimization*. Engineering Optimization, Vol. 11, 195 - 216

Powell, M.J.D. (1978): *A fast algorithm for nonlinearly constrained optimization calculations*. in: Numerical Analysis, G.A. Watson ed., Lecture Notes in Mathematics, Vol. 630, Springer

Powell, M.J.D. (1983): *On the quadratic programming algorithm of Goldfarb and Idnani*. Report DAMTP 1983/Na 19, University of Cambridge, Cambridge

Schittkowski, K. (1981): *The Nonlinear Programming method of Wilson, Han* and *Powell with an augmented Lagrangian type line search function.* Numerische Mathematik, Vol. 38, 83 - 114

Schittkowski, K. (1983): *On the convergence of a Sequential Quadratic Programming method with an augmented Lagrangian line search function.* Optimization, Vol. 14, 197 - 216

Schittkowski, K. (1985/86): *NLPQL: A FORTRAN subroutine solving constrained nonlinear programming problems.* Annals of Operations Research, Vol. 5, 485-500

Schittkowski, K. (1992): *Solving nonlinear programming problems with very many constraints.* Optimization, Vol. 25, 179-196

Schittkowski, K.; Zillober, Ch.; Zotemantel, R. (1993): *Numerical Comparison of Nonlinear Programming Algorithms for Structural Optimization.* Structural Optimization, Vol. 7, 1-19

Sömer, M. (1987): *Aufstellen und Testen eines hybriden SQP-GRG Optimierungsalgorithmus.* Studienarbeit, Institut für Mechanik und Regelungstechnik, Universität-GHS Siegen, Siegen

Svanberg, K. (1987): *The Method of Moving Asymptotes - a new method for Structural Optimization.* International Journal for Numerical Methods in Engineering, Vol. 24, 359 - 373

Svanberg, K. (1993): *The Method of Moving Asymptotes (MMA) with some extensions.* In: Rozvany, G. (ed.) Lecture Notes for the NATO/DFG ASI "Optimization of Large Structural Systems", Vol. 1, 55 - 66, Berchtesgaden, Germany

Zillober, Ch. (1992): *Eine global konvergente Methode zur Lösung von Problemen aus der Strukturoptimierung.* Dissertation, Technische Universität München

Zillober, Ch. (1993a): *A globally convergent version of the Method of Moving Asymptotes.* Structural Optimization, Vol. 6, 166-174

Zillober, Ch. (1993b): *SCP - An implementation of a sequential convex programming algorithm for nonlinear programming.* DFG-Report No. , Schwerpunkt Anwendungsbezogene Optimierung und Steuerung

Zotemantel, R. (1993): *MBB-LAGRANGE: A computer aided structural design system.* in: Software Systems for Structural Optimization, H. Hörnlein, K. Schittkowski eds., Birkhäuser, Basel, Boston, Berlin

TARGET COSTING: THE DATA PROBLEM

Peter Abel[1], Stefan Niemand[2] and Markus Wolbold[2]

[1] *USU Softwarehaus Unternehmensberatung GmbH, Spitalhof,*
D-71693 Möglingen, Germany.
[2] *University of Stuttgart, Department of Controlling, Keplerstr. 17,*
D-70174 Stuttgart, Germany.

ABSTRACT:

In this article we introduce target costing as a management concept which receives considerable attention in the current business literature and practice. It has acquired major importance in shaping and structuring product costs in the early phases of the development process. The dominant role of market information makes this concept a major instrument to align modern companies with market requirements.

In section 2 of this article we describe how target costing works and how it is integrated into existing organizational and management structures. This is illustrated by a case study using real life data. Finally, a survey on the use of target costing in Japanese companies shows the relevance of this concept for industrial application.

Market data and the definition and setting of target costs are two main areas of the target costing approach where stochastic optimization decision models can help to solve the problems based on the required availabilty of perfect information about the input data for target costing models. In section 3 and 4 we discuss modifications of the traditional target costing tools and models to allow the use of random variables with known or only partial known probability distribution as input data.

1. Introduction

Target costing is a cost management concept that has been developed and practised in Japanese companies since the 1970s and has been described in English mainly by Japanese authors like *Hiromoto* [1989], *Sakurai* [1989], *Tanaka* [1989] and others. Yet, the concept is not entirely new. It combines some elements of existing cost management and cost engineering tools (e.g. cost engineering, design to cost) with the necessity to bring the whole company in line with market demands.

In western industrial companies target costing is known for about half a decade. It widens the (traditional) orientation on only technically possible solutions by the demands of (subjective) customer wishes. Using the target costing approach one has to find out how much a customer is willing to pay for a product and what kind of quality the customer is expecting from this product. Typical application areas for target costing are

- product design and development,
- cost reduction of existing products,
- efficiency improvement in all business areas.

For a detailed introduction to target costing see e.g. *Horváth, Niemand, Wolbold* [1993a and b], *Horváth* [Ed. 1993], *Horváth, Seidenschwarz* [1992] and *Seidenschwarz* [1993].

In general, target costing approaches today work with "hard" customer and production data as input. The knowlegde of the perfect information about the input data is necessary for using the known target costing tools. The first experiences in German real-life projects show how difficult it is to describe e.g. "soft" customer wishes with "hard" data. In section 3 we discuss this problem using a real-life example from a German automotive company. To avoid this difficulty it is possible to use random variables as input data. Stochastic optimization models like "stochastic programming with recourse" and "chance constrained" are well-accepted tools to solve this kind of decision problem.

2. Target Costing: Features and Application

2.1 Definition of Target Costing

There is no generally accepted definition of target costing. It has mainly been developed by Japanese companies like *NEC, Sony, Nissan* and especially *Toyota* as an instrument for cost planning. Consequently, it is often referred to by Japanese authors as "cost planning" or "cost projection" (in Japanese "genka kikaku"). Target costing is essentially a management tool and not an accounting routine, so that the term "Zielkosten*rechnung*", which is sometimes used in German literature, is not correct.

The spectrum of definitions ranges from rather narrow ones, e.g. *Cooper* [1992]: "The object of target costing is to identify the production cost of a proposed product so that, when sold, it generates the desired profit margin", to fairly comprehensive concepts. According to *Sakurai* [1989] " ... target costing can be defined as a cost management tool for reducing the overall cost of a product over its entire life cycle with the help of the production, engineering, R&D (research and development), marketing, and accounting departments."

German authors like *Horváth* and *Seidenschwarz* view target costing as an instrument of strategic cost management which is capable of linking products, markets and resources on a strategic basis and transforming this information into quantitative operational measures.

Target costing, as we understand it, is built on a comprehensive set of cost planning, cost management and cost control instruments which are aimed primarily at the early stages of product and process design in order to influence product cost structures resulting from the market-derived requirements. The target costing process requires the cost-orientated coordination of all product-related functions.

Thus target costing requires the definition of the highest acceptable market price of a product, the specification of a target profit, and the decomposition of the resulting target cost to the parts and component level. This process has to be entirely driven by market information, i.e. customer requirements (see also Fig. 2.1).

In general, companies want to achieve the following main objectives with the help of target costing:
- Market orientation of the whole company and especially of cost management.
- Improvement of the strategic position of a company through market-oriented R&D.
- Support of cost management in the early design phases of a product.
- Motivational aspects because behaviour is directly influenced by market-based requirements and not by abstract company goals.

2.2 Information Structures for Target Costing

Target costing is a complex process which requires an effective organizational structure to control and evaluate the flow of information between management systems and the people involved in the target costing process. The ultimate purpose is to use the information for goal-attaining action.

Target costing is essentially a multi-functional activity. A basic requirement is the use of interdisciplinary teams. These are generally composed of members of the following functions: production technology, design, purchasing, development, marketing, manufacturing, product planning, and accounting. These teams are overlapped by "heavyweight" product managers who are responsible for the overall development of a specific product. This kind of matrix organization (ad hoc and usually for a limited period of time) combines the specialist knowledge of the corporate functions (which is necessary to drive target costs) with the total view of a product (which ensures that the market requirements are effectively answered in a comprehensive product concept).

This organization is a prerequisite for *concurrent engineering*. From the beginning of the design process, the product manager is able to optimize the achievement of product functions as components are developed concurrently. In this case the only limitation is to meet all the functional requirements as well as the target cost, as opposed to the sequential approach to development where additional limitations arise from the fact that fully developed components influence and determine the technological choices for components which are still to be developed, i.e. later components are already constrained in design and cost by earlier decisions. An additional effect of concurrent engineering is that the development process is accelerated and the increasingly important "time to market" is reduced.

2.3 Identifying and Setting the Target Cost

Target costing starts at the product planning stage. After the marketability of a certain product is assured, the *target price* is formulated by evaluating market research, competitive research and users' needs. This process is called *market into company* and is the valid approach to target costing. However, other forms of setting the target price attempt to avoid the difficulties of market research by replacing direct market orientation by secondary information. But in the following we shall concentrate on *market into company* as the essential way to set target costs.

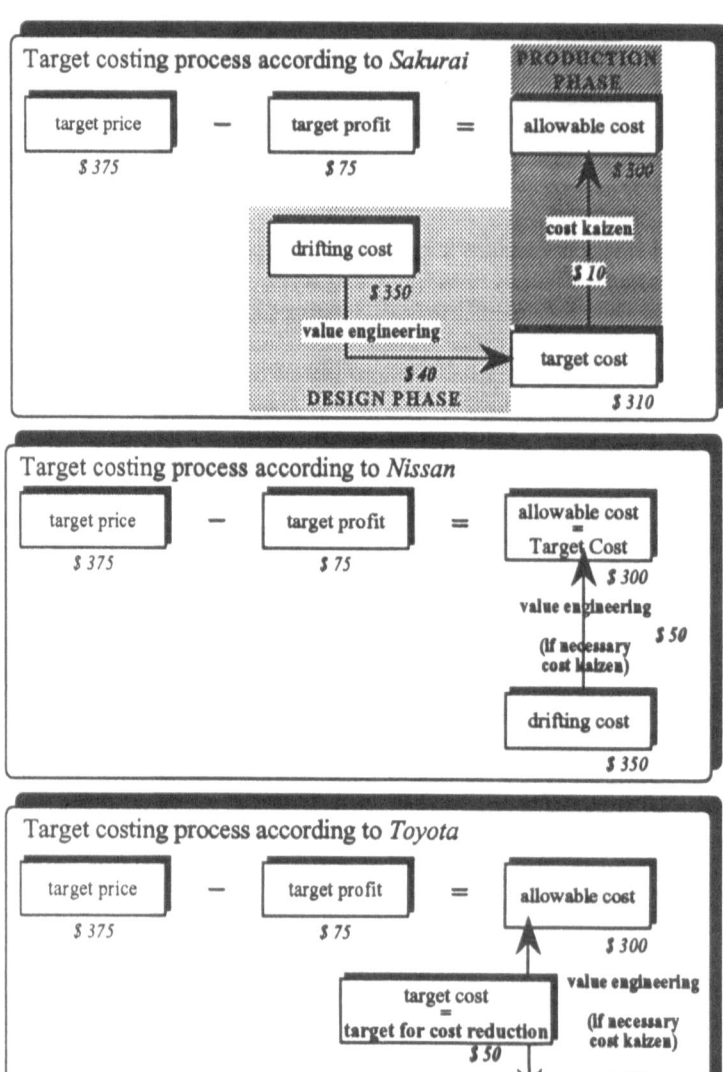

Figure 2.1: *Different ways of defining the target cost.*

The desired *target profit* is derived from profit planning. The result is the *allowable cost* (target price - target profit), defined as the highest acceptable cost of a product for the quality level defined by customer requirements and competitive constraints. In the next step it is compared to the *drifting cost*; this is the estimated cost of the product that can be reached by using existing technology and process standards. The difference between these two costs is the goal of the target costing process. The target cost itself can be defined in different ways as shown in figure 2.1.

Action is needed to close the gap between allowable and drifting cost. However, the target cost as a single figure is too aggregated for direct cost reduction. It requires a further breakdown, usually to the level of components.

To realize this breakdown, *Tanaka* describes a model called *functional area method* to evaluate components and parts based on market information. This views a product as a combination of functions and evaluates each product component with regard to the contribution it makes to the performance of each function. This ensures that a consistent market orientation is retained even when working at the target costs of components and parts. This method is explained in chapter 2.6 by means of a real life example.

2.4 Target Costing and Market Research

Target costing starts with the procurement of market information. To ensure that this information is fully used in target costing, an organizational link is needed between market research and the other departments involved in the development process. The flow of information is not limited to one direction or to a certain point in time in this process. There will be feedback and feedforward loops - e.g. design alternatives will have to be evaluated with regard to customer requirements - and a constant exchange of information during the whole development process. Thus market research is needed to ensure the market orientation of the whole development process.

To support the development process with suitable information, market research has to explore customer needs and wishes and has to transform this aggregated information into specific requirements of product functions. Within the concept of the new product, these functions have to be weighted according to their degree of importance to the customer. One has to distinguish between hard functions (mechanical functions which are necessary to fulfil technical requirements) and soft functions (which describe the convenience and personal values of the customer) (see figure 2.2).

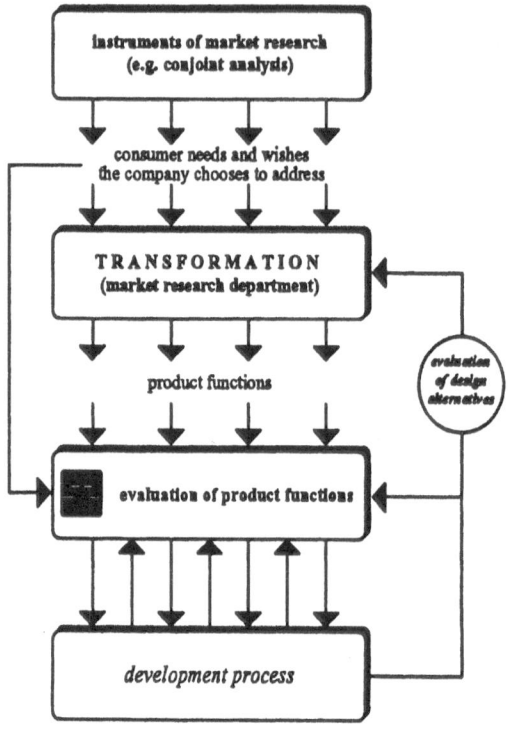

Figure 2.2: *Target costing and market research.*

To derive the required product functions from general customer wishes and needs and from the total product concept, market research has to apply instruments for the decomposition of information on customers' perceptions and needs. An instrument which has acquired major importance in this context is *conjoint analysis*. Conjoint analysis or conjoint measurement is a method for measuring psychological value judgements. It is the combination of a number of methods of multivariate data analysis which attempt to define the relation between the total view of an object and the features combining to this view.

In the process of measuring a complete product is evaluated by a potential customer. The global judgement is decomposed and so called part worth utilities are derived by decompositional methods. Thus the contribution of each component to total utility of the product can be specified. It is also possible to estimate how the subjectively perceived value changes if product features are modified.

2.5 Practical Experiences with Target Costing

The concept of target costing was mainly developed by Japanese companies (especially assembly oriented industries) in the early 1970s. Since then, the use of target costing has spread to other industries and is widely used by all kinds of businesses and industries in Japan. European and American experiences with target costing are still very limited. The concept became known in the late 1980s and companies are still evaluating and testing target costing for practical application. Therefore, empirical evidence on the use of target costing can only be found in Japan. The findings described below originate from a field study which was carried out in Japan in 1991/92 and was published in Japanese starting May 1992 (*Kobayashi et al* [1992]).

The study shows that target costing is mainly applied in the automobile (100%) and electrical industry (88,5%) as well as machinery (82,8%) and precision equipment (75%). It is hardly or not at all used in chemical/pharmaceutical industry (31,3%), food (28,6%), steel (23,1%), and paper/pulp (0%). The major departments involved in target costing are design (22,0%), accounting and product planning (both 17,6%), production technology (14,3%) and development (12,1%).

When adopted, the main objective of target costing is cost reduction. Other objectives like *"satisfying customer needs"*, *"quality"*, and *"time to market"* become increasingly important the longer target costing is applied.

As described, target costing is essentially a team-oriented, cross-functional activity. The members of the target costing team are from the following functional areas: production technology (73,6%), design (70,8%), purchasing (67,9%), development (54,7%), marketing (46,2%), manufacturing (45,3%), product planning (41,5%), and accounting (37,7%). In general suppliers are not represented in the team (6,6%). In most cases, there is no special person responsible for the target costing process.

At present, material cost (98,1%), purchased parts (97,2%), and direct processing costs (99,1%) are the object of target costing in almost all surveyed companies using "Genka Kikaku". Indirect costs become more and more important (indirect processing costs 80,7%, logistics costs 69,4%). If they are not yet included, their integration is at least under consideration.

Industry Application of Target Costing
1992 Study of Japanese Companies

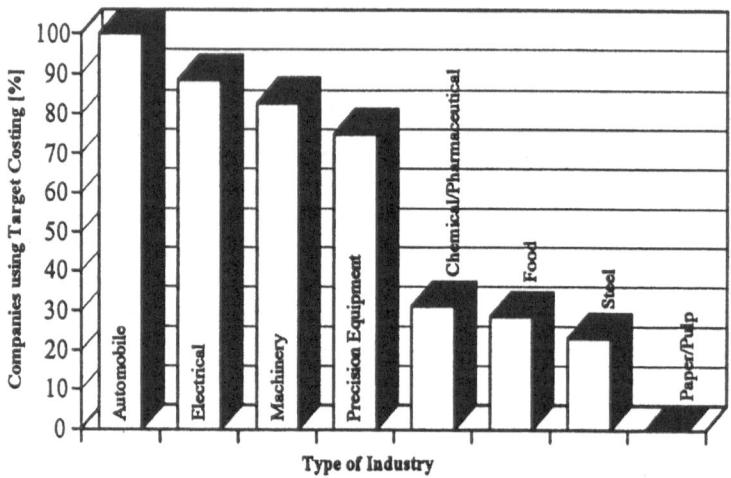

Departments Involved in Target Costing
1992 Study of Japanese Companies

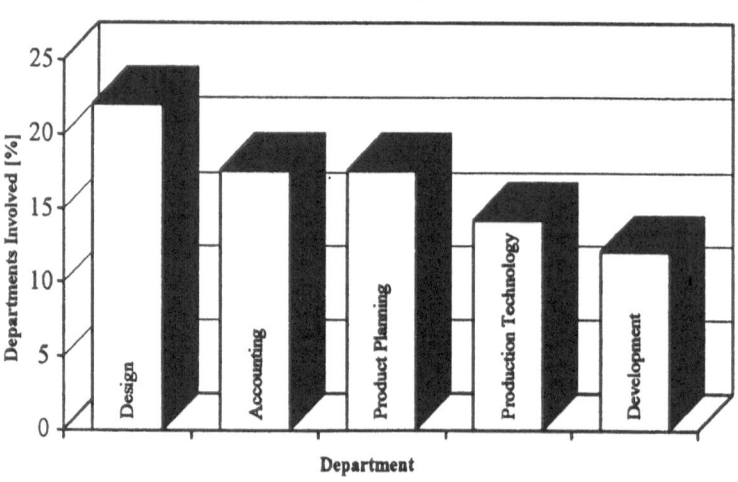

Figure 2.3: *Results of the empirical study, part I*

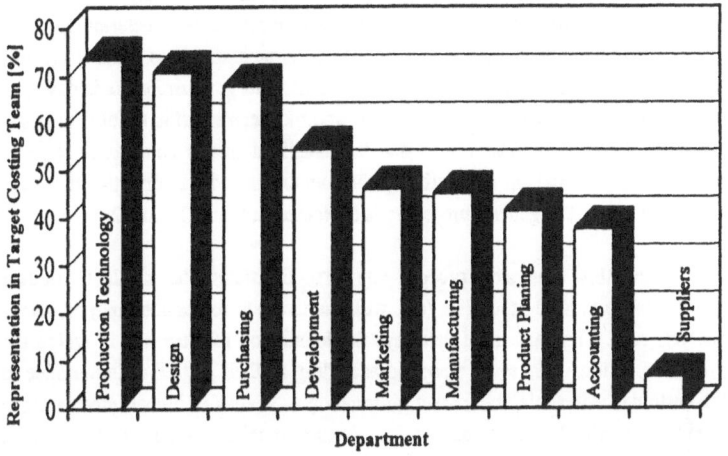

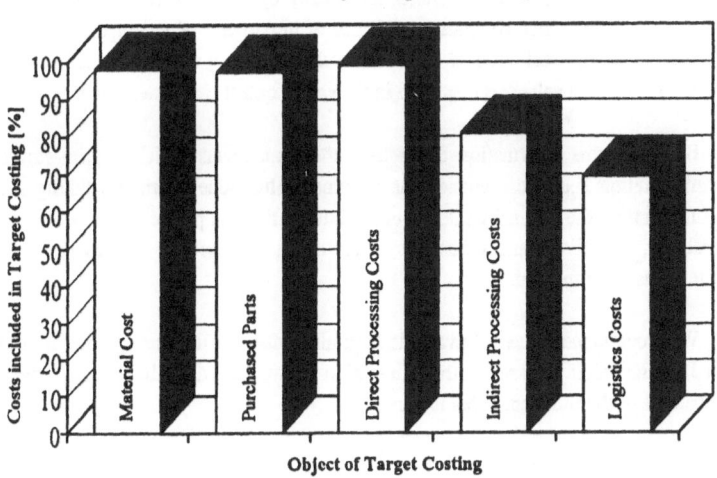

Figure 2.4: *Results of the empirical study, part II*

2.6 Example for the Application of the Functional Area Method

The following (real-life) example shall give an idea of how target costing can work in practice. After the target cost of a new product has been determined based on market research and internal cost analyses, the next step is to break down this aggregated figure into parts and component costs. A *target cost matrix* is designed to support this step by relating components and their contribution to fulfil the market derived functions. The results can be visualized in the *value control chart* which is also a visual aid to identifying the components where corrective measures are most urgent. The general procedure is explained in figure 2.5 and 2.6.

To enter this four step process one first translates the available market information into hard input data (e.g. with multivariate regression analysis, factor analysis etc.). The hard input data lays the foundation for step one in which the product functions have to be weighted according to their degree of importance to the customer. After this, the market information as well as technical information is used to determine how the product's components combine to realize and fulfil each function which is assumed to be realized 100 % by existing or future components.

In step two the degree of importance of each component is determined by multiplying the components' contributions to the product functions and the relative importance of the product functions, and finally by summing the results for each component. Step three compares the degree of importance of a component and its actual share of total product cost. The value index, giving a measure of the deviation of both values, is displayed in the value control chart (step four).

Based on this information one selects the components which require further consideration (i.e. those outside the "optimal value zone"), without taking into consideration the original weakness of the information. So problems can occur with stability, sensitivity, and robustness of the decisions based on the results of the functional area method.

We recommend to use all available hard and soft/weak information (not translated) in the functional area method. In section 3.2 we shall describe necessary modifications of the functional area method.

The target costing process can be devided into three phases in which the different departments and functions of a company are involved with different levels of intensity (see figure 2.7). The marketing function is of major importance in defining the product functions based on customer wishes or demands.

In a first step, the product functions, F1 to F8, which the customers require, are weighted according to their degree of importance. This is the external perspective. Second, it has to be determined to what extent (in %) the product's components, C1 to C7, contribute to the realization of each product function. This is the internal perspective.

Functions / Components	F 1	F 2	F 3	F 4	F 5	F 6	F 7	F 8	Total
	0.26	0.24	0.15	0.10	0.08	0.10	0.02	0.05	1.00
C 1	13.4	38.7	19.0	23.0	28.8	28.2	20.0	33.6	
C 2	32.7		18.7	18.0	2.5	10.4		11.3	
C 3	50.6	6.4	20.3	25.6	23.7	30.7	45.0	24.6	
C 4		9.2	5.3	6.7	5.0	9.7		2.3	
C 5		4.2	13.7	7.0	6.2	4.8	5.0	15.6	
C 6	3.3	27.8	18.7	17.0	25.0	9.7	20.0	7.7	
C 7		13.7	4.3	2.7	8.8	6.5	10.0	4.9	
Total	100	100	100	100	100	100	100	100	

In the next step, the degree of importance of each component is determined. First, the degree of importance of each function is multiplied by the percentages the components contribute to the realization o f each function. Second, the lines are added up. The total of each line shows the degree of importance of each component (in %).

Functions / Components	F 1	F 2	F 3	F 4	F 5	F 6	F 7	F 8	Total
	0.26	0.24	0.15	0.10	0.08	0.10	0.02	0.05	1.00
C 1	3.5	9.2	2.9	2.3	2.3	2.8	0.4	1.7	25.1
C 2	8.5		2.8	1.8	0.2	1.0		0.6	14.9
C 3	13.2	1.5	3.0	2.6	1.9	3.0	0.9	1.2	27.3
C 4		2.2	0.8	0.7	0.4	1.0		0.1	5.2
C 5		1.0	2.1	0.7	0.5	0.5	0.1	0.8	5.7
C 6	0.9	6.6	2.8	1.6	2.0	1.0	0.4	0.4	15.7
C 7		3.3	0.6	0.3	0.7	0.7	0.2	0.3	6.1
Total									100.0

Figure 2.5: *Building up the target cost matrix.*

154

The degree of importance of each component is compared to the actual percentage of cost a component consumes. Dividing the degree of importance by the percentage of cost of each component gives the value index, which is a measure for the deviation of importance (which the component should be given) and component cost.

Components	Percentage of Cost	Degree of Importance	Value Index
C 1	13.9	25.1	1.8
C 2	12.0	14.9	1.2
C 3	17.6	27.3	1.6
C 4	7.6	5.2	0.7
C 5	8.6	5.7	0.7
C 6	16.8	15.7	0.9
C 7	23.5	6.1	0.3

The value control chart (below) relates degrees of importance and percentage of cost of the components. In the ideal case, the value index is 1 (i.e. degree of importance = percentage of cost). The deviation of both measures is accepted within a certain range, called optimal value zone. Value points outside this zone require further consideration.

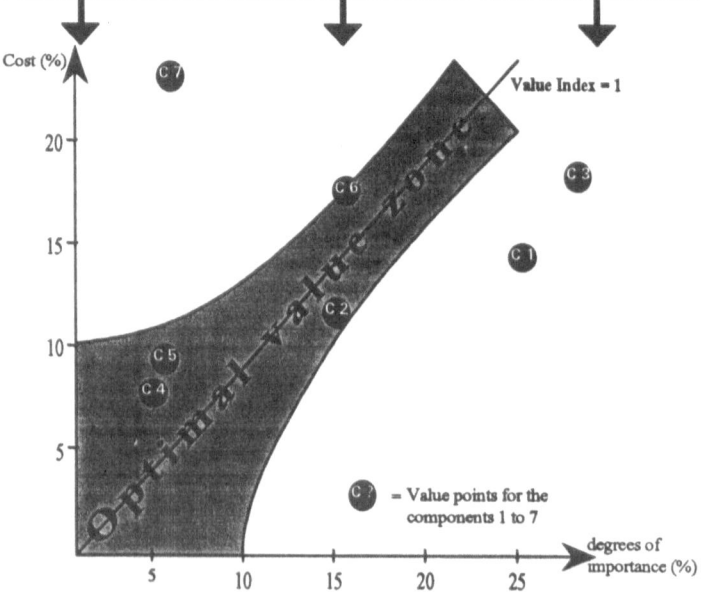

Figure 2.6: *Building up the value control chart.*

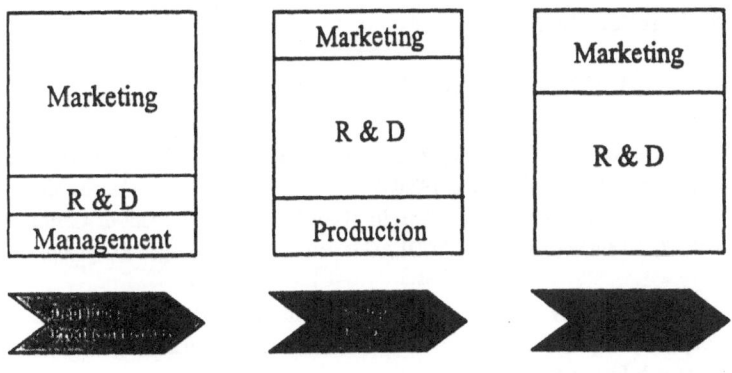

Figure 2.7: *Importance of corporate functions in different phases of the target costing process.*

In phase 2, the R&D department has to develop a basic concept of the new product which aims at fulfilling the market requirements. At this stage a close cooperation with marketing and production people is needed to ensure market orientation as well as manufacturability of the concept design. Finally, the determined product functions have to be transformed into specific technical solutions for product components by the R&D department. Again, the market information on functional requirements and target costs for components is most important.

The different phases, with different disciplines dominating the information processing process, also indicate which kind of data problem we encounter. We therefore selected two areas, market data and the setting of target costs, to develop our ideas.

3. Market Data

3.1 Real-Life Example - the Case of AUDI AG

AUDI is one of Germany's major car manufacturers. Sales in 1992 were 16.7 bill. DM and production volume was 492,000 cars. The company has production sites in Ingolstadt (head office) and Neckarsulm (both located in the south of Germany) with 37,160 employees (1992). AUDI AG is part of the Volkswagen AG group.

In this case study target costing was applied to the AUDI B-series (AUDI 80, Coupé, Cabriolet). The assumption was made that all data (including soft factors) are exactly known ("hard data").

In the past, planning for a new model was mainly based on the following information:
- estimated production volume, depending on assumptions on the market development;
- desired target profit depending on internal criteria like general profitability and growth targets;
- general requirements on technology and schedule of the development process; target price, based on the price of the old model plus a certain increase.

Based on this information the target cost of the new model was determined. Yet, to guideline the development process more differentiated cost targets for parts and components were needed. The required process of breaking down the target cost was mainly achieved by comparison to the old model. Thus the production cost of main components was estimated, targets for cost reduction were set and the development cost as well as the required capital investment was estimated. So, one main objective of the target costing implementation was to decompose the target cost of the new model along market requirements.

Another objective was to have more market information in the first place. The problem of providing adequate market data proved to be one of the main difficulties of the study.

The determination of the required product functions is probably the most difficult part of the target costing process. Different customers view a product in different ways. What they perceive is a mixture of objective product functions which are mainly of technical nature and which can be directly influenced by the producer (so called "hard functions"), of subjectively determined benefits which are corresponding to the utility created by the product, and of subjectively perceived product features which convey a certain style or image (so called "soft functions"). Depending on the specific product and market, these criteria are of different importance and lead to a more or less technically seizable product profile. In order to support this transformation process, market research applies different methods to determine the relevant product functions.

In general, market information can be procured by questioning (potential) customers or by observation. This primary data is analysed with uni- or bivariate methods (e.g. correlation or regression analysis) or with multivariate methods (e.g. multiple regression, variance, factor and cluster analysis, multidimensional scoring). The importance of conjoint measurement for analysing customers' perceptions of products and their features has already been mentioned.

The degree of importance of the product functions was derived from existing market studies concerning product image and new-car buyer attitudes. The data was based on questioning 32,000 new-car buyers, there was no differentiation between car types and brands. In this special case a combination of data on "satisfaction with car features or functions" and "importance of the functions to the customers" was used and similar features were combined to main functions. The results are shown in figure 3.1.

The transformation of "the importance of functions" to "the importance of components" was realized with the target cost matrix (see figure 3.2). The extent to which the main components fulfil the required functions was determined in discussions with representatives of the departments involved in the development process, as well as marketing representatives.

A comparison of these results to the estimated standard manufacturing cost of the five main components showed that cost reductions are required for the body (-16%) and the chassis (-28%), the engine (+5%), the electric (+100%) and the equipment (+18%) might be more expensive in manufacturing (see figure 3.3). The comparison of allowed cost and estimated standard cost was also displayed in the value control chart with two different optimal value zones: q = 10 and q = 15 (see figure 3.4). This was the basis of a discussion on further steps concerning the development of components and measures for cost management.

Using the results from uni- or multivariate analysis methods as hard input data for the selection of components which require further consideration, without taking into consideration the underlying mathematical assuptions and statistical results, no information will be available for example about the "quality of decision" and stability of this selection.

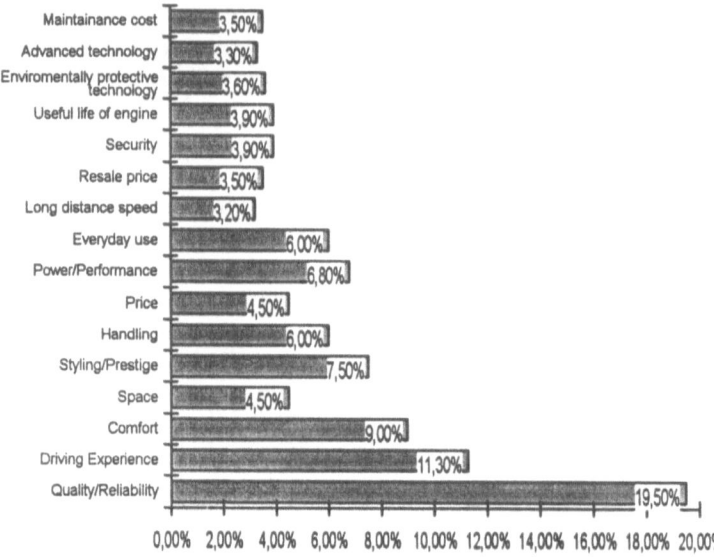

Figure 3.1: *Degrees of importance of the car functions.*

%	Car Features	Engine		Electric		Body		Chassis		Equipping	
19,50%	Quality/Reliability	20%	3,9	18%	3,5	30%	5,9	15%	3	17%	3,3
11,30%	Driving Experience	21%	2,4	9%	1	12%	1,4	51%	5,7	7%	0,8
9,00%	Comfort	8%	0,7	8%	0,8	17%	1,5	5%	0,5	62%	5,6
4,50%	Space	5%	0,2	5%	0,2	58%	2,6	20%	0,9	13%	0,6
7,50%	Styling/Prestige	8%	0,6	11%	0,9	44%	3,3	15%	1,2	21%	1,6
6,00%	Handling	-	-	51%	3	3%	0,2	10%	0,6	36%	2,2
4,50%	Price	15%	0,7	25%	1,1	23%	1	13%	0,6	25%	1,1
6,80%	Power/Performance	45%	3,1	13%	0,9	18%	1,2	15%	1	10%	0,7
6,00%	Everyday use	27%	1,6	4%	0,2	39%	2,3	24%	1,4	7%	0,4
3,20%	Long distance speed	20%	0,6	20%	0,6	20%	0,6	20%	0,6	20%	0,6
3,50%	Resale price	10%	0,4	5%	0,2	50%	1,8	5%	0,2	30%	1,1
3,90%	Security	5%	0,2	5%	0,2	50%	2	10%	0,4	30%	1,2
3,90%	Useful life of engine	95%	3,7	5%	0,2	-	-	-	-	-	-
3,60%	Envirom. techn.	30%	1,1	15%	0,5	20%	0,7	20%	0,7	15%	0,5
3,30%	Advanced technology	20%	0,7	20%	0,7	20%	0,7	20%	0,7	20%	0,7
3,50%	Maintenance cost	15%	0,5	15%	0,5	45%	1,6	20%	0,7	5%	0,2
100,00%	Sum in %		20		14		27		18		20

Figure 3.2: *Target cost matrix.*

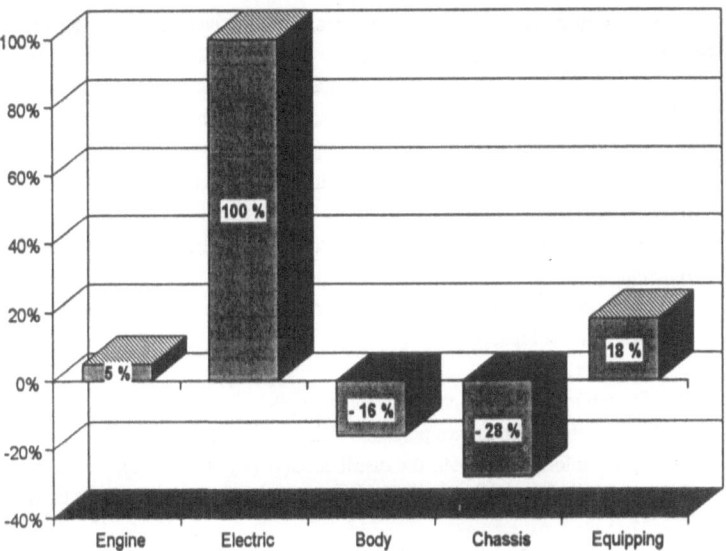

Figure 3.3: *Comparison of allowed cost and estimated standard cost.*

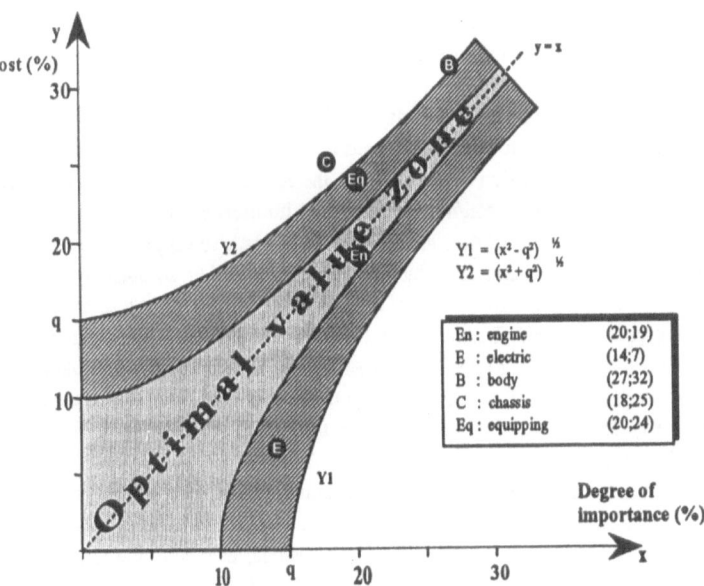

Figure 3.4: *Value control chart.*

A typical way of consolidating contradictory (subjective) customer statements about the same (soft) product function today is the use of mean values, without taking into consideration any other additional information as input data for the functional area method. So we have to discuss how we can use weak information about

- the degree of importance of the product's functions especially for the soft functions,
- the contribution of the product's components to the realization of each product function,
- and which modification of the functional area method is necessary to work with this kind of weak input data.

Weak information about the data can be for example
- random variables with known probability distribution(s),
- incomplete information about the distribution(s) (e.g. information about mean and variances of the random variables or so-called linear partial information).

Typical weak information about the degree of importance of the functions are
- finite (discrete) distributions as a result of market research,
- uniform or normal distributions,
- results from multivariate regresssion or factor analysis (conjoint analysis), including statistical information .

Using weak information as input data for the functional area method the selection of components which require further consideration is based on all available information about the decision model, without the necessity to add mathematical assumptions to describe soft/non-deterministic information with hard numbers. Changes in the functional area method have to be made to support the use of non-deterministic information as input data. In the following sections we describe necessary modifcations for the use of random variables with known or only partial known probability distribution as input data for the degree of importance of the components and the target cost matrix coefficients. We propose two possible ways:

- Replace the optimal value zone by a so-called α-optimal value zone (see section 3.2.1). In this case the classification of the product components is based on a given probability α.
- Modify the computing of the degree of importance of components using a stochastic programming approach (see section 3.2.2). In this case we can use the other components of the functional area method, especially the well-accepted optimal value zone chart as (management) presentation tool.

3.2 Exactly Known Distribution

We assume that the "degree of importance" $f := (f_j)_{j=1,2,...,n}$ of the functions F_j ($j=1,2,...,n$), which the customers require, is a random vector

$$f: \Omega_f \to \mathfrak{R}^n,$$

with $\Sigma_j f_j(\omega) = 1$ for every $\omega \in \Omega_f$ and probability distribution P_f. For each function F_j we assume, that the contribution of each component C_i ($i=1,2,...,m$) to the realization of product function F_j (in %) is a random vector $c_j := (c_{ij})_{i=1,2,...,m}$:

$$c_j: \Omega_j \to \mathfrak{R}^m,$$

with $\Sigma_i c_{ij}(\omega) = 100$ for every $\omega \in \Omega_j$, $j=1,2,...,n$ and probability distribution P_j. For $j=1,2,...,n$ random variable c_j represents a column of the target cost matrix.

3.2.1 α-optimal Value Zone

One way of using random variables as input data for the functional area method is the replacement of the optimal value zone by a so-called α-optimal value zone, because the definition of the optimal value zone is not usable with random variables as input data.

Definition 3.1: For a given probability $0 \leq \alpha \leq 1$ the value point of component C_i is element of an *α-optimal value zone*, if

$$\text{prob}(\,|\,\text{imp}_i^2(\omega) - \text{poc}_i^2\,| \leq q^2) \geq \alpha, \tag{3.1}$$

for a given q, with random variable ("degree of importance")

$$\text{imp}_i(\omega) := \Sigma_j f_j(\omega_f) c_{ij}(\omega_j).$$
$$\text{imp}_i: \quad \Omega_f \times \Omega_1 \times \Omega_2 \times \ldots \times \Omega_m \to \mathfrak{R}.$$

"Percentage of cost" poc_i normally is coming from cost tables (databases of detailed cost information based on a number of manufacturing variables).

Components with their value points outside the α-optimal value zone require further consideration. For restriction (3.1) we used the boundary functions from the definition of the optimal value zone (see section 2.6). Other definitions for the inequality condition $|\text{imp}_i^2(\omega) - \text{poc}_i^2| \leq q^2$ in (3.1) are also possible, using e.g. another distance measure .

In some real-life situations parameter q was used to handle the uncertainty in the input data. But in fact q is a target costing model parameter. With definition 3.1 of an α-optimal value zone it is possible to separate the management of the influence coming from the uncertainty in the input data material (using parameter α) from the definition of the target costing model parameter q itself.

Example 3.1: We assume, that only the degree of importance of the functions f_j is random and take the numbers from section 2.6. The expected value of element f_j from random vector f is the degree of importance from this deterministic example. Mean values for example were used from AUDI (see section 3.1 and *Deisenhofer* [1993]) to consolidate contradictory (subjective) customer statements about the same product function:

number of realizations of f: $N := 2$,

probabilities: $p_1 := P_f(\omega_1) := 0,8$,

$p_2 := P_f(\omega_2) := 0,2$

In case of a finite (discrete) distribution from random vector f with N realizations of f and finite set $\Omega_f := \{\omega_1, \ \omega_2,..., \ \omega_N\}$

$$\text{prob}(\ |\ \text{imp}_i{}^2(\omega) - \text{poc}_i{}^2\ | \leq q^2) \ = \ \Sigma_k \ P(\omega_k) \ 1_A(\omega_k),$$

with $A := \{\omega \in \Omega_f \ | \ (\ |\ \text{imp}_i{}^2(\omega) - \text{poc}_i{}^2\ |\ ) \leq q^2\}$.

The following steps are necessary for the computation of the components inside and outside the α-optimal value zone:

Step 1: For every realization $f(\omega_k)$ of the random variable "degree of importance" f (see table 3.1) compute for every component C_i, whether this component is inside or outside the optimal value zone, using the functional aera method, which is described in section 2.6. The results are shown in table 3.2 For every realization of the random variable you have to solve a non-stochastic problem.

Step 2: For every component C_i determine the realizations $f(\omega_k)$ of random vector f with component C_i inside the optimal value zone (for $k = 1,2,...,N$) and compute the sum of the corresponding probabilities $P(\omega_k)$. If the sum of these probabilities is greater or equal α, than C_i is inside the α-optimal value zone, otherwise C_i is outside (see table 3.3).

f	$f(\omega_1)$	$f(\omega_2)$	Ef
f_1	0,16	0,66	0,26
f_2	0,3	0	0,24
f_3	0,1875	0	0,15
f_4	0,11	0,06	0,10
f_5	0,1	0	0,08
f_6	0,08	0,18	0,10
f_7	0	0,1	0,02
f_8	0,0625	0	0,05

Table 3.1: *Values for random vector "degree of importance" f.*

c	poc	imp(ω_1)	imp(ω_2)	imp(Ef)	value index for imp(ω_1)	imp(ω_2)	imp(Ef)
C_1	13,9	27,1	17,3	25,1	1.9	1.2	1.8
C_2	12,0	12,5	24,5	14,9	1,0	2,0	1,2
C_3	17,6	23,0	45,0	27,3	1,3	2,6	1,55
C_4	7,6	5,9	2,2	5,2	0,8	0,3	0,7
C_5	8,6	6,6	1,8	5,7	0,8	0,2	0,7
C_6	16,8	18,0	6,9	15,7	1,1	0,4	0,9
C_7	23,5	6,9	2,3	6,1	0,3	0,1	0,3

poc: percentage of cost of components
imp: degree of importance of components

Table 3.2: *Value index for example 3.1.*

For $\alpha=0,75$ and $q=15$ the following restrictions (see (3.1)) for the α-optimal value zone have to be fulfilled

$$\text{prob}(\mid \text{imp}_i{}^2(\omega) - \text{poc}_i{}^2 \mid \leq 225) \geq 0,75,$$

for i=1,2,...,7. Every component C_i which fulfils the restriction above, is with*in* the α-optimal value zone, if the restriction is not fulfilled, the component is *out*side (see table 3.3). In table 3.3 the results for the α-optimal value zone ("0,75-optimal value zone") are compared with the determinsitic examples ("optimal value zone") with input data $f(\omega_1)$, $f(\omega_2)$ (= realizations of random vector f), and Ef (= results from section 2.6) for the degree of importance of the functions.

In our (non-deterministic) example component C_3 is lying within the α-optimal value zone, but is outside the optimal value zone in the corresponding deterministic example, which is defined based on the expected value (Ef) as input data. For input data $f(\omega_1)$ and $f(\omega_2)$ we have different results in comparison to the results from the deterministic example from section 2.6, only for component C_4, C_5 and C_7 we have the same results ("in(side) or out(side) the (α-)optimal value zone) with the different deterministic and non-deterministic input data. In our example the results for the components in/out the α-optimal value zone highly depend upon the results for $f(\omega_1)$, because probability $p(\omega_1) = 0,8$.

Component	C_1	C_2	C_3	C_4	C_5	C_6	C_7
prob($|imp_i(\omega)^2-poc_i^2|\leq225$)	0,2	0,8	0,8	1	1	0,8	0
α-optimal value zone	out	in	in	in	in	in	out
optimal value zone for $f(\omega_1)$	out	in	in	in	in	in	out
optimal value zone for $f(\omega_2)$	in	out	out	in	in	out	out
optimal value zone for Ef[1])	out	in	out	in	in	in	out

[1]) deterministic example from section 2.6

Table 3.3: *(α-)optimal value zone for example 3.1.*

We assume furthermore, that only the degree of importance of the functions f is a random vector. Random variables f with finite (discrete) distribution are e.g. results from interviewing a certain number of people. If one only knows that the elements of vector f are within an interval without any further information about the distribution, the following example for the degree of importance of the functions can be an approach for this situation:

$$f: \Omega_f \rightarrow \{[a_1,b_1] \times [a_2,b_2] \times ... \times [a_n,b_n]\} \cap \{\xi \in \Re^n | \Sigma_j \, \xi_j = 1\},$$

and f uniformly distributed with fixed $a_j, b_j \in \Re$ for $j=1,2,...,n$.

3.2.2 Degree of Importance: A Stochastic Optimization Model

Replacing the computing of the degree of importance of the components by a stochastic optimization model in the non-deterministic case allows us to use the optimal value zone chart as (management) presentation tool for the results of target costing models. To find optimal deterministic solutions x_i^* for the (stochastic) degree of importance, we have to solve the following optimization problem with m objective functions:

$$\min \{(| x_i^2 - poc_i^2 |^{1/2})_{1 \leq i \leq m}\} \qquad (3.2)$$
with restrictions: $x_i = \Sigma_j \, f_j(\omega_f) c_{ij}(\omega_j),$
$\Sigma_i \, x_i = 100,$
$x_i \geq 0$ for $i=1,2,...,m$

and all $\omega_f \in \Omega_f$, $\omega_j \in \Omega_j$ for $j=1,2,...,n$.

The restriction $\Sigma_i \, x_i = 100$ is included to show the importance of this restriction, but is automatically fulfilled, because $\Sigma_i \, x_i = \Sigma_i \Sigma_j \, f_j(\omega_f) c_{ij}(\omega_j) = 100$ for all $\omega_f \in \Omega_f$, $\omega_j \in \Omega_j$ for $j=1,2,...,n$.

Other definitions of the objective functions from (3.2) are also possible, e.g.

$$\min \{((d(x_i, poc_i))_{1 \leq i \leq m}\}$$

with distance d. Objective is to find the degree of importance of the components close to the percentage of cost. The important components for decisions are components with high percentage of cost (poc_i) values. We can not solve the m optimization problems (3.2) independently because of the (joint) distribution of the right-hand side of the restrictions. To solve (3.2) we can generate a corresponding optimization problem, using the weighted sum of the m objective functions (with weights poc_i) as objective function:

$$\min \{\Sigma_i \, poc_i | x_i^2 - poc_i^2 |^{1/2}) \ | \quad x_i = \Sigma_j \, f_j(\omega_f) c_{ij}(\omega_j), \qquad (3.3)$$
$\Sigma_i \, x_i = 100,$
$x_i \geq 0$ for $i=1,2,...,m\},$

and all $\omega_f \in \Omega_f$, $\omega_j \in \Omega_j$ for $j=1,2,...,n$

Point-of-interest are solutions x_i^* of (3.3), with minimal difference in the restrictions after the realization of the random variables especially for components C_i with large poc_i values. Feasibility and optimality for a solution of (3.3) are not well-defined. We use the *two-stage programming* (*stochastic programming with recourse*) approach to define feasibility and optimality for (3.3). The following deterministic equivalent for (3.3) has to be solved to find optimal solutions "degree of importance" x^*:

$$\min E\{\Sigma_i\, poc_i \,|\, x_i^2 - poc_i^2 \,|^{1/2}\} + \min\{\Sigma_i\, poc_i^2(y_i^+(\omega)+y_i^-(\omega))\,| \quad (3.4)$$
$$y_i^+(\omega) - y_i^-(\omega) = \Sigma_j\, f_j(\omega_f)c_{ij}(\omega_j)\text{-}x_i,$$
$$y_i^+(\omega),\ \ y_i^-(\omega) \geq 0,\ i=1,2,...,m\}|$$
$$\Sigma_i\, x_i = 100,$$
$$x \geq 0\ \},$$

for all $\omega_f \in \Omega_f$, $\omega_j \in \Omega_j$ for $j=1,2,...,n$ and $\omega \in \Omega$. The important situation for practice are random vectors with discrete distributions. In this case (3.4) can be solved with well-known numerical methods.

The expected value E in (3.4) can be replaced by E - kV (with parameter k) to keep the differences in the objective function (after realization of the random variables) sufficiently small.

Example 3.2: We assume that only the degree of importance of the functions f is a random vector with finite discrete distribution P_f from f with N realizations of the random variable f and $\Omega_f := \{\omega_1, \omega_2,..., \omega_N\}$ is a finite set. In this case we have to solve the following (deterministic) optimization problem (3.4):

$$\min\{\Sigma_i\, poc_i \,|\, x_i^2 - poc_i^2\,|^{1/2}\}\ +\ \Sigma_k\, P(\omega_k)\, \Sigma_i\, poc_i^2(y_i^+(\omega_k) + y_i^-(\omega_k))\,|$$
$$x_i + y_i^+(\omega_k) - y_i^-(\omega_k) = \Sigma_j\, f_j(\omega_k)c_{ij}, \qquad (3.5)$$
$$\Sigma_i\, x_i = 100,$$
$$x_i,\ y_i^+(\omega_k),\ y_i^-(\omega_k) \geq 0,\ \text{for } 1\leq i\leq m,\ 1\leq k\leq N\}.$$

Using the numbers from example 3.1 we have to solve the following nonlinear optimization model with linear restrictions:

$$\min\{13,9|x_1^2 - 13,9^2|^{1/2} + 12,0|x_2^2 - 12,0^2|^{1/2} + 17,6|x_3^2 - 17,6^2|^{1/2}$$
$$+ 7,6|x_4^2 - 7,6^2|^{1/2} + 8,6|x_5^2 - 8,6^2|^{1/2} + 16,8|x_6^2 - 16,8^2|^{1/2}$$
$$+ 23,5|x_7^2 - 23,5^2|^{1/2}$$
$$+ 0,8(13,9^2y_{11}^+ + 13,9^2y_{11}^- + 12,0^2y_{21}^+ + 12,0^2y_{21}^- + 17,6^2y_{31}^+$$
$$+ 17,6^2y_{31}^- + 7,6^2y_{41}^+ + 7,6^2y_{41}^- + 8,6^2y_{51}^+ + 8,6^2y_{51}^-$$
$$+ 16,8^2y_{61}^+ + 16,8^2y_{61}^- + 23,5^2y_{71}^+ + 23,5^2y_{71}^-)$$

$$+ 0,2(13,9^2y_{12}^+ + 13,9^2y_{12}^- + 12,0^2y_{22}^+ + 12,0^2y_{22}^- + 17,6^2y_{32}^+$$
$$+ 17,6^2y_{32}^- + 7,6^2y_{42}^- + 7,6^2y_{42}^- + 8,6^2y_{52}^+ + 8,6^2y_{52}^-$$
$$+ 16,8^2y_{62}^+ + 16,8^2y_{62}^- + 23,5^2y_{72}^+ + 23,5^2y_{72}^-) \quad |$$

$$
\begin{aligned}
x_1 + y_{11}^+ - y_{11}^- &= 27,1 \\
x_2 + y_{21}^+ - y_{21}^- &= 12,5 \\
x_3 + y_{31}^+ - y_{31}^- &= 23,0 \\
x_4 + y_{41}^+ - y_{41}^- &= 5,9 \\
x_5 + y_{51}^+ - y_{51}^- &= 6,6 \\
x_6 + y_{61}^+ - y_{61}^- &= 18,0 \\
x_7 + y_{71}^+ - y_{71}^- &= 6,9 \\
x_1 + y_{12}^+ - y_{12}^- &= 17,3 \\
x_2 + y_{22}^+ - y_{22}^- &= 24,5 \\
x_3 + y_{32}^+ - y_{32}^- &= 45,0 \\
x_4 + y_{42}^+ - y_{42}^- &= 2,2 \\
x_5 + y_{52}^+ - y_{52}^- &= 1,8 \\
x_6 + y_{62}^+ - y_{62}^- &= 6,9 \\
x_7 + y_{72}^+ - y_{72}^- &= 2,3 \\
\Sigma_i x_i &= 100 \\
x, \quad y^+, \quad y^- &\geq 0 \}.
\end{aligned}
$$

Remarks: This example cannot be solved with the well-known standard algorithms for nonlinear programming, because it is not possible to compute the gradients of the objective function. An earlier version of this optimization problem (with other weights) was solved from F. Schwefer (Federal Armed Forces University, Munich), using some modifications of existing computer algorithms. Based on this results we modified the weights from the y-variables in model (3.4) and used poc_i^2 as weights, which gives a more realistic balance between the x- and y-variables. For models with large m-values it is important for the solution effort, that we can avoid the problems, which come from the structure of the objective function. The well-known solution methods can be used, when we replace the m-objective functions from model (3.2) with

$$((x_i^2 - poc_i^2)^2)_{1 \leq i \leq m}. \tag{3.2a}$$

In this case the following modifications are necessary for the objective functions from model (3.3.)-(3.5) .

For model (3.3): $\qquad \Sigma_i poc_i (x_i^2 - poc_i^2)^2,$ $\qquad\qquad$ (3.3a)

for model (3.4): $\quad E\{\min_y \Sigma_i (poc_i ((x_i^2 - poc_i^2)^2 + (poc_i(y_i^+(\omega) + y_i^-(\omega)))^4))\},$ $\quad$ (3.4a) and for model (3.5):

$$\Sigma_i \left(poc_i((x_i^2 - poc_i^2)^2 + \Sigma_k P(\omega_k)(poc_i(y_i^+(\omega_k) + y_i^-(\omega_k)))^4) \right) = \quad (3.5a)$$
$$\Sigma_i \left(poc_i((x_i^2 - poc_i^2)^2 + \Sigma_k P(\omega_k)poc_i^4(y_i^+(\omega_k)^4 + y_i^-(\omega_k)^4)) \right),$$

because $y_i^+(\omega_k) \cdot y_i^-(\omega_k) = 0$ for every $\omega_k \in \Omega_f$ for $k=1,2,...,N$.

The modified objective function in example 3.2 is:

$$13,9(x_1^2 - 13,9^2)^2 \;+\; 12,0(x_2^2 - 12,0^2)^2 \;+\; 17,6(x_3^2 - 17,6^2)^2 \;+\; 7,6(x_4^2 - 7,6^2)^2$$
$$+\; 8,6(x_5^2 - 8,6^2)^2 \;+\; 16,8(x_6^2 - 16,8^2)^2 \;+\; 23,5(x_7^2 - 23,5^2)^2$$
$$+\, 0,8\, (\; 13,9^5(y_{11}^+)^4 + 13,9^5(y_{11}^-)^4 + 12,0^5(y_{21}^+)^4 + 12,0^5(y_{21}^-)^4 + 17,6^5(y_{31}^+)^4$$
$$+\; 17,6^5(y_{31}^-)^4 + 7,6^5(y_{41}^+)^4 + 7,6^5(y_{41}^-)^4 + 8,6^5(y_{51}^+)^4 + 8,6^5(y_{51}^-)^4$$
$$+\; 16,8^5(y_{61}^+)^4 + 16,8^5(y_{61}^-)^4 + 23,5^5(y_{71}^+)^4 + 23,5^5(y_{71}^-)^4\;)$$
$$+\, 0,2\, (\; 13,9^5(y_{12}^+)^4 + 13,9^5(y_{12}^-)^4 + 12,0^5(y_{22}^+)^4 + 12,0^5(y_{22}^-)^4 + 17,6^5(y_{32}^+)^4$$
$$+\; 17,6^5(y_{32}^-)^4 + 7,6^5(y_{42}^+)^4 + 7,6^5(y_{42}^-)^4 + 8,6^5(y_{52}^+)^4 + 8,6^5(y_{52}^-)^4$$
$$+\; 16,8^5(y_{62}^+)^4 + 16,8^5(y_{62}^-)^4 + 23,5^5(y_{72}^+)^4 + 23,5^5(y_{72}^-)^4\;).$$

3.3 Partial Information about the Distribution

Distribution P_f in example 3.1 can be e.g. the result of the work of two different market research institutes. The computing of the exact values of p_1 and p_2 can become a major problem in such a situation.

We assume that only vector f is a random variable with finite (discrete) distribution P_f, $\Omega_f := \{\omega_1, \omega_2, ..., \omega_N\}$ is a finite set, and we have only partial information Θ_f about $P_f := (p_1, p_2,..., p_N)$, where Θ_f is a so-called *Linear Partial Information* (LPI)

$$\Theta_f := \{ p \in \Re^N | \; Gp \geq h, \; 0 \leq p_i \leq 1, \; \Sigma_i\, p_i = 1 \}, \quad (3.6)$$

with constant matrx G, and constant vector h (see e.g. *Kofler et. al.* [1980]). Typical examples for LPI are lower and upper bounds for the probabilities of example 3.3.

3.3.1 α-optimal Value Zone

In case of an LPI definition 3.1 for an α-optimal value zone cannot be used anymore, because the probability distribution P_f is not known exactly. Using α-optimal value zones one looks for components C_i with a degree of importance (imp_i) close to the percentage of cost (poc_i). Under the assuption of the availability of an LPI Θ_f as information about the distribution we select a distribution $P^* \in \Theta_f$ as "optimal estimation" for the only partial known distribution as solution of the optimization problem

$$\max_{P \in \Theta_f} \Sigma_i E_P((|imp_i(\omega)^2 - poc_i^2|)^{1/2}) \qquad (3.7)$$

taking into consideration the available information about the decision problem. We look for a distribution with the highest expected values, which is a pessimistic view of the whole problem. Other principles are also possible, e.g. an more optimistic oriented approach can be

$$\min_{P \in \Theta_f} \Sigma_i E_P((|imp_i(\omega)^2 - poc_i^2|)^{1/2}).$$

In this case we look for a distribution with minimal expected values.

Weighted sums with e.g. weights poc_i are also possible.

For solution $P^* \in \Theta_f$ of (3.7) we can use definition 3.1 for the α-optimal value zone and can continue with the target costing process as described in section 3.2.1.

Summary: In case of partial information about the probability distribution two steps are necessary to find the components inside and outside the α-optimal value zone:
- Step 1: Selection of a distribution P^* as element of the partial information. For this selection all available information about the decision problem should be used.
- Step 2: Using definition 3.1 for an α-optimal value zone for the distribution P^* as solution from step 1, all components C_i inside and outside the α-optimal value zone can be computed.

Example 3.3: We use the numbers from example 3.1, but assume that we have only partial information Θ_f (LPI) about probability distribution P_f available with

$$\Theta_f := \{p \in \Re^2 \mid 0 \leq p_2 \leq p_1 \leq 1,\ p_1+p_2 = 1\}.$$

For example distribution p with $p_1=0,8$, $p_2=0,2$ is an element of Θ_f.

<u>Step 1:</u> *Computation of an optimal distribution P^** as solution of optimization problem (3.7), using the numbers from table 3.2. We have to solve the following linear program:

$$\max_P\{80,8p_1 + 127,5p_2 \mid p_1 + p_2 = 1,\ p_1 - p_2 \geq 0,\ p_1,\ p_2 \geq 0\}.$$

Optimal solution $P^* = (0,5\ ,\ 0,5)$.

<u>Step 2:</u> *α-optimal value zone for P^**:

Using definition (3.1) we have the same linear restriction for the α-optimal value zone as in example 3.1. As only difference in comparison to example 3.1 we have to replace $p_1=0,8$, $p_2=0,2$ with $p_1{}^*=p_2{}^*=0,5$.

Component	C_1	C_2	C_3	C_4	C_5	C_6	C_7
prob($\mid$imp$_i(\omega)^2$-poc$_i{}^2\mid \leq 225$)	0,5	0,5	0,5	1	1	0,5	0
α-optimal value zone	out	out	out	in	in	out	out

Table 3.4: *Optimal value zone for example 3.2.*

Under a pessimistic oriented point-of-view only components C_4 and C_5 lie within the α-optimal value zone with $\alpha=0,5$. This result is not surprising because C_4 and C_5 are the only components, which always lie within the optimal value zone in example 3.1. If one can accept a small $\alpha \leq 0,5$, then C_1, C_2, C_3, and C_6 also lie within the α-optimal value zone.

3.3.2 Degree of Importance

In case of partial information about the distribution, the stochastic programming approach also can be used to compute the degree of importance similar to model (3.4). The necessary modifications of the model are described in detail in *Abel* [1984].

4. Setting the Target Cost

The setting of the target cost can be done using some modifcations of well-known production models. In this section we describe the necessary modifications of existing (linear) production models.

As described in section 2, the term "target cost" is used as target for cost reduction (see e.g. *Tanaka* [1993]),

> target cost = drifting cost - allowable cost,
> allowable cost = target price - target profit,

where the target price is a fixed parameter coming from the market.

The target cost (tc) of the product is the sum of the target cost (tc_i) of the m components of the product:

$$tc := \Sigma_i tc_i, \tag{4.1}$$

tc_i are output data of the functional area method (see pos_i in section 3).

$$tc_i \leq costold_i\text{-}cost_i, \tag{4.2}$$

where $costold_i$ are the production cost (today) for component $C_i^{(old)}$, $cost_i$ are the (future) drifting cost for the production of component C_i as solution of e.g. a linear production model

$$cost_i := \min\{c^{(i)}x \mid A^{(i)}x = b^{(i)}, x \geq 0\}, \tag{4.3}$$

$C_i^{(old)}$ is the old version of component C_i.

For a detailed discussion of production models with random variables as input data see e.g. *Abel and Thiel* [1981].

The values of tc_i and $costold_i$ in general are given parameter.

The coefficients of $A^{(i)}$, $b^{(i)}$, $c^{(i)}$ contain information about the future production. At this stage of the product planning process (where we talk about the targets) in general not every necessary information about the production processes is available. So we have to assume, that some of the coefficients are random variables. In this case optimality of a solution of (4.3) is not well-defined, but this problem can also be solved with stochatic programming approaches (see e.g. *Kall* [1976]). Problems like (4.3) are typical examples, where two-stage programming models (or stochastic programming with recourse) can be used to define feasability and optimality of a solution. The deterministic equivalent for (4.3) as *two-stage programming* model for component C_i has the following form:

$$\min E\{c^{(i)}(\omega)'x + \min\{q^{(i)}(\omega)'y(\omega)| \qquad (4.4)$$
$$W^{(i)}(\omega)y(\omega) = b^{(i)}(\omega) - A^{(i)}(\omega)x, \; y(\omega)\geq 0\}, \; x\geq 0\},$$

with random variables $A^{(i)}(\omega)$, $b^{(i)}(\omega)$, $c^{(i)}(\omega)$, $q^{(i)}(\omega)$, and $W^{(i)}(\omega)$ and known probability distribution. We assume, that $W^{(i)}(\omega)$ (for all $i=1,2,..,m$ and $\omega\in\Omega$) is a complete recourse matrix (see e.g. *Abel*[1984], p.47), the recourse problem has a finite solution for every feasable x, and the expectation in (4.4) exists.

The solution of (4.4) has to fulfil restriction (4.2) to reach the target. So we have to add restriction (4.2) in production model (4.3). Using the chance-constrained approach from *Charnes and Cooper* [1959] the complete deterministic equivalent for a *production model with target cost restriction* for component C_i has the following structure:

$$E(cost_i(\omega)) = \quad \min_x E\{c^{(i)'}(\omega)x + \min_y\{q^{(i)'}(\omega)y(\omega) | \qquad (4.5)$$
$$W^{(i)}(\omega)y(\omega) = b^{(i)}(\omega) - A^{(i)}(\omega)x,$$
$$prob(q^{(i)'}(\omega)y(\omega) \leq costold_i - tc_i - c^{(i)'}(\omega)x) \geq \alpha_i,$$
$$y(\omega) \geq 0\} |$$
$$x \geq 0\},$$

with given $0\leq\alpha_i\leq 1$.

Remarks:

- Model (4.5) contains random variables, so we can compute only the expected value for $cost_i(\omega)$ as optimal value of (4.5), because at this stage of the product planning process we don't know the realizations of the random variables.

- With restriction

$$prob(q^{(i)'}(\omega)y(\omega) \leq costold_i - tc_i - c^{(i)'}(\omega)x) \geq \alpha_i$$

we ensure, that with given probability α_i the target cost goal tc_i for component C_i will be reached.
- For models with only partial information about the probability distribution see e.g. *Abel* [1987].
- This model can be solved numerically e.g. in case of (finite) discrete distributions of the random variables.

The optimal solutions of (4.5) can be used to set the target cost.

It is not necessary that every tc_i will be reached, but you have to reach the target $tc := \Sigma_i tc_i$. In this case the following deterministic equivalent from a production model for the whole product has to be solved:

$$E(cost(\omega)) = \min_x E \Sigma_i \{c^{(i)'}(\omega)x^{(i)} + \min_y \{q^{(i)'}(\omega)y^{(i)}(\omega) \mid \qquad (4.6)$$
$$W^{(i)}(\omega)y^{(i)}(\omega) = b^{(i)}(\omega) - A^{(i)}(\omega)x^{(i)},$$
$$prob(\Sigma_i q^{(i)'}(\omega)y^{(i)}(\omega) \leq \Sigma_i costold_i - tc - \Sigma_i c^{(i)'}(\omega)x^{(i)}) \geq \alpha$$
$$y^{(i)}(\omega) \geq 0\} \mid$$
$$x^{(i)} \geq 0\},$$

with given $0 \leq \alpha \leq 1$.

Remarks:
- For solutions from model (4.6) the probability is at least α, that the target cost goal tc for the complete product will be reached, but it is not necessary that all target cost goals for the components C_i have to be reached with probability α.
- The expected values for $cost_i$ for the components can be computed directly from the solutions of model (4.6), Σ $E(cost_i(\omega))$ is the optimal value from model (4.6).
- A set of solutions from model (4.4) for i=1,2,..,m is a feasible solution for model (4.6) if $\alpha_i \geq \alpha$ for all i, under the assumption that we have no stochastic dependencies between the probability distributions of the random variables in the models of the components.
- Dependencies between the production of different components of the complete product can be included in model (4.6) using restrictions of structure Wy=b-Ax.

5. Summary

A successful use of target costing approaches depends upon the quality of the available information/data material. Target costing goals with the resulting cost reduction effects can only be reached if it is possible for Controlling to manage high quality data and not trying to control low quality information during the whole process chain. The first German target costing attempts e.g. in automotive industry show that the availability of high quality "hard" data as a necessary input for target cost management is one of the major problems in target costing projects. Possible solutions are modifications of the models and tools in a way that random variables as input data are also possible, e.g. the use of an α-optimal value zone instead of the "traditional" deterministic optimal value zone definition, or the use of stochastic optimization models to compute the degree of importance of the components based on stochastic input data.

In the corresponding production models for the components and the whole product, restrictions should be included to ensure that with the optimal solutions of the production models the target costing goals (tc, tc_i) can be reached. If some of the input data are random variables stochastic programming approaches like two-stage and chance-constrained programming are possible solution approaches.

6. Acknowledgement

The authors wish to thank Prof. K. Marti (Federal Armed Forces University Munich) for beneficial discussions of the stochastic optimization model in section 3. Also we want to express our gratitude to Dipl.-Stat. F.W. Schwefer (Federal Armed Forces University Munich), who computed the numerical solution of the nonlinear optimization problem from the older version of example 3.2.

REFERENCES

ABEL, P. [1984]: Stochastische Optimierung bei partieller Information. Athenäum, Hain, Hanstein, Königstein, 1984.

ABEL, P. [1987]: Stochastic Linear Programming with Recourse under Partial Information. In: VIERTL, R. [Ed. 1987], pp. 1 - 6.

ABEL, P.; THIEL, R. [1981]: Mehrstufige stochastische Produktionsmodelle. Fischer Verlag, Frankfurt a. Main, 1981.

CHARNES, A.; COOPER, W.W. [1959]: Chance-constrained programming. In: Management Science [1959] 6, pp. 73 - 79.

DEISENHOFER, T. [1993]: Marktorientierte Kostenplanung auf Basis von Erkenntnissen der Marktforschung. In: HORVÁTH, P. [Ed. 1993], pp. 93 - 117.

HORVÁTH, P. [Ed. 1993]: Target Costing - Marktorientierte Zielkosten in der deutschen Praxis. Schäffer-Poeschel Verlag, Stuttgart, 1993.

HORVÁTH, P.; NIEMAND, S.; WOLBOLD, M. [1993 a]: Target Costing - State-of-the-Art Review. Research Paper No. 35, Institute for Business Administration, Department of Controlling, University of Stuttgart, 1993.

HORVÁTH, P.; NIEMAND, S.; WOLBOLD, M. [1993 b]: Target Costing - A State-of-the-Art Review. IFS International Ltd., Bedford 1993

HORVÁTH, P.; SEIDENSCHWARZ, W. [1992]: Zielkostenmanagement. In: Controlling 4 [1992] 3, pp. 142 - 150.

KALL, P. [1976]: Stochastic Linear Programming. Springer Verlag, Berlin, Heidelberg, New York, 1976.

KOBAYASHI, T.; TANI, T.; KATO, Y.; OKANO, H.: SHIMIZU, N.; IWABUCHI, Y.; FUKUDA, J.; COORAY, S. [1992]: Genka-Kikaku in Japanese Firms: Current State of the Art (1), (2) and (3). In: Kigyo Kaikei (Business Accounting) 44 (1992) 5, pp. 86 - 91; 6, pp. 74 - 79; 7, pp. 84 - 89.

KOFLER, E.; MENGES, G.; FAHRION, R.; HUSCHENS, S.; KUß, U. [1980]: Stochastische Partielle Information (SPI). In: Statistische Hefte [1980] 21, pp. 160 - 167.

MONDEN, Y.; SAKURAI, M. [Ed. 1989]: Japanese Management Accounting: A World Class Approach to Profit Management. Productivity Press, Cambridge/ Mass., 1989.

SEIDENSCHWARZ, W. [1991]: Target Costing, ein japanischer Ansatz für das Kostenmanagement. In: Controlling 3 [1991] 4, pp. 198 - 203.

SEIDENSCHWARZ, W. [1993]: Target Costing. Vahlen Verlag, München, 1993.

TANAKA, M. [1989]: Cost Planning and Control Systems in the Design Phase of a New Product. In: MONDEN, Y.; SAKURAI, M. [Ed. 1989], pp. 49 - 71.

TANAKA, M. [1993]: Target Costing at Toyota. In: Journal of Cost Management 7 [1993] 1, pp. 4 - 11.

VIERTL, R. [Ed. 1987]: Probability and Bayesian Statistics. Plenum Press, New York, London, 1987.

STATISTICAL CHARACTERIZATION OF GRANULAR ASSEMBLIES

K. Bagi and A. Vásárhelyi

Technical University of Budapest

ABSTRACT

On the basis of experimental results a numerical
model was elaborated to follow the changes of the
microstructure of granular assemblies. The
distributions of internal forces and displacements were
not possible to determine so we tried to characterize
the assemblies with the angles of neighbouring
contacts.

Keywords: granual assembly, material quality,
statistics, potimization

1. Introduction to the mechanical problem

The mechanical description of granular materials is
an ancient, and surprisingly stubborn problem: in spite
of the long practical tradition, and in spite of its
enormous economical importance, engineers are still not
able to reliably predict the load bearing ability of
the sand under a house, the stability of an embankment,
or, for example, the behavior of corn in a silo.

Soil mechanics is doubtlessly the most important application of the mechanics of granular materials. Recently the engineers in the design of geotechnical objects have to confine themselves with experimentally estimated, very rough predictions. The problem is not only that the design cannot be economical: in worse cases the lack of exact knowledge of material behavior can lead to accidents like sinking of buildings etc. Beside geotechnical engineering several other fields like powder metallurgy, blood research, oceanography etc. are strongly interested in the micromechanical research.

The basic aim of these researches is to predict the behavior of the assemblies. More accurately, the engineers need the answer for two questions:
→ If loads are put on the material how will it deform?
→ How much load can be put on the material without damaging it?

Traditionally the continuum mechanical variables stress and strain tensors have been used as the basic variables that uniquely describe the state of the material in any point. Relation between stress and strain are given by the so called constitutive equations which, through their experimentally measured parameters, are supposed to contain all the necessary information on the material properties.

This concept usually works well in case of 'continuous' materials like metals or plastics. Granular materials, however, are totally different. They consist of separate grains supporting each other with the help of forces acting at the contact points of neighboring grains ('contact forces'); the grains can slide or roll on each other, more or less independently; even the whole internal structure of the assembly can rearrange under the effect of external loads and the changing of this internal geometry is a fundamental characteristic of the material. Continuum mechanical theories based on stress and strain tensors are simply not able to contain enough information to reflect these phenomena.

Micromechanical analysis of granular materials is a relatively new area. Researches became intensive about 15-20 years ago when the inefficiency of the traditional continuum mechanical approach became obvious. At the beginning several researchers concentrated on trying to find additional state variables beside stress and strain tensors (the different fabric tensors for example, see [1], [2] etc.) that could describe the geometrical state of the material. In this concept the three fundamental variables, stress, strain and fabric tensors, would have provided the basis of a theory that was expected to be able to predict the behavior. Unfortunately these trials have not led to success yet. Its reason seems to be hidden in the fact that since these variable are defined as the **averages** of micro variables (contact forces, position vectors of contact points, etc), they cannot contain enough information: significant amount of important data gets lost during the averaging.

To overcome this difficulty an other method starts to develop recently: the statistical analysis of the microstructure. In this concept the frequency diagrams of different microvariables are considered instead of the averages only. The aim is to find the type of these distributions, to identify the parameters determining them, and finally, to find what kind of relations exist between them. Initial steps have been done by a few researchers (see [3], [4] for instance) and our work is also a part of this research.

Chapter 2. of this paper summarizes the most important micro level properties of the material, and the necessary micro variables will be introduced. In Chapter 3. we shall show how we tried to find the distributions of the variables we measured, and how the load bearing capacity of the material was formulated as an optimization problem.

2. Micro-level analysis of granular materials

A typical stress-strain curve of granular assemblies is shown in Figure 2.1. An assembly of round river stones was subjected to axial compression.(The curve was measured in a compression test by M. G"los, TJ Budapest)

We shall focus our attention on the stress jumps as a typical phenomena in case of granular materials.

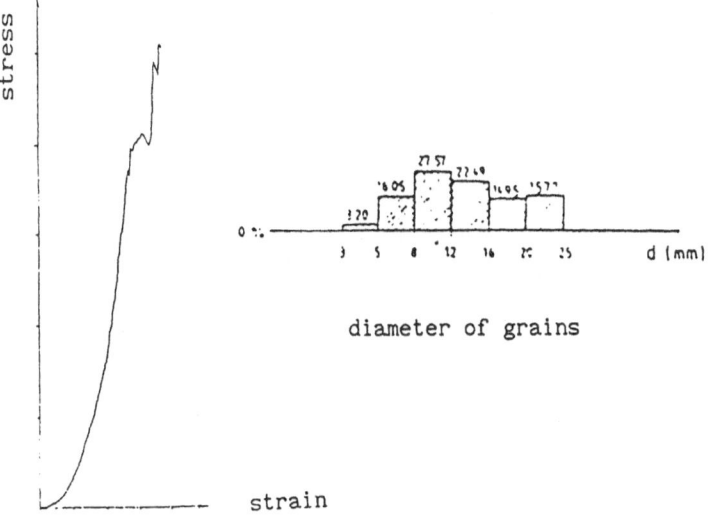

diameter of grains

Figure 2.1.

The explanation of the existence of these jumps is illustrated by Figure 2.2. As the load acting on the assembly increases, the grains move relatively to each other and the internal geometry starts to change. With the load increasing, these changes become more and more significant and finally unstable parts will form in the assembly. These unstable parts are not able any more to carry the loads: they will collapse, and the grains fall into a new position where they may form a new structure being stable under the actual load.

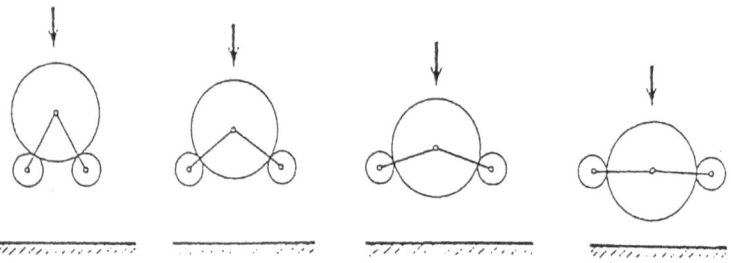

Figure 2.2.

Obviously, the state and changes of the micro structure have a fundamental effect on the global stress-strain behavior of the material. To follow and understand these micro level processes we use numerical simulations (run by a computer code) instead of the real physical experiments. The application of numerical simulations has several advantages from our point of view. The most important one is that detailed information (exact position of each individual grain; grain displacements; forces acting between the grains; etc.) is available at any time during the experiment - while most of these data cannot be registrated in usual physical tests. Besides, numerical tests are easy to reproduce; the initial conditions and the properties of the grains can exactly be prescribed; easy-to-interpret output (like graphs or statistical parameters) can be asked from the computer; the costs of a numerical test cannot even be compared with the costs of a real experiment etc. No wonder that the application of numerical simulations is a widely used method all over the world. We shall also follow this route. For simplicity, the analysis will be restricted to two-dimensional assemblies of perfectly circular discs.

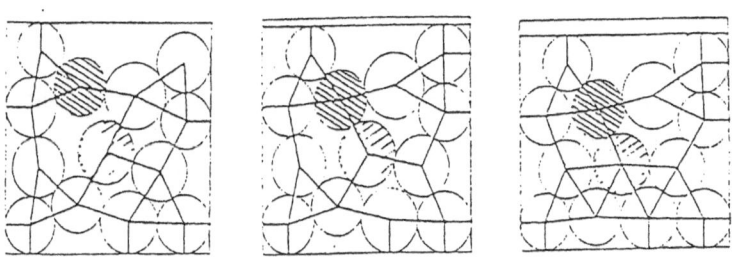

Figure 2.3.

Figure 2.3. is an illustration given by the numerical model to show the changing of microstructure under vertical compression process.

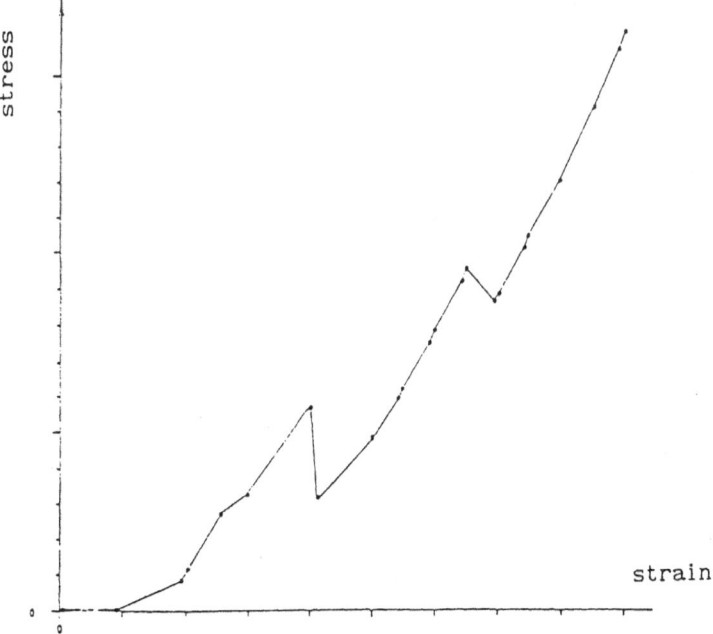

Figure 2.4.

Figure 2.4. is the corresponding stress-strain curve; obviously, its character is rather similar to that of the real experiment.

Now the most important <u>micro-variables</u> of granular materials will be introduced.

1. Coordinates of grain centers

Usually a Descartes coordinate system is used. Position of the grains is characterized by the coordinate vector pointing to the grain centers (Fig. 2.5.).

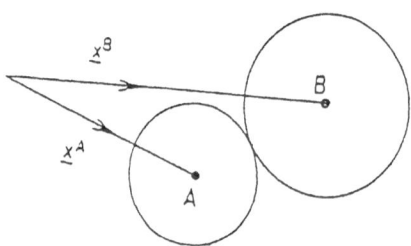

Figure 2.5.

2. Displacements

The displacements are the variables that describe the motion of each individual grain from the original position to the new, displaced state. <u>Translations</u> and <u>rotations</u> belong to each grain; independently from each other. The translations ($\underline{u}$) are simply the difference

between the new and the original coordinates of grain
centers; rotations (ω) are considered as scalars in 2D
(see Figure 2.6.)

It has to be underlined that the displacements serve
for the <u>comparison</u> of two states of the assembly.

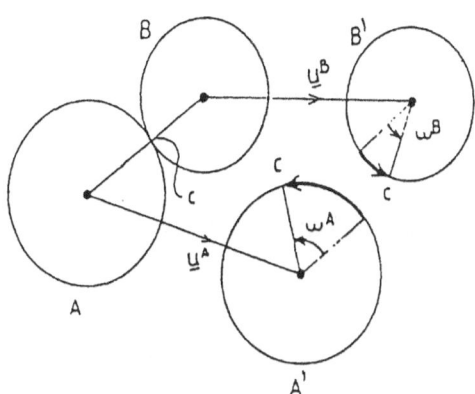

Figure 2.6.

3. Contact forces

If two grains are pressed against each other, they
will deform in the neighborhood of the contact, and
their material will resist the further deformation. In
other words the grains express forces on each other and
these forces hold the grains against getting even
closer. These forces are called 'contact forces'. Of
course, the contact force acting on grain B by grain A,
and the force acting on grain A by grain B, are just
the opposite of each other (Newton's law). The contact
forces can be separated into a 'normal component', $\underline{N}^c$
(normal to the common tangent of the touching grains)

and a 'tangential' or 'shear component', $\underline{T}^c$ (parallel with the common tangent) as shown by Figure 2.7. (index 'c' denotes that the force belongs to contact 'c').

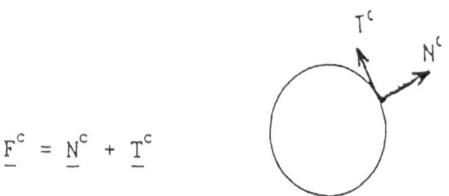

$$\underline{F}^c = \underline{N}^c + \underline{T}^c$$

Figure 2.7.

5. Angle of neighboring contacts

This geometrical variable, α, measures the angle of any two neighboring contacts on a given grain, in the sense shown in Figure 2.8. (The contacts of the grain are denoted by the corresponding radius in the Fig. 2.8.)

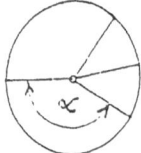   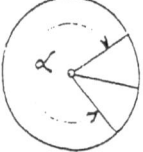

Figure 2.8.   Figure 2.9.

It should be noticed that if there exists an α larger than π, then it means that the grain is not supported against translation in certain directions so it can move away (see Figure 2.9.). Moreover, if there are relatively large angles (even being smaller than π), the situation is somewhat dangerous because the grain can relatively easily be translated in those directions near to the middle of the large angles. So α is a rather good measure of the stability of individual grains.

3. Statistical analysis

The following numerical experiment was performed in order to get the frequency diagrams of the micro variables:

Initial assembly:
→ 467 grains with an average diameter of 0.2355 mm (uniform distribution function was used to generate the diameters of the grains between D=0.215 mm and D=0.256 mm)
→ the assembly is under isotropic compression

Load:
The assembly is compressed equally in the x and y directions; magnitude of isotropic strain: 0.8‰ .

Frequency diagrams of translations, rotations, contact forces, and of the angles of neighboring contacts were measured.

To determine what type of distribution functions we have in case of contact forces and displacements the Kolmogorov-Smirnov, von Mises and the χ^2 tests were used. The patterns were done with Gaussian, lognormal and Weibull distribution functions. The statistical patterns were done by S. Fegyverneki [6],[7].

Our sample contained 467 grains and 1087 contact places. In the following tables the results of the pattern to the Gaussian distribution in case of the displacements in x and y directions and the rotations are shown, respectively. The other patterns gave worse results.

u_x	Kolm.-Smirnov		von Mises		χ^2	
M: 0.0001 V: 0.0151	1.8072	98%	0.9865	99%	76.67	76%
Robustus P_p: 1.006 S_p: 0.0117	1.006	73%	0.1072	0%	38.12	38%

Table 1.

u_y	Kolm.-Smirnov		von Mises		χ^2	
M: 0.0001 V: 0.0159	1.5349	97%	0.7455	98%	49.64	50%
Robustus P_p: 0.006 S_p: 0.0127	0.723	32%	0.0844	31%	38.90	39%

Table 2.

φ	Kolm.-Smirnov	von Mises	χ^2
M: 0.0018 V: 0.0776	1.4318 96%	0.4117 93%	41.67 41%
Robustus P_p : 0.0008 S_p : 0.0117	0.8291 48%	0.1229 51%	28.82 27%

Table 3.

where M and V are the expected value and the variance respectively, in robustus tests P_p and S_p denote the place and scale parameters respectively.

As an example the distribution function of rotation can be seen in Figure 3.1.

Gaussian net

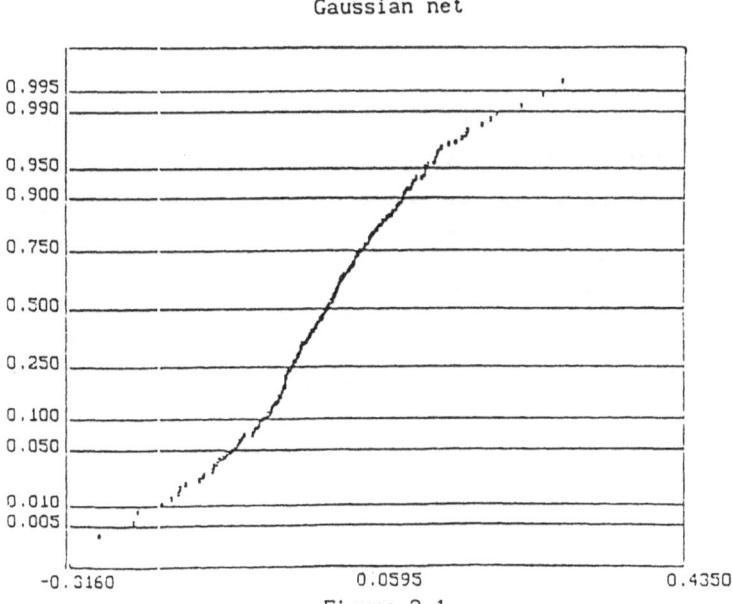

Figure 3.1.

On base of Kolmogorov-Smirnov and von Mises tests it may be accepted to have Gaussian distribution functions for the displacements.

We tried to pattern the contact normal and shear forces as well but the distribution functions of them were not possible to determine. Besides, we could not find any connection between the distribution functions of contact forces and displacements. It is in contradiction with the well known deterministic relation between the contact forces and displacements. The explanation may be given by the fact that the order of magnitude of displacements are very small so the truncation errors strongly influenced the pattern.

After this unsuccessful trial we were looking for some other quantity to characterize the physical attitude of the assembly. The large motion of the grains happened in the domains where the angles between neighboring contacts were relatively large, larger than approximately $\pi/2$. First we tried to pattern all the angles but they didn't show any type of distribution functions. After that, the blunt angels were patterned (Fig. 3.2.) and the result was that they had a Weibull distribution function:

$$F(\xi) = 1 - \exp\left[-V\left(\frac{\xi-a}{b}\right)^c \right]$$

where V was the volume of the assembly. The values of the parameters in our pattern were: a = 0.0, b = 0.003, c = 0.9135.

The result of the Kolmogorov-Smirnov test was 1.25 which corresponds to 91%, of the Mises test was 0.04 that is 93% .

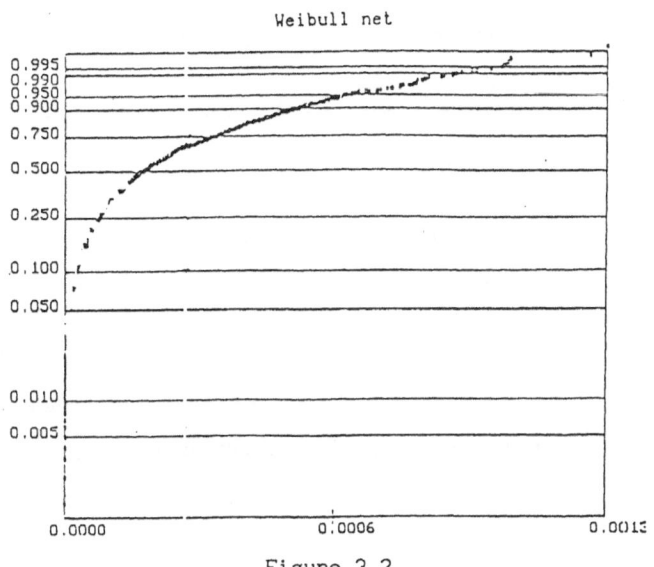

Figure 3.2.

So it was between the normal and the Laplace distributions.

4. Stochastic optimization problem

To check whether the blunt angles really characterize the material quality of the assembly, we did a collapse load analysis at the first external load caused collapse in the microstructure of the assembly. In the collapse load analysis the load is supposed to be one-parametric and we look for the value of the load parameter where the equilibrium equations for the contact forces and the inequalities for tne displacements are satisfied. In the inequalities the stochastic character of the blunt angles are taken into consideration weighting the limit displacements by them. This way the problem has very large dimensions:

$$F_{ij} + m \hat{F}_{ij} = 0, \qquad j = 1, \ldots, s, \quad i = 1, \ldots, n,$$

boundary conditions,

$$P\left(\arcsin\left(\frac{F_{ij}}{\varepsilon_{Lj} K r_i} \right) \le \alpha_i \right) \ge p, \quad i = 1, \ldots, z,$$

$$m \Rightarrow \max.$$

where F_{ij} is the contact forces of the i-th grain in the j-th direction, n is the number of grains, $\hat{F}_{ij}$ is the external load of the i-th grain in the j-th direction, m is the load parameter, s is the number of directions, z is the number of blunt angles, α_i is the i-th blunt angle, ε_L is the limit strain in the j-th direction, K is the Young modulus of the grain, r_i is the radius of the i-th grain.

In case of our sample there were 467 grains and 1087 contact points. Taking the normal components and shear components of the contact forces as unknowns, the number of them is 2174. The equilibrium of the grains is expressed by 3*467=1401 equality constraints. In our sample 26 grains touched the walls; in these cases the directions of the supports were in normal to the walls, that is we had 26 equalities as boundary conditions. There are 83 blunt angles, they gave 83 inequality constraints. Blunt angles were supposed to have no correlation betwen them.

We had no success to solve this problem, the results were numerically instable and they depended on the starting points very strongly.

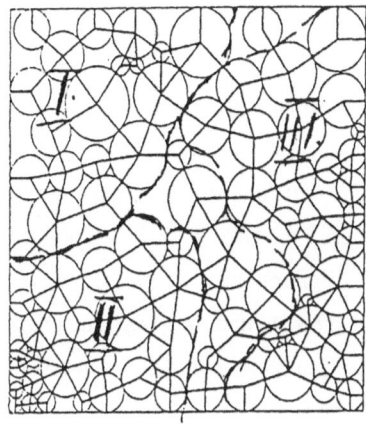

Figure 4.1.

The next step was to restrict the dimensions of the problem. By hand we chose three domains which seemed to be more stable and most blunt angles were in the regions of the domains.(As an illustration see the Fig. 4.1.) In this way the collapse load analysis problem was decreased: it contained 3 domains, 127 unknown contact forces, 63 equalities and 26 inequalities.

This problem was solved by a reduced gradient method. The results show a material harder by cca 15% than it was found in the numerical experiments.

Probably this deviation was caused when the stable domains were formed

SUMMARY

A numerical model was elaborated to examine the character of granular assemblies. According to the statistical patterns of the contact forces and displacements it was obvious that the usual way of characterisation was not applicable here. We tried to find some other quantity to express the special properties of granular assemblies. Perhaps the distribution of the blunt angles in the assembly will be good for this purpose but it is necessary to do more numerical experiments.

ACKNOWLEDGEMENTS

This research has been supported by the OTKA program of the Hungarian Academy of Sciences, under project numbers 5313 and F-7641.

REFERENCES

[1] Oda, M.; Nemat-Nasser, S.; Mehrabadi, M.M.: A statistical study of fabric in a random assembly of spherical granules. Int. J. for Numerical and Analytical Methods in Geomechanics Vol.6, 1982

[2] Satake, M.: Fundamental quantities in the graph approach to granular materials. Mechanics of Granul·r Materials, eds. J.T.Jenkins and M.Satake, Elsevier, 1983

[3] Sidoroff, F.; Cambou, B.; Mahboubi, A.: Contact force dist- ribution in granular media. Proceedings of US-Japan Seminar on Micromechanics of Granular Materials, 1992

[4] Rothenburg, L.: Micromechanics of idealised granular systems. PhD Dissertation, Carleton University, Ottawa, 1980

[5] Bagi, K.: On the definition of stress and strain in granular assemblies. Powders and Grains, ed. C. Thornton, Balkema, 1993

[6] Fegyverneki, S.: Robustus estimation of space- and scale- parameters according to the model distribution function. Procs. MICROCAD'92, Miskolc, 1992 (in Hungarian)

[7] Fegyverneki, S.: Robustus estimations according to the model distribution, and their applications. Doctorial thesis, Kossuth Lajos University, Debrecen, 1993 (in Hungarian)

[8] A.Vásárhelyi: Limit Analysis and Optimal Plastic Design by Stochastic Programming with Uncertainties of Material Quality. Journal of Structural Mechanics Vol15. No.2. 1987.

ON STOCHASTIC ASPECTS
OF A METAL CUTTING PROBLEM*

Jitka Dupačová[1], Pavel Charamza[1] and Jan Mádl[2]

[1]Department of Statistics, Charles University, Sokolovská 83, 186 00 Prague, Czech Republic
[2]Department of Machining, Technical University, Technická 4, 166 07 Prague, Czech Republic

Abstract. Taylor's tool life equation that relates the tool life and the machining conditions plays an important role in designing optimization schemes for various machining processes. This equation results from statistical regression analysis therefore, in the contrast to common approaches, we suggest to treat the corresponding input values of the coefficients in the related geometric programming problems as statistical estimates of the true coefficients. The proposed statistical sensitivity analysis of the optimal machining conditions and of the minimal value of the total machining costs provides an additional statistical information about their precision. Details concerning the suggested method and a numerical example are given for the case of optimization of the single pass, single tool turning operation.

Keywords. Optimal cutting conditions, tool life, Taylor's equation, stochastic geometric programming, sensitivity analysis, estimated coefficients

1. Introduction

Modern cutting processes require cost effective working conditions and high reliability of the performance. It is machining that contributes essentially to the final production costs and for the purposes of designing the process, various mathematical models for optimization of machining conditions have been applied. Modern monitoring and control systems contribute to further improvement of economic parameters of complex expensive machine tools, e.g., by reduction of the non-productive time and they are necessary for obtaining satisfactory results for new cutting materials or for machining processes running under extreme conditions, e.g., at very high speed and using a lot of coolant. Another type of problems is connected with reliability of the machining process that depends on tool life variation, tool wear, tool breakage and tool chipping; besides of monitoring (see e.g. Mádl 1992a,b), exploitation of suitable statistical techniques can be of great help (see e.g. Szántai et al. 1993).

The first attempts to optimization of machining processes date probably from the very beginning of this century and in this context, paper of Taylor (1907) is reported. Simple low dimensional problems often allow for an analytical solution. Nevertheless, various techniques of mathematical programming, namely, linear and geometric programming have been considered and applied already in the seventies, cf.

Research supported in part by grants from Charles University (GAUK-357) and from the Grant Agency of the Czech Republic (402/93/0631, 101/93/2425).

for instance Ermer (1971), Philipson and Ravindran (1979), Chakrabarti (1986), Mukherjee and Pal (1986), Gopalakrishnan and Al-Khayyal (1991). The most frequent optimization criterion is that of minimization of machining cost; other criteria used in this context are maximization of production rate, maximization of profit rate, maximization of metal removal rate, maximization of number of parts between tool changes or maximum tool life. The main control variables are the cutting speed v or the number of revolutions r, depth h and feed f, the constraints are mostly given by the machine tool specification (the available maximum and minimum cutting speed and feed), by the machine tool dynamics (the allowable maximum cutting force and power consumption, etc.), by tool specification, by workpiece material and by component specification (the maximum surface roughness and the required accuracy). For single tool, single pass turning operation, these constraints can be usually linearized. This, however, does not apply to more practical cutting processes that use several tools operating on the workpiece in sequence or for extensions to automatic lines. Also the already mentioned most favorable objective function does not allow for linearization; hence, the increasing interest in geometric programming.

The machining process takes place in uncertain environment that influences the output. The tool life T that enters the objective is known to be related to the machining conditions (cutting speed, feed, etc.). The famous Taylor's equation (1907) as well as its updates (cf. e.g. Colding and Konig 1971 or Wagner and Barash 1971) that are used in this context are based on statistical regression analysis. Nevertheless, they have mostly entered the related optimization problems as postulated functional relationships with presumably known "standard" values of coefficients. A similar situation can be discovered in the case of "empirical" coefficients in some of the above mentioned constraints. The papers that take into account the probabilistic character of these coefficients are rare. For instance, Iwata et al. (1972) use individual probabilistic constraints to cope with uncertainties in the linearized constraints on machine tool dynamics; they assume uncorrelated normal coefficients.

The random nature of the "standard" values and of "empirical" constants in the relatively simple posynomials of the underlying geometric programming formulation of the metal cutting problem concerns both multipliers c_i and exponents α_{ij} and neither their randomness nor their statistical dependence can be excluded by ad hoc arguments. Therefore, the known results from the field of stochastic geometric programming, cf. Avriel and Wilde (1970), Wiebking (1977), Ellner and Stark (1980), Jagannathan (1990), cannot be applied in their original form.

In our formulation of the metal cutting problem we shall work with *the Taylor's equation considered as a regression relationship* and with the coefficients obtained as *statistical estimates of the true values*. Accordingly, the optimal cutting conditions and the optimal value of the objective are regarded as estimates of the true ones. Their approximate distribution can be derived and used subsequently to complement the obtained numerical values of optimal machining conditions and of the corresponding costs by an additional information about their confidence regions, etc.

The suggested approach is based on ideas of Dupačová (1984) and on results by Kyparisis (1988) concerning differentiability of the optimal solutions of the geometric programming problems with respect to parameters. The details will be given for optimization of the single pass, single tool turning operation. We shall concentrate to the case of random parameters only in the objective function assuming thus that the only source of uncertainties stems from the Taylor's equation. The way how to treat more complicated cases is clear, though quite demanding as to the required technicalities. The proposed method is illustrated on a numerical example (Section 5) and some of possible extensions are outlined in the last Section.

2. Optimization of single pass single tool turning operation

The most frequently used optimality criterion is that of minimizing the total machining cost (without material and nonproductive handling costs) per component, cf. Philipson and Ravindran (1979), Gopalahrishnan and Al-Khayyal (1991). It equals to the sum of machining time cost, cost of tool changing time per component and tool cost per component:

$$y_c = x\, T_c + x\, T_d \frac{T_{ac}}{T} + y\, \frac{T_{ac}}{T} \tag{1}$$

where
x is the labor plus overhead cost rate
y is the tool cost
T_c machining time
T_{ac} actual cutting time
T_d tool changing time
T tool life.

Whereas x, y, T_d are understood as a fixed input, the tool life, the cutting and machining time depend on the cutting conditions, such as speed v or number of revolutions r, feed f and depth h. The variables v and r are related by a simple linear equation

$$r = \frac{10^3\, v}{\pi\, D} \tag{2}$$

where D denotes the workpiece diameter. The used variables are given in the following unites: v (m/minute), r (revolutions/minute), h (mm), f (mm/revolution), D in mm, the tool life and all kind of times are given in minutes.

The relationship among T, r, and f reads

$$T_c = L\, r^{-1}\, f^{-1} \tag{3}$$

with L (in mm) the length of the feed motion between the tool and the workpiece. A frequent assumption is

$$\tau := \frac{T_{oc}}{T_c} \approx 1 \tag{4}$$

The tool life is assumed to fulfil the following form of Taylor's equation

$$v\, T^n\, f^m\, h^p = A \tag{5}$$

with n, m, p, A positive empirical constants related to the tool and workpiece material. Substituting (2) - (5) into (1), we get the objective function

$$c_1\, v^{-1}\, f^{-1} + c_2\, v^{\frac{1}{n}-1}\, f^{\frac{m}{n}-1}\, h^{\frac{\ell}{n}} \tag{6}$$

or

$$k_1\, r^{-1}\, f^{-1} + k_2\, A^{-\frac{1}{n}}\, r^{\frac{1}{n}-1}\, f^{\frac{m}{n}-1}\, h^{\frac{\ell}{n}} \tag{7}$$

to be minimized. The coefficients c_1, c_2 and k_1, k_2 are positive and they can be easily obtained from the fixed parameters and empirical constants that appear in (1) - (5).

The constraints on machine tool dynamics can be also expressed in terms of v, r, f and h; for instance *the machine power constraint* has the form

$$k_{F_c}\, f^{y_{F_c}}\, h^{x_{F_c}}\, v \le 60 P_{eff} \tag{8}$$

where P_{eff} (in W) denotes the effective power and k_{F_c}, x_{F_c} and y_{F_c} are the empirical constants that appear in the cutting force function. Prescribed upper and lower bounds on the product fr restrain the feed per minute, the upper and lower bounds on v (or r) and on f are of a complex nature being given not only by the maximal and minimal available cutting speed and feed but also by the allowable cutting force, etc. Further constraints of a similar nature can be related to *surface finish requirements*. For details see, e.g., Philipson and Ravindran (1979).

The resulting mathematical program is evidently of the form of the following simple geometric program

$$\text{minimize } g_0(t) \text{ subject to } g_k(t) \le 1, \ k=1,...,K, t > 0 \tag{9}$$

with $t_1 = v$, $t_2 = f$, $t_3 = h$ and

$$\begin{aligned}
g_0(t) &= c_1\, t_1^{a_{11}}\, t_2^{a_{12}} + c_2\, t_1^{a_{21}}\, t_2^{a_{22}}\, t_3^{a_{23}} \\
g_k(t) &= c_{2+k}\, t_1^{a_{2+k,1}}\, t_2^{a_{2+k,2}}\, t_3^{a_{2+k,3}}, \ k=1,...,K.
\end{aligned} \tag{10}$$

Even if the empirical relationships are believed to be precise enough in the considered region of h, v and f, so that one can use "standard" values of empirical constants in the Taylor's equation, in the machine power constraint, etc., with values specified according to the given cutting conditions, machine and material, the obtained optimal specification of cutting feed, speed and depth does not provide in general the best expected performance as to the tool life and the overall costs. The reason is attributed to Taylor's equation and to the observed large volatility of A. A lot of effort has been spent to get more precise relation for the tool life. In this paper, we shall instead return to the origin of Taylor's equation relating it to a regression relationship that explains the tool life by the cutting conditions out of which speed, feed and depth can be taken as the control variables.

3. The Taylor's equation

We shall examine Taylor's tool life equation (5) from the point of view of regression analysis that is definitely the source of the "standard" coefficient values. It is natural to take T as the dependent variable and v, f, h as regressors and to linearize:

$$\lg T = \lg c_0 - \frac{1}{n} \lg v - \frac{m}{n} \lg f - \frac{p}{n} \lg h + e \tag{11}$$

with $c_0 = A^{1/n}$. If e in (11) is independent on $\lg T$ with zero mean value and a fixed variance σ^2 we can use the least squares method to estimate the true coefficient values. We put (11) into the standard form of linear regression model

$$y = \beta_0 + \beta_1 x_1 + \beta_2 x_2 + \beta_3 x_3 + e \tag{12}$$

with $y = \lg T$, $x_1 = \lg v$, $x_2 = \lg f$, $x_3 = \lg h$ and with $\beta_0 = \lg c_0$, $\beta_1 = -1/n$, $\beta_2 = -m/n$, $\beta_3 = -p/n$ and σ^2 the parameters to be estimated. Let X denote the $N \times 4$ design matrix whose first column consists of fixed values of $x_{i0} = 1$, the second one of values x_{i1}, $i = 1, ..., N$, the third one of x_{i2}, $i = 1, ..., N$ and the fourth one of x_{i3}, $i = 1, ..., N$. If the matrix X has a full column rank, the best linear unbiased estimate of the true vector β^* of regression coefficients is

$$\hat{\beta} = (X^T X)^{-1} X^T y,$$

and

$$s^2 = \frac{1}{(N-4)} [y^T y - \hat{\beta}^T X^T y]$$

is an unbiased estimate of σ^2 that is independent of $\hat{\beta}$. For our purposes, the most important properties are *consistence* of $\hat{\beta}$ and its asymptotic normality that hold true under weak conditions on the design matrix X and on the errors e (see e.g. Dhrymes 1974, Section 4.2):

$$\sqrt{(N)}(\hat{\beta} - \beta^*) \sim N[0, \sigma^2] \quad \text{with} \quad M = \lim_{N \to \infty} \left(\frac{X^T X}{N} \right)^{-1} \tag{13}$$

Moreover, some of empirical results (see also Section 5) support hypothesis that e is a vector of i. i. d. random variables with normal $N(0, \sigma^2)$ distribution independently on T so that (12) is approximately a linear normal regression model. This is certainly useful for small sample properties of the estimates and also for constructing confidence bounds and the confidence band around regression hyperplane (12).

4. Sensitivity for the geometric program

Parametric and statistical sensitivity results will be derived for the following *transformed* geometric program that can be obtained from (9), (10) by means of transformation $z_1 = \lg t_1 \ (= \lg v)$, $z_2 = \lg t_2 \ (= \lg f)$, $z_3 = \lg t_3 \ (= \lg h)$:

$$\text{minimize} \quad c_1 \exp(\alpha_{11} z_1 + \alpha_{12} z_2) + c_2 \exp(\alpha_{21} z_1 + \alpha_{22} z_2 + \alpha_{23} z_3) \quad (14)$$
$$\text{subject to} \quad \lg c_{k \cdot 2} + \alpha_{k \cdot 2,1} z_1 + \alpha_{k \cdot 2,2} z_2 + \alpha_{k \cdot 2,3} z_3 \le 0, \quad k=1,...,K.$$

The objective function in (14) is convex and its coefficients c_1, α_{21}, α_{22}, α_{23} depend on parameters of the Taylor's equation (5), i.e., on parameters β_0, β_1, β_2, β_3 in regression (12); we have

$$c_2 = \lambda \exp(-\beta_0), \ \alpha_{21} = -\beta_1 - 1, \ \alpha_{22} = -\beta_2 - 1, \ \alpha_{23} = -\beta_3 \quad (15)$$

The multiplier λ in (15) and all remaining coefficients in nonlinear program (14) will be treated as fixed known numbers. Sensitivity results for the optimal solution of program (14) parametrized with respect to $\beta = [\beta_0, \beta_1, \beta_2, \beta_3]^T$

$$\text{minimize} \quad G_0(z, \beta) :=$$
$$c_1 \exp(\alpha_{11} z_1 + \alpha_{12} z_2) + c_2(\beta) \exp(\alpha_{21}(\beta) z_1 + \alpha_{22}(\beta) z_2 + \alpha_{23}(\beta) z_3) \quad (16)$$
$$\text{subject to} \quad \lg c_{k \cdot 2} + \alpha_{k \cdot 2,1} z_1 + \alpha_{k \cdot 2,2} z_2 + \alpha_{k \cdot 2,3} z_3 \le 0, \quad k=1,...,K.$$

follow from application of general sensitivity results for nonlinear programs: The basic assumptions for differentiability of the optimal solution $z_1(\beta)$, $z_2(\beta)$, $z_3(\beta)$ (see e.g. Fiacco 1983) are the second order sufficient condition, the strict complementarity condition and the linear independence condition. These conditions guarantee, inter alia, that the set of the active constraints does not change for small changes of parameter values and that the derivatives can be obtained from the 1st order necessary conditions by the implicit function theorem. For geometric programs, these assumptions can be put into a simpler form and formulas for derivatives can be received, cf. Kyparisis (1988). We shall apply the sensitivity results directly to program (16) with the *fixed polyhedral set*

$$Z = \{z \in \Re^3: \lg c_{k \cdot 2} + \alpha_{k \cdot 2,1} z_1 + \alpha_{k \cdot 2,2} z_2 + \alpha_{k \cdot 2,3} z_3 \le 0, \quad k=1,...,K\} \quad (17)$$

of feasible solutions. We are interested in sensitivity analysis on a neighborhood of the true solution, i.e., on a neighborhood of $[z_1(\beta^*), z_2(\beta^*), z_3(\beta^*), \beta^*]$. We shall denote by A^* the matrix of coefficients $\alpha_{k \cdot 2, j}$, $j=1,2,3$, of the active constraints for which the corresponding Lagrange multipliers μ^*_k are nonzero. Furhter we denote by C^*, B^* matrices of the second order derivatives

$$C^* = \nabla^2_\alpha G_0(z(\beta^*), \beta^*), \quad B^* = \nabla^2_{z\beta} G_0(z(\beta^*), \beta^*).$$

Now, the matrices of the first partial derivatives of the true optimal solution $z_1(\beta^*)$,

$z_2(\beta^*)$, $z_3(\beta^*)$ and of the corresponding multipliers μ_k^* with respect to $\beta_0, \beta_1, \beta_2, \beta_3$ at $\beta^*_0, \beta^*_1, \beta^*_2, \beta^*_3$ written symbolically as $\dfrac{\partial z}{\partial \beta}$ and $\dfrac{\partial \mu}{\partial \beta}$, can be obtained by solving the following matrix equation

$$
\begin{pmatrix} C^* & A^{*T} \\ A^* & 0 \end{pmatrix} \begin{pmatrix} \dfrac{\partial z}{\partial \beta} \\ \dfrac{\partial \mu}{\partial \beta} \end{pmatrix} = -\begin{pmatrix} B^* \\ 0 \end{pmatrix} \tag{18}
$$

In our case, it is not difficult to spell out explicit conditions under which the matrix

$$
\begin{pmatrix} C^* & A^{*T} \\ A^* & 0 \end{pmatrix}
$$

is nonsingular. Namely, the matrix C^* can be written as

$$
C^* = A_0^{*T} H^* A_0^*
$$

with H^* diagonal matrix with positive entries

$$
c_1 \exp(\alpha_{11} z_1(\beta^*) + \alpha_{12} z_2(\beta^*)) \ ,
$$
$$
c_2(\beta) \exp(\alpha_{21}(\beta) z_1(\beta^*) + \alpha_{22}(\beta) z_2(\beta^*) + \alpha_{23}(\beta) z_3(\beta^*))
$$

and

$$
A_0^* = \begin{pmatrix} \alpha_{11} & \alpha_{12} & 0 \\ \alpha_{21}(\beta^*) & \alpha_{22}(\beta^*) & \alpha_{23}(\beta^*) \end{pmatrix}
$$

Nonsingularity of A_0^* implies nonsingularity of C^*. The linear independence condition implies A^* of full row rank. Notice that in our case, non degeneracy of vertices of Z is sufficient for linear independence condition and that no more than three constraints can be active. The only essential assumption for differentiability of the optimal solution that cannot be easily secured for the true program is the strict complementarity condition. If it does not hold true, we still can get *differentiability of the optimal value*

$$
\varphi(\beta^*) = G_0(z(\beta^*), \beta^*).
$$

$$
\nabla_\beta \varphi(\beta^*)^T = \nabla_\beta(z(\beta^*), \beta^*)^T =
$$
$$
c_2(\beta^*) \exp(\sum_{j=1}^{3} \alpha_{2j}(\beta^*) z_j(\beta^*)) [-1, -z_1(\beta^*), -z_2(\beta^*), -z_3(\beta^*)] \tag{19}
$$

At this moment, we can follow the approach suggested in Dupačová (1984) and to apply theorems on transformed asymptotically normal vectors (see, e.g., Serfling 1980) to get the main result.

Proposition: Let β^* be the vector of the true parameter values and assume that for the optimal solution $z(\beta^*)$ of the parametric program

$$\text{minimize} \quad G_\theta(z,\beta^*) \quad \text{on the set } Z \tag{20}$$

the strict complementarity condition holds true, the matrix A^* is of full row rank and the matrix A^*_θ is nonsingular. Let β be the asymptotically normal estimate of β^* given by (13). Then

(i) The optimal solution $z(\beta)$ of the estimated parametric program

$$\text{minimize} \quad G_\theta(z,\beta) \quad \text{on the set } Z \tag{21}$$

is asymptotically normal

$$\sqrt{N}(z(\beta)-z(\beta^*)) \sim N[0,\sigma^2(\frac{\partial z}{\partial \beta})M(\frac{\partial z}{\partial \beta})^T] \tag{22}$$

provided that $\dfrac{\partial z}{\partial \beta}$ given by (18) is a non zero matrix. The rank of distribution (22) equals to the rank of

$$(\frac{\partial z}{\partial \beta})M(\frac{\partial z}{\partial \beta})^T.$$

(ii) The optimal value $\varphi(\beta)$ of the estimated program (21) is asymptotically normal

$$\sqrt{N}(\varphi(\beta)-\varphi(\beta^*)) \sim N[0,\sigma^2\nabla_\beta\varphi(\beta^*)^TM\nabla_\beta\varphi(\beta^*)] \tag{23}$$

with $\nabla_\beta\varphi(\beta)$ given by (19) even if the strict complementarity conditions for (21) are not fulfilled.

Whereas the distribution (23) can be used immediately to get an asymptotic confidence interval for the minimal cost per component, (22) applies to *logarithms* of the optimal speed, feed and depth in the original formulation of the metal cutting problem. For application of these results, it is essential that for N large enough, the covariance matrix in (22) and the variance in (23) can be evaluated at the estimated values of parameters β and σ^2; for a general discussion on this point see Dupačová (1984).

5. Example

A single carbon steel part (85 mm diameter, 100 mm length) is to be rough turned by sintered carbide tool using optimal cutting conditions that minimize the total machining cost. The lathe has an effective power of 8 kW. The maximum speed capacity is 2000 revolutions per minute, the minimum speed available is 50 revolutions per minute, the maximum available feed per minute is 1000 mm/minute, the minimum available feed per minute is 5 mm/minute.

From the point of view of cutting tool from sintered carbide, the main constraints are: The maximum feed (chip formation) is 0.45 mm/revolution, the minimum feed (chip formation) is 0.05 mm/revolution, the maximum depth of the cut is 5 mm (strength of the tool), the minimum depth of the cut is 1.0 mm (chip formation). The machine power constraint has the form (8).

The coefficients of the cost function (7) and of the constraint (8) are $k_1 = 283$, $k_2 = 807$, $x_{F_z} = 1$, $y_{F_z} = 0.75$, $k_{F_z} = 1066$. The standard values of parameters in the Taylor's equation (5) are $A = 293$, $n = 0.36$, $m = 0.39$, $p = 0.11$. The used variables are given in the following unites: v (m/minute), r (revolutions/minute), h (mm), f (mm/revolution).

At first we give some statistical results. The collected data are given in the following table:

T	h	f	v	T	h	f	v	T	h	f	v
69.8	2	0.1	151	56.7	2	0.1	151	62.3	2	0.1	151
25.7	2	0.1	203	27.8	2	0.1	203	27.7	2	0.1	203
16.3	2	0.1	242	21.3	2	0.1	242	11.7	2	0.1	242
46.2	2	0.2	123	56.5	2	0.2	123	54.3	2	0.2	123
30.2	2	0.2	149	26.7	2	0.2	149	25.4	2	0.2	149
11.6	2	0.2	196	17.8	2	0.2	196	14.9	2	0.2	196
28.9	2	0.4	111	24.5	2	0.4	111	36.7	2	0.4	111
24.5	2	0.4	129	21.5	2	0.4	129	18.7	2	0.4	129
8.9	2	0.4	162	12.4	2	0.4	162	13.5	2	0.4	162
35.7	1	0.2	149	38.6	1	0.2	149	41.5	1	0.2	149
8.3	1	0.2	218	12.6	1	0.2	218	15.1	1	0.2	218
7.8	1	0.2	243	9.8	1	0.2	243	12.1	1	0.2	243
39.8	4	0.2	121	44.1	4	0.2	121	45.2	4	0.2	121
27.8	4	0.2	134	34.4	4	0.2	134	37.0	4	0.2	134
9.8	4	0.2	186	16.2	4	0.2	186	13.5	4	0.2	186

The following regression estimates and statistics were computed using Famulus modeling system (Famulus 1992).

```
Dependent variable: ln(T)

              b      std.dev.      t       signif.

  ln(c₀)   16.0462    0.6061    26.475*** 0.000
  ln(h)    -0.3121    0.0616    -5.067*** 0.000
  ln(s)    -1.0805    0.0646   -16.719*** 0.000
  ln(v)    -2.8308    0.1245   -22.729*** 0.000

residual sum squares            1.051205
residual std. dev.              0.160122
degrees of freedom                    41
coeff. of determination          93.44%
```

The one-to-one inverse transformation of the regression estimates provides the estimates of the coefficients in the Taylor equation (5). We get $n = 0.3532$, $m = 0.3816$, $p = 0.1103$, $A = 289.3155$. The differences between these transformed estimates and the tabelized coefficients as they were given in the beginning of this section are nonsignificant.

As was mentioned in the section 3 the normality of residuals can be often observed. This is also our case as the following statistical normality tests prove:

```
type of the test    test value    signif.

skewness test         -1.551        0.121
Cramer-von Mises       0.097       ~0.121
kurtosis test         -0.530        0.596
Watson                 0.088       ~0.129
Anderson-Darling       0.488       >0.150
omnibus                2.876        0.237
Shapiro-Wilk           0.968        0.358
```

Also residuals diagrams which are not shown here for the sake of brevity prove a good model choice. According to the previous results we can consider the normal distribution in our further analysis.

The variance matrix $\sigma^2 M$ was estimated as:

$$\sigma^2 M = \begin{pmatrix} \ln(v) & \ln(f) & \ln(h) & \ln(A) \\ 0.016 & 0.004 & 0.004 & -0.075 \\ 0.004 & 0.004 & 0.001 & -0.016 \\ 0.004 & 0.001 & 0.004 & -0.019 \\ -0.075 & -0.016 & -0.019 & 0.367 \end{pmatrix}$$

The optimization program provides results (GAMS 1992) as follows:

```
OPTIMAL PRICE = 1.185 (first summand = 0.765, second summand = 0.419
Lg(r) =  6.710  (r=820.571)
Lg(v) =  5.389  (v=218.984)
Lg(f) = -0.799  (f=0.45)
Lg(h) =  0.000  (h=1)

Lg(T) =  1.673  (T = 5.333).
```

The following constraints were active: lower bound for the logarithm of depth (Lagrange multiplier = 0.489), and upper bound for the logarithm of feed (Lagr. mult. = 2.744). From here it follows that the largest possible unit improvement of the price can be achieved by relaxation the upper feed constraint. Solving the system of equations (18) we get the solution as

$$\left(\frac{\partial z}{\partial \beta}\right) = \begin{pmatrix} -0.353 & -2.095 & 0.282 & 0.000 \\ 0 & 0 & 0 & 0 \\ 0 & 0 & 0 & 0 \end{pmatrix}$$

Since the only active constraints were of the simple form mentioned above the two latest rows of the previous matrix are zero. Hence the second and the third coordinate of the optimal solution are not sensitive to small changes of the parameters β. However, if the active constraint would not be of the simple form all the coordinates could be sensitive with respect to the estimated parameters.

Using the Proposition we get the variance matrix of the optimal vector z as

$$\begin{pmatrix} 0.00208 & 0 & 0 \\ 0 & 0 & 0 \\ 0 & 0 & 0 \end{pmatrix}$$

and the variance of the optimal price reads 0.006. Hence the 2σ interval for z_1, i.e., for the optimal value of logarithm of speed, can be given as (5.300,5.480) and for the optimal price we get the interval (1.030,1.400).

For a better performance we give also optimality results computed for the tabelized coefficients in the Taylor's equation.

```
OPTIMAL PRICE = 1.177 (first summand = 0.754, second summand = 0.424
Lg(r)  =   6.727  (r=834.640)
Lg(v)  =   5.407  (v=222.962)
Lg(f)  =  -0.799  (f=0.45)
Lg(h)  =   0.000  (h=1)

Lg(T)  =   1.624  (T = 5.073).
```

The optimal values are fully covered by the mentioned intervals given by the stochastic approach.

Confidence interval width information is of a great value for analysis of economic features of the solved cutting problem. It provides bounds on the optimal cost of rough cutting of one part of a large series, say several thousands of pieces. This kind of results can be also used for decisions on using a tool life monitoring system.

In practical problems the depth of the cut is usually fixed for a particular pass in turning regarding the requirements on the final diameter of the part. Hence our results are the same as if we took $h=1$ mm.

6. Extensions

The approach developed in Section 4 can be generalized in many ways. The first generalization is to allow for *random coefficients in the constraints* of geometric program (9),(10), for instance in the machine power constraint (8). In this case, we can use formulas for derivatives in Kyparisis (1988) but we can arrive at difficulties connected with obtaining data suitable for a proper treatment of the corresponding *multiple* regression model. A similar problem can be discovered in course of application of a simplified approach based on individual probabilistic constraints for (14) (cf. Iwata et al. 1972) along with the proposed treatment of the estimated coefficients in the objective function.

Additional variables and constraints appear in connection with *fine turning* (cf. Chakrabarti 1986): Surface finish equation and requirements on accuracy can be considered, both of them often based on a sum of hyperbolic terms, and the nose radius is included as an additional decision variable. The resulting optimization problem is of the form of geometric program again, the derivatives can be obtained and the approach of Section 4 can be applied.

Optimization of multiple tools turning operation is based on a time diagram according to which various tools are supposed to be active at a given time. It means that the constraints on available power consumption (generally, in the form of a sum of hyperbolic terms each of which corresponds to one active tool) have to be considered separately in all particular time points in which the pattern changes. Moreover, the number of decision variables increases (cutting speed, feed and depth for each of considered tools) and it is necessary to evaluate the contribution of all tools to the total machining cost per component including the various tool lives. The resulting optimization problem is of the form of geometric program and, theoretically, the suggested approach can be applied again. Another possibility could be to take the tool lives of individual tools as decision variables, too, and to use additional constraints based on individual "relaxed" Taylor equations: The idea is simple - for instance, instead of substituting for T into the objective function according to (5) one uses constraints (compare with (11))

$$L_\alpha c_0 v^{-\frac{1}{n}} f^{-\frac{m}{n}} h^{-\frac{\ell}{n}} \leq T \leq U_\alpha c_0 v^{-\frac{1}{n}} f^{-\frac{m}{n}} h^{-\frac{\ell}{n}}$$

with prior bounds L_α, U_α on the precision of the Taylor's equation that can be related to the $(1-\alpha)$ confidence band around the corresponding linearized regression model.

In the case of *multipass turning operation* (cf. Mukerjee and Pal 1986), additional constraints have to be considered again. The new feature is that there are *differences of polynomials* in the objective function so that we get a *generalized* geometric programming problem with estimated coefficients. At the first instance, generalization of the parametric sensitivity results to generalized geometric programs is needed as a starting point for application of our approach.

References

M. Avriel and D. J. Wilde: Stochastic geometric programming. In: H. W. Kuhn (ed.), Proc. of the Princeton Symp. of Math. Progr. (Princeton Univ. Press, 1970)

K. K. Chakrabarti: Cost of surface finish - A general optimisation approach. In: Proc. 12th AIMTDR Conference, Delhi (Tata McGraw-Hill Publ., New Delhi, 1986), p. 504-507.

B. Colding and W. Konig: Validity of the Taylor equation in metal cutting. Ann. of CIRP XVIV (1971) 793-812.

P. J. Dhrymes: Econometrics (Springer, Berlin, 1974).

J. Dupačová: Stability in stochastic programming with recourse - Estimated parameters. Math. Progr. 28 (1984), pp. 72-83.

P. M. Ellner, R.M. Stark: On the distribution of the optimal value for a class of stochastic geometric programs. Naval Res. Log. Quart. Q7 (1980), pp. 549=571.

D. S. Ermer: Optimization of the constrained machining economics problem by geometric programming. J. of Engineering for Industry Trans. of ASME Vol.93 (1971) 1067-1072.

Famulus and FamStat software, release 3.5, English version. Famulus Corp., FamStat Corp. 1993.

GAMS software, release 2.25, GAMS Development Corp. 1992.

A. V. Fiacco: Introduction to Sensitivity and Stability Analysis in Nonlinear Programming (Academic Press, New York, 1983).

B. Gopalakrishnan and Faiz Al-Khayyal: Machine parameter selection for turnihg with constraints: An analytical approach based on geometric programming. Int. J. Prod. Res. 29 (1991) 1897-1908.

K. Iwata et al.: A probabilistic approach to the determination of the optimum cutting conditions. J. of Engineering for Industry Trans. of ASME Vol.94 (1972) 1099-1107.

R. Jagannathan: A stochastic geometric programming problem with multiplicative recourse. Oper. Res. Letters 9 (1990) 99-104.

J. Kyparisis: Sensitivity analysis in posynomial geometric programming. J. Optim. Theory Appl. 57 (1988) 85-121.

J. Mádl: Tool wear and breakage monitoring in machining. VDI BERICHTE 940 (1992) 125-132.

J. Mádl: Monitoring systems and machining costs. In: Proc. 6th International Conference on the Theory of Machines and Mechanisms, Liberec 1992, p. 149-154.

S. K. Mukherjee and M. N. Pal: Application of complimentary geometric programming technique in optimisation of a multipass turning operation. In: Proc. 12th AIMTDR Conference, Delhi (Tata McGraw-Hill Publ., New Delhi, 1986), p.487-489.

R. H. Philipson and A. Ravindran: Application of mathematical programming to metal cutting. Math. Progr. Study 11 (1979) 116-134.

R. J. Serfling: Approximation Theorems of Mathematical Statistics (Wiley, New York, 1980).

T. Szántai, I. Mészáros, J. Völgyi: Stochastic optimization models for machinig optimization. Lecture presented at 2nd GAMM/IFIP Workshop on Stochastic Optimization: Numerical Methods and Technical Applications, UniBw Munich-Neubiberg, June 15-17, 1993.

F. W. Taylor: On the art of cutting metals. Trans. of ASME 28 (1907) 31-35.

J. G. Wagner and M. M. Barash: Study of the distribution of the life od HSS tools. Trans. of ASME 93 (1971) 1044-1050.

R. D. Wiebking: Optimal engineering design under uncertainty by geometric programming. Manag. Sci. 6 (1977) 644-651.

Consideration of Stochastic Effects for Finding Optimal Layouts of Mechanical Structures

H.A. Eschenauer and T. Vietor

Research Center for Multidisziplinary Analysis
and Applied Structural Optimization (FOMAAS)
Institute of Mechanics and Control Engineering
University of Siegen, 57068 Siegen, Germany

Abstract. The use of advanced materials will increasingly gain importance in future developments of constructions in different disciplines. For this reason, the material behaviour in particular has to be considered when finding optimal layouts for components. Here, the different failure mechanisms of the applied materials must be taken into consideration. This paper presents a comparison between conventional, ductile materials and brittle ceramics as an example of an advanced material. In order to find a failure criterion which is characteristic for the material, stochastic models of the defects determining the failure of ceramic materials have been included. Because of the stochastic nature of the material parameters the classical deterministic optimization model is not sufficient. For this reason an augmented optimization procedure for shape optimization is introduced and tested for two examples.

Keywords. Advanced materials, brittle materials, cantilever disk, ceramics, failure of brittle materials, optimization procedure, shape optimization, turbine disk

1 Introduction

The optimal layout of structures using advanced materials (e.g. ceramics, fiber-reinforced materials) calls for the augmentation of existing optimization procedures as well as for an multidisciplinary cooperation of mechanics, material sciences and design. Apart from considering the sometimes substantially diverging material characteristics, it becomes necessary to precisely describe the failure mechanisms of the different materials. An increased range of application is characteristic of one advanced material, namely ceramics which belong to the group of brittle materials. So far, results have been obtained in the field of structural optimization predominantly for the use of ductile materials. This paper presents first the optimal layout of a cantilever disk using the example of a specific ceramic (Al_2O_3). A comparison is made between the results for this brittle material and ductile materials (steel, aluminium). The goal is to find some basic effects of brittle materials on the optimal shape and compare these to the effects of ductile materials. In the following part the developed optimization procedure is used for the optimization of a simple turbine disk.

Shape optimization of structures is a well known problem which LAGRANGE (1736-1813) and CLAUSEN (1801-1885) already applied to bending beams using variational principles. But only the introduction of efficient and flexible analysis procedures like the Finite-Element (FE) Analysis allowed the application of the shape optimization to a wide range of problems.

Freudenthal (1968) first dealt in his paper with a probabilistic procedure for deriving a failure criterion for brittle materials. Here, the term *failure probability* is introduced, and basic physical phenomena of brittle materials are described mathematically. A large number of papers based upon this paper, i.g. Evans (1978) and Batdorf, Crose (1974) describe ways of calculating failure criteria for ceramic materials. Koski, Silvennoinen (1990) show the result of a shape optimization using brittle materials.

2 Failure of Brittle Materials

2.1 Comparison of Ductile and Brittle Materials

In order to classify ductile and brittle materials, the value of the *critical stress intensity factor* K_{IC} which is a real material parameter proved useful. For a Griffith-crack in the one-dimensional stress state and the so-called crack opening mode I the stress intensity factor is given by the equation

$$K_{IC} = \sigma_c \sqrt{\pi x} \tag{1}$$

with σ_c as the critical stress value and x as the half length of the Griffith-crack. For brittle materials a range of $K_{IC} \leq 10MN/m^{3/2}$ and for ductile materials a range of $K_{IC} \geq 25MN/m^{3/2}$ is assumed. Correspondingly, the material behaviour lies between brittle and ductile for $10MN/m^{3/2} \leq K_{IC} \leq 25MN/m^{3/2}$.

Conventional Failure Criteria
For the conventional layout of components failure criteria are established by means of strength hypotheses which transform a multi-axial stress state into an equivalent one-axial stress state. With the obtained equivalent stress a statement concerning failure can be made by comparison with characteristic values generally determined under a one-axial stress. The type of hypothesis to be used depends on the failure mechanism which ultimately depends on the material.

a) Ductile materials
In the case of yielding before failure of a component the distortion energy hypothesis according to LEVY, HUBER, VON MISES leads to the following expression

for the equivalent stress:

$$\sigma_e = \frac{1}{\sqrt{2}} \sqrt{\left[(\sigma_{xx} - \sigma_{yy})^2 + (\sigma_{yy} - \sigma_{zz})^2 + (\sigma_{zz} - \sigma_{xx})^2 + 6(\tau_{xy}^2 + \tau_{yz}^2 + \tau_{zx}^2) \right]}.$$
(2)

b) Brittle materials

In the case of material failure caused by rupture perpendicular to the direction of the principal stress (e.g. cast materials), the normal stress hypothesis according to Rancine and Lame is valid. Here, the highest principal normal stress determines the failure:

$$\sigma_e = \sigma_1, \; \sigma_1 = \max_i \sigma_i.$$
(3)

This hypothesis can be applied only in parts to ceramics as it does not consider the stochastic distribution of defects which determine the failure.

2.2 Failure Criteria for Brittle Materials

For ductile materials an equivalent stress criterion is applied in many cases. From the given multi-axial stress state an uniaxial equivalent stress σ_e is calculated. A comparison with the feasible value σ_{ef} supplies information about the failure. For brittle materials with a critical stress intensity factor $K_{Ic} \leq 10MN/m^3$ this procedure is impossible. Because of the stochastic distribution of the flaws determining the failure, it is necessary to derive a stochastic failure criterion. For this purpose, there exist different approaches with different ranges of application and effort. Two basic approaches are briefly introduced in the following.

2.2.1 Weibull-Approach

This approach is based on the following assumptions:

Assumption 1
The material behaviour is isotropic and all flaws are oriented perpendicular to the applied loads (crack opening mode I).

Assumption 2
The failure stress of a local crack can be calculated with the GRIFFITH-equation $\sigma_c x^{1/2} = const$. The volume can be subdivided into parts so that every volume element only contains one crack to be considered.

Assumption 3
The stiffness of the total volume is determined by the weakest-link-theory. As a result one obtains an equation for calculating the failure probability

$$P_f = 1 - \exp(-X)$$
(4)

with

$$X = \left(\frac{1}{m}!\right)^m \left(\frac{1}{\sigma_c}\right)^m \frac{1}{V_c} \int\limits_V (\sigma_1^m + \sigma_2^m + \sigma_3^m) dV, \qquad (5)$$

where the material parameters σ_c, V_c and the Weibull-modulus m are given.

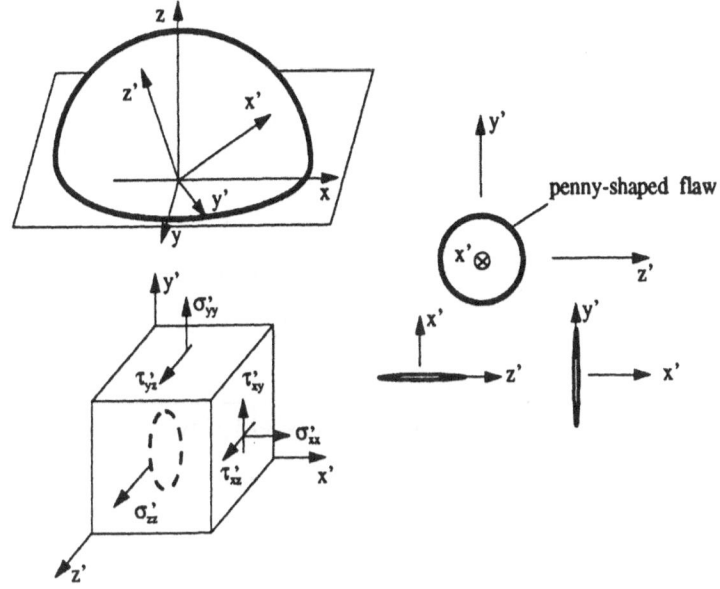

Fig. 1. Simple flaw model and critical stresses.

2.2.2 Batdorf-Approach

The assumptions used here read:

Assumption 1
The material behaviour is isotropic and all flaws are arranged with different orientations.

Assumption 2
The influence of the shape and the orientations of the defects is described by different failure models.

Herewith, we obtain

$$P_f = 1 - \exp[-X]$$

with

$$X = \int\limits_{V} \int\limits_{0}^{2\pi} \int\limits_{0}^{\frac{\pi}{2}} N(\sigma_{Ieq} = \sigma_{Ic}) \sin \varphi \, d\varphi \, d\psi \, dV. \tag{6}$$

For the function $N(\sigma_{Ieq} = \sigma_{Ic})$ in (6) exist different approaches. In many cases the assumption of a Weibull-distribution is valid for the critical defects and thus

$$N(\sigma_{Ic}) = \left(\frac{\sigma_{Ic}}{\sigma_{I0}}\right)^{m} \tag{7}$$

with the Weibull-modulus m. The parameter σ_{I0} is calculated from the distribution function of the strength

$$P_s = \exp[-(\frac{\sigma_{Ieq}}{\sigma_0})^m]. \tag{8}$$

For a penny-shaped flaw acc. to Fig. 1 we obtain the following stresses

$$\begin{aligned}
\sigma_{Ieq} &= f(\sigma_n, \tau), \\
\sigma_n(\varphi, \psi) &= \sigma_{xx}(\varphi, \psi), \\
\tau(\varphi, \psi) &= \sqrt{\tau_{xy}^2(\varphi, \psi) + \tau_{xz}^2(\varphi, \psi)}
\end{aligned} \tag{9}$$

with the failure model f and the normal and shear stresses σ_n, τ that cause the crack.

2.2.3 Comparison of the Approaches

The different orientations of the defects remain unconsidered in the Weibull-approach; in the Batdorf-approach all orientations are included in the calculation. Strictly speaking, the validity of the Weibull-approach is restricted to uniaxial stress states. The Weibull-approach can be extended to multi-axial stress states by forming the product of the single reliabilities. This is not motivated physically. Efficient approximations of the failure probability can be obtained though in stress states that are not too complex. The Batdorf-approach allows the integration of different failure criteria. In practice, however, only the normal stress criterion or the criterion of non-coplanar strain energy release rate are used. In the following, the latter is integrated into the post-processing of a finite element (FE) program and is applied to the example of a turbine disk.

2.2.4 Augmentation of the post-processing

In order to analyse the integral acc. to eq. (6), the structural analysis used in each case has to be augmented by a post-processing. From the structural analysis we obtain the stress state σ, from which $f(\sigma_n, \tau)$ can be calculated by means of a failure model. In the present paper, the development of a post-processor for the finite element program system ANSYS (NN, 1992) is developed. With that it is possible to calculate the failure probability of a material for numerous structures in the post-processing. For carrying out a numerical integration, the integral acc. to eq. (6) is solved numerically. Basically, it is possible to deal with uniaxial, biaxial and triaxial stress states. Here, however, we limit ourselves to biaxial stress states. The integral to be analysed reads:

$$
\begin{aligned}
X &= \left[\int_V \int_0^{2\pi} \int_0^{\frac{\pi}{2}} N(x, y, z, \varphi, \psi) \sin\varphi \, d\varphi \, d\psi \, dV \right], \\
&\approx \int_V \sum_{i=1}^{n} \sum_{j=1}^{n} w_{ij} \, N(x, y, z, \varphi_i, \psi_j) \sin\varphi_i \, dV, \qquad (10) \\
&\approx t \sum_{i=1}^{n} \sum_{j=1}^{n} \sum_{k=1}^{n} \sum_{l=1}^{n} w_{ijkl} \, N(x_i, y_j, \varphi_k, \psi_l),
\end{aligned}
$$

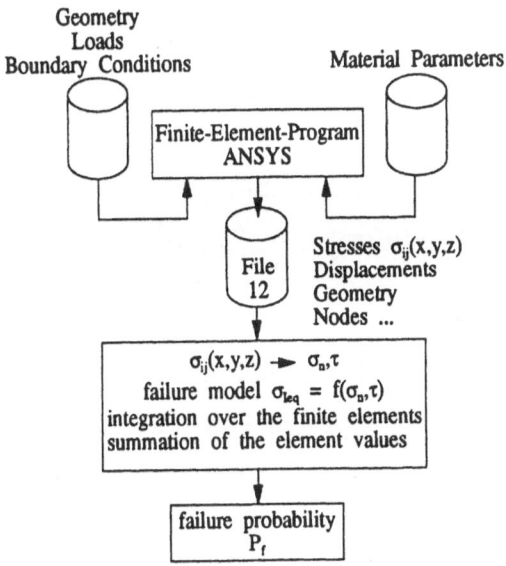

Fig. 2. Post-processing for calculating the failure probability with the finite element program ANSYS.

abbreviated this means

$$X \approx tS = t \sum_{i=1}^{n_e} S_{ei} \tag{11}$$

with the thickness t, the weighting factors w_{ijkl}, the number of integration points n that are assumed to be equal for all directions, the number of elements n_e and the integral over one element S_{ei}. The numerical integration is realized for a 2-D 8-node stress solid element and an axial harmonic 2-D 8-node stress solid element as a GAUSSian quadrature. With varying stresses in the direction of the thicknesses, the multiplication by t in eq. (10) is substituted by a further integration. The character of the integrand makes the use of a corresponding high number of integration points necessary. With $n = 8$ integration points, a numerically sufficient result is obtained in the case of the example given in the following. Thus, the integration effort is high, because 8 integration points in each direction have to be analysed. But for each analysis of the integral only one structural analysis is necessary. The functional values of the integrand can be determined by means of the approach functions of the isoparametric elements used at the integration points. The structure of the post-data file is advantageous since it can be read externally and the calculation of the integral only then becomes possible. The structure of the post-processing is given in Fig. 2.

3 The Shape Optimization Problem

The definition of the shape optimization problem taking the stochastic constraints into consideration reads as follows:

$$F^*\big[R^k(\xi^\alpha)\big] = \underset{R^k}{\text{Min}}\bigg\{ \quad F\big[R^k(\xi^\alpha), p\big] \ \bigg| \\ R^k(\xi^\alpha) \in G_R \bigg\}, \tag{12}$$

$$G_R = \bigg\{ \quad R^k(\xi^\alpha) \in \mathbb{R}^3 \ \bigg| \\ H_i\big[R^k(\xi^\alpha)\big] = 0 \ \ \forall i = 1, \ldots, n_h; \\ G_j\big[R^k(\xi^\alpha)\big] \geq 0 \ \ \forall j = 1, \ldots, n_{G1}; \\ G_k\big[R^k(\xi^\alpha, p\big] \geq 0 \ \ \forall k = 1, \ldots, n_{G2}; \\ R^{ml} \leq R^m \leq R^{mu} \ \ \forall m = 1, \ldots, n_R \bigg\} \tag{13}$$

with

F :   objective functional vector,

H_i, G_j, G_k : vector of equality and inequality constraint functionals,
R^k : shape functions,
ξ^α : GAUSSians surface parameters,
R^{kl}, R^{ku} : lower and upper bounds of the shape functions,
G_R : set of feasible shape functions

and the assumption for the shape functions

$$R^k(\xi^\alpha) \approx \hat{R}^k(\xi^\alpha, z^k),$$

The splitting of the functional vector of the inequality constraints

$$G^T = (G_1, G_2, \ldots, G_{n_{G_1}}, G_{n_{G_1}+1}, \ldots, \qquad (14)$$
$$G_{n_{G_1}+n_{G_2}})$$

into deterministic and stochastic constraints. The shape optimization problems is solved by a direct procedure where the problem with approach functions is

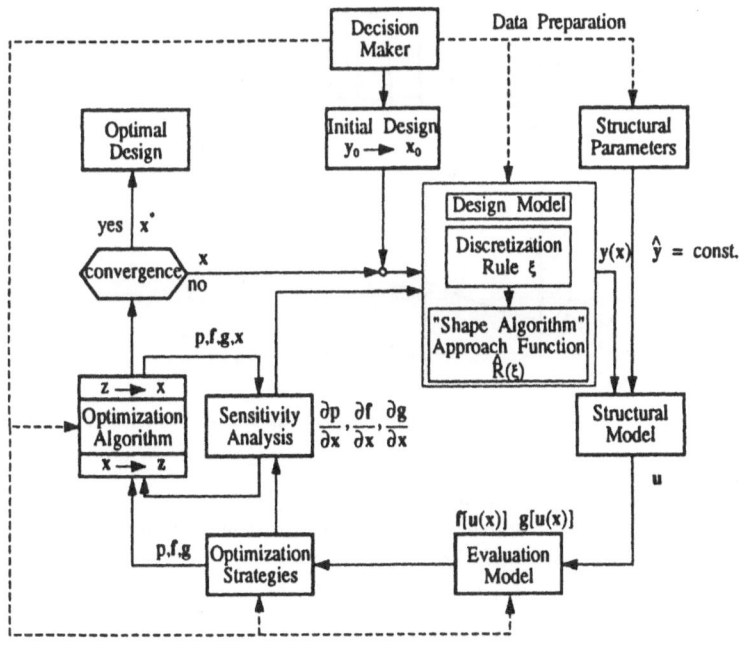

Fig. 3. Optimization loop for shape optimization.

transformed into a vector optimization problem which is transformed into a scalar optimization problem by means of a preference strategy. The design variable vector contains the independent shape parameters. The solution of the scalar optimization problem is carried out by procedures of mathematical programming and the optimization procedure SAPOP (Eschenauer et.al., 1993). Here, B-splines and Bezier curves are used as approach functions. The optimization loop for shape optimization is shown in Fig. 3.

4 Shape Optimization of a Cantilever Disk

Fig. 4 shows the initial design for a cantilever disk. The constant thickness of the disk is t = 10 mm. The initial design is a rectangular disk. The lower contour is described by B-splines and the coordinates of the control points are chosen as the design variables of the optimization model. The optimization model is used in accordance with the Three-Columns-Concept due to Eschenauer (1992).

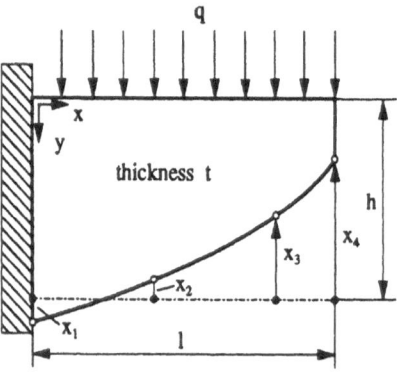

Fig. 4. Initial design, control points and design variables of the cantilever disk.

a) **Column 1: Structural Model**
The FE-program system ANSYS is used for structural analysis. The 8-nodes isoparametric shell element is used.

b) **Column 2: Optimization Model**
Fig. 4 shows the definition of the design variables as control points of the B-spline function. The following items are used as objective functions: mass of the disk, failure probability (in the case of brittle material) or the maximum equivalent stress σ_e (in the case of ductile material).

c) Column 3: Optimization Algorithms

Two different optimization algorithms are used. First a Generalized Reduced Gradient Algorithm with an efficient strategy for finding feasible design points and second a Sequential Linearization Strategy.

The formulation of the objective functions reads as follows:

$$f_1(x) \triangleq \rho t \sum_{i=1}^{n_e} A_i(x) \tag{15}$$

for the weight of the disk with ρ as the density of the material, A_i the area of the i.th finite element, the number of finite elements n_e and t the thickness of the disk.

$$f_2(x) \triangleq \begin{cases} P_f & | \quad \text{brittle material} \\ \sigma_e & | \quad \text{ductile material} \end{cases} \tag{16}$$

as the failure criteria for the specified material. Both objective functions create a multicriteria optimization problem which in the present case is transformed into a scalar optimization problem by applying a constrained oriented transformation. For that purpose, one objective function is transformed into a constraint by determining a demand level. So it is necessary to define maximum feasible values $P_{ffeas}, \sigma_{efeas}$.

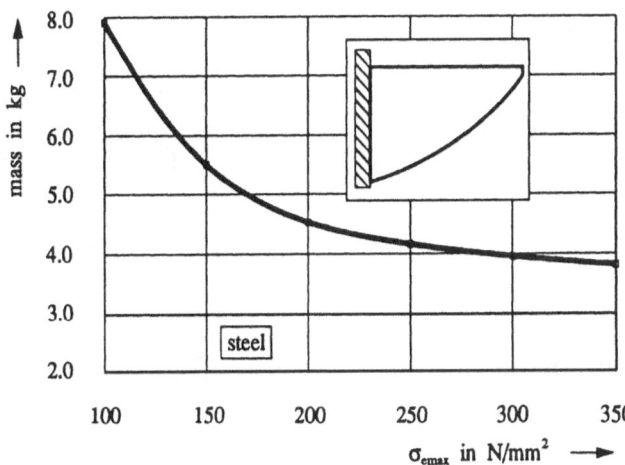

Fig. 5. Functional-efficient boundary of the disk out of steel and one optimal design.

Figs. 5 and 6 show the optimization results in the form of functional-efficient boundaries and the optimal design of one point. Each point on the boundary corresponds to a best compromise design. For the feasible failure probabilities in Fig. 6 a logarithmic scale is used. For the brittle material the optimal boundary shape of the disk is concave in contrast to the optimal shape using the ductile material. This results from the volume effect for ceramics. According to this effect the failure probability increases with constant stress and growing volume. It pre-dominates the stress reducing effect of a convex boundary for the ductile material. In Fig. 6 the influence of different WEIBULL-moduls m is shown.

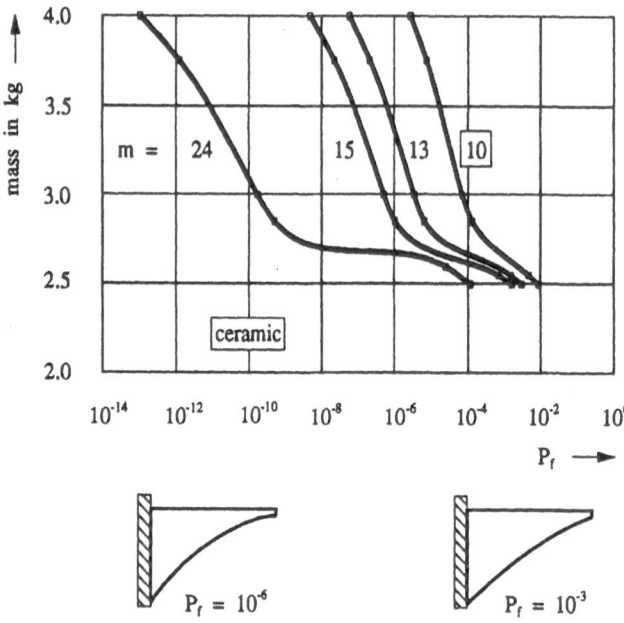

Fig. 6. Functional-efficient boundary of the disk out of ceramic and typical optimal designs.

5 Optimization of a Turbine Disk

With the results a comparison of ductile and brittle materials is given. In order to ensure comparability, the optimization is carried out with constrained equivalent stresses. The results of the optimization using ceramics and taking the special failure criteria of them into consideration are given then.

5.1 Problem definition

For the idealized turbine disk illustrated in Fig. 7 following is valid for the geometry, the loads and the material data:

	geometry		loads	
a	60.0 mm	p	10	N/mm^2
b	7.5 mm	ω	5000	$1/s$
h	10.0 mm	$\Theta(x = a)$	80	K

material parameters Al$_2$O$_3$
E	297000 N/mm^2
ν	0.23
ρ	$0.37 * 10^{-5}\, kg/mm^3$
m	24
α_T	$0.8 * 10^{-5}\, 1/K$
σ_{I0}	370 N/mm^2

material parameters steel		material parameters aluminium	
E	210000 N/mm^2	E	70000 N/mm^2
ν	0.30	ν	0.30
ρ	$0.78 * 10^{-5}\, kg/mm^3$	ρ	$0.28 * 10^{-5}\, kg/mm^3$
α_T	$0.16 * 10^{-4}\, 1/K$	α_T	$0.24 * 10^{-4}\, 1/K$

A special aluminium oxide ceramic Al$_2$O$_3$ is used with low density and low thermal expansion coefficient values but with a high YOUNG's modulus. The disk rotates with ω and the load p results from the centrifugal forces of the turbine blades that are not given in the structural model. The temperature load given as a simple fourth power function

$$\Theta(x) = \Theta^{(0)}\left(\frac{x}{b}\right)^4 \tag{17}$$

results from the heated steam or gas and describes the difference between the real and the stress free temperature of the structure. The parameter $\Theta^{(0)}$ is determined by means of the boundary conditions.

5.2 Structural Analysis

For the structural analysis, the FE-program system ANSYS is used. The structure is discretized by an axial harmonic 2-D 8-node stress solid element (element number 82) for which the augmentation of the post-processing for calculating the failure probability of the ceramic material has been carried out. The disk is discretized by up to 39 elements. The number of elements can be varied during the optimization. The adaptation to the current variables is carried out by the pre-processing of the FE-program. The elements used are in as far advantageous as only the cross-sectional area has to be discretized. Here, one can furthermore take advantage of the symmetry of the area. The integration acc. to eq. (6) can be carried out easily in circumferential direction so that the integration direction φ, ψ for determining the crack orientations and x, y in the cross-sectional plane remain. The contour of the disk is described by B-spline approach functions by means of 4 control points, the y-coordinates of which are the design variables $x_1, \ldots, x_4$.

In a parameter study preceding the optimization, the range of application of a ceramic turbine disk with regard to the loads is defined. Fig. 8 illustrates the dependence of the failure probability P_f of the ceramic material on the loads. If a value of $P_f < P_{ffeas}$ is demanded for practical applications, the curves given in Fig. 8 can be used to decide whether the ceramic material is suitable. For

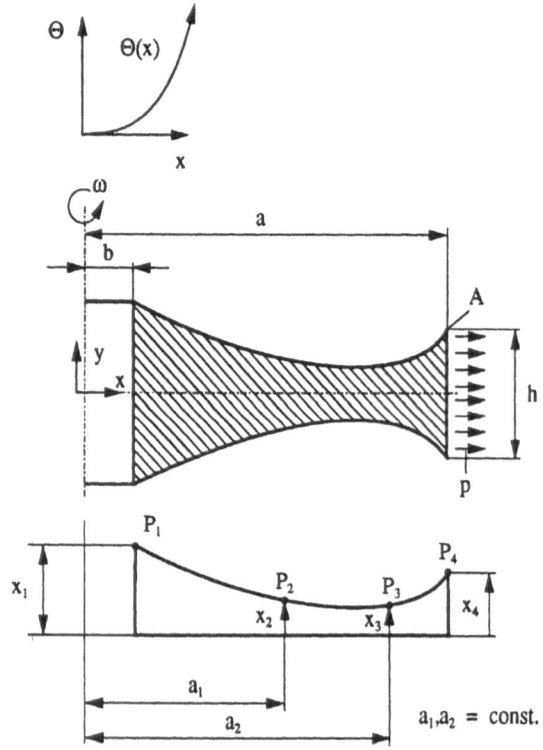

Fig. 7. Turbine disk under different loads.

other ceramics with different material parameters the curves are either shifted
to higher or lower values of P_f. Here, the Weibull-modulus m and the stress σ_{I0}
have a strong influence.

5.3 Optimization Model

For the optimization, two optimization models are defined. In both models, a
geometrical design model is used for the shape optimization, and a constraint-
oriented transformation for the vector optimization. The contour of the disk is
described using B-splines with the given control points. The y-coordinates of
these points are used as design variables. There is a difference in the definition
of the objective function vectors. In model 1, the objective function vector
contains the mass of the disk and the maximum radial displacement of point A
(see Fig. 7). In model 2 it additionally contains the failure probability of the
material as a further objective. Thus, the objective functions read

$$
\begin{aligned}
&\text{Model 1:}\\
&\boldsymbol{f}_1^T(\boldsymbol{x}) \;=\; (f_1, f_2) \qquad \text{vector of objectives,}\\[4pt]
&f_1 \;=\; M \qquad\qquad \text{mass,}\\[4pt]
&f_2 \;=\; u_{mA} \qquad\quad \text{displacement.}\\[4pt]
&\text{Model 2:}\\
&\boldsymbol{f}_2^T(\boldsymbol{x}) \;=\; (f_1, f_2, f_3) \quad \text{vector of objectives,}\\[4pt]
&f_1 \;=\; M \qquad\qquad \text{mass,}\\[4pt]
&f_2 \;=\; u_{mA} \qquad\quad \text{displacement,}\\[4pt]
&f_3 \;=\; P_f \qquad\qquad \text{material failure.}
\end{aligned}
\tag{18}
$$

As equality constraints the system equations $h(\boldsymbol{x}) = 0$ have to be fulfilled in
both models. In model 1 the following inequality constraints are given

$$
\boldsymbol{g}^T(\boldsymbol{x}) = (g_1, g_2, \ldots, g_{n_e})
\tag{19}
$$

with

$$
g_i = 1 - \frac{\sigma_{ei}}{\sigma_{e\,max}} \quad \forall i = 1, \ldots, n_e
\tag{20}
$$

and

$$
\sigma_{ei} =
\begin{cases}
\text{acc. to v. Mises} & |\quad \text{ductile}\\
\text{max. equivalent stress} & |\quad \text{brittle}
\end{cases}
\tag{21}
$$

and the number of finite elements n_e. It has to be emphasized here that the
layout of the disk made of ceramics with a comparative stress criterion is insuf-
ficient because of the extremely brittle character of the material, and it is only
carried out to make a comparison with ductile materials possible.

224

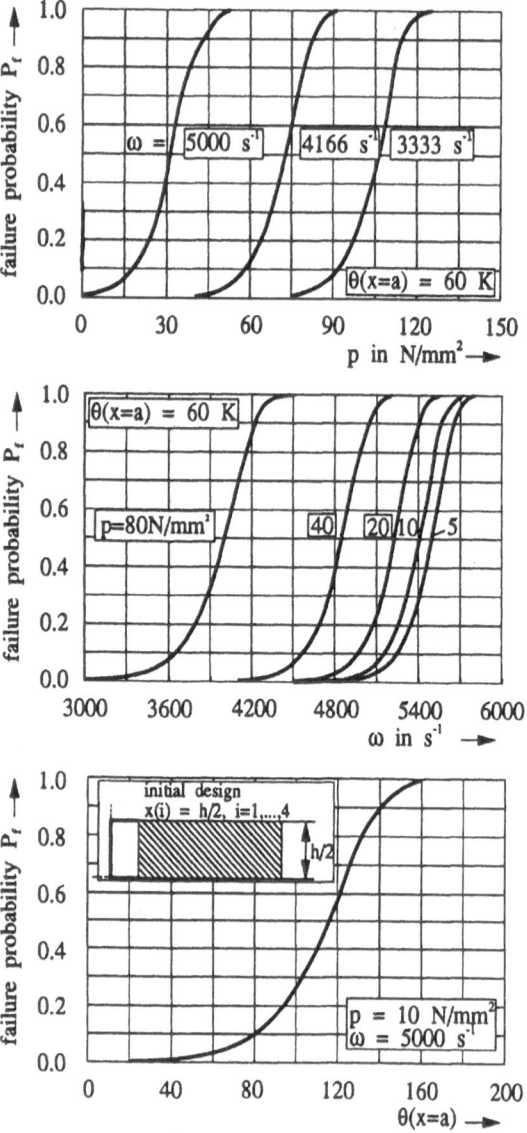

Fig. 8. Failure probability P_f for the initial design of a rectangular disk.

5.4   Optimization Algorithms

The choice of an appropriate optimization algorithm extremely depends on the problem. The optimization program system SAPOP used here offers the possibility of applying different algorithms individually. For the example given, the application of the sequential linearization procedure (SLP) seems promising if the starting point x_0 is extremely unfeasible. With a low unfeasibility or a feasible starting point, an algorithm of the generalized reduced gradient (GRG) proves to be advantageous. In combination with a quadratic approximation procedure (SQP), the optimal point can be calculated accurately.

5.5   Optimization Results

Fig. 9 gives the results of the optimization using optimization model 1 as functional-efficient boundaries.

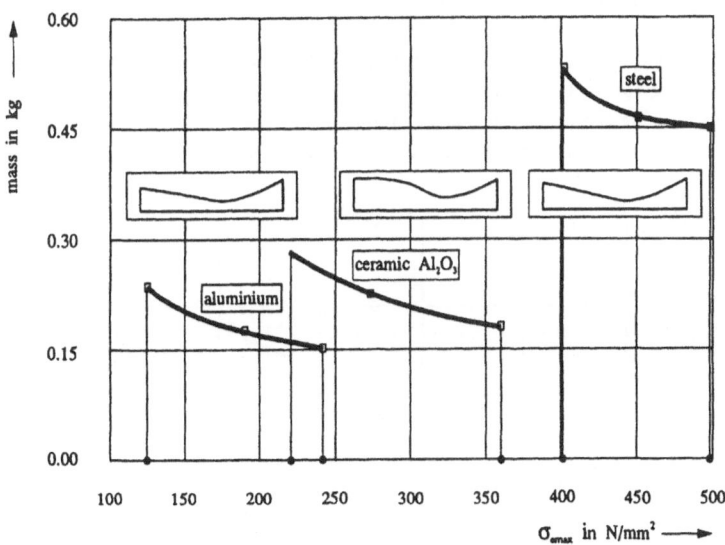

Fig. 9. Functional-efficient boundaries as a solution of the optimization problem for the optimization model 1.

If one compares the results for the three materials used, one recognizes that at different stress levels nearly the same optimal masses can be obtained for ceramics and aluminium. When using the latter, the low density is of essential importance for the good results. With $\rho = 0.28 * 10^{-5} kg/mm^3$ the density is lower than the density of the ceramic, and with the high speed of the plate it is the decisive load for the construction. The deformation constraint is inactive for all designs and it only influences the optimization results if the admissible value is assumed to be very small when determining the demand level. The optimal masses for steel are higher compared to the other two materials. Thus, with regard to the weight, steel is unsuitable for the optimal layout in this special case. The points on the stress axis in Fig. 9 mark the limits of the highly efficient solutions. The solutions are weak efficient towards to higher stresses whereas no feasible solutions can be found for lower stresses. In this example, the range of the efficient solutions is very small. Thus, the range in which a certain material can be optimally used for a turbine disk is also small.

Fig. 10 illustrates the optimization results of ceramic materials. The optimal failure probability is given above the thermal load. If one compares these optimization results to the failure probabilities of the initial design given in Fig. 8, it becomes obvious that the values are reduced decisively. This improvement, however, should not be overestimated since the initial designs are relatively arbitrary. The failure probabilities can be reduced to the extent that the application temperature range defined in Fig. 8 can be extended to temperature differences exceeding 100 K. Compared to aluminium, the ceramic material has the advantage of being thermostable. Above a temperature of 150 K in Fig. 10 it is shown that it does not make much sense to use the disk because of the great values of

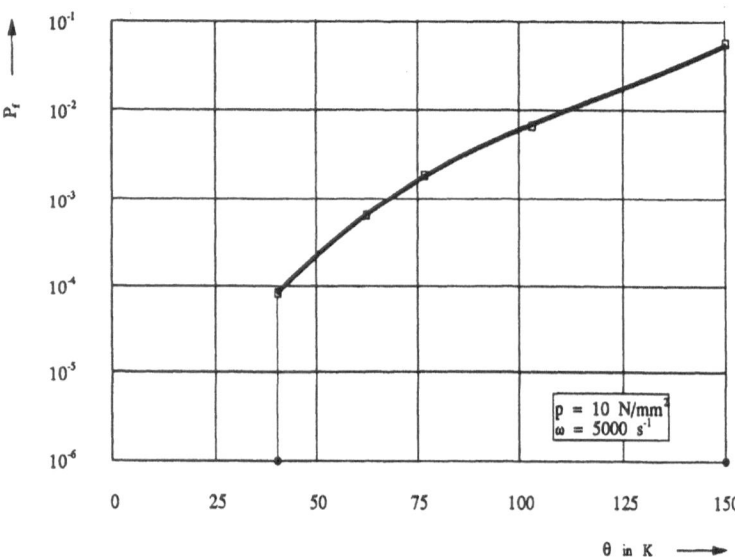

Fig. 10. Optimal failure probabilities for the ceramic turbine disk.

the failure probability. Below 40 K we did not calculate optimal solutions.

Fig. 11 gives the results of the optimization using optimization model 2 as functional efficient boundaries. The optimal mass for the cantilever disk is given above the values of the failure probabilities. The difference in the optimal shapes using an equivalent stress criterion or the failure probability is small. But with these small variations in the contour it is possible to reduce the failure probality from $2.0 * 10^{-1}$ to $4.8 * 10^{-4}$.

6 Conclusion

In the present paper, failure models are introduced for ceramic materials and integrated into the optimization model. The basic influence of the different failure criteria is presented at the example of a simple cantilever disk. Here the great influence of the Weibull-modulus to the optimal designs is shown. For a turbine disk subjected to pressure and temperature loads at a high speed, a special ceramic seems especially suitable compared to steel. As regards the weight, aluminium can be used as well. In a parameter study, the load range of the use of the plate made of ceramics can be determined. In the optimization calculations the failure probability of the material is reduced enormously compared to the initial design, and the application range that was defined in the parameter study can be extended. The time-dependent growth of the cracks, and the changing failure behaviour of ceramic materials subjected to high temperatures are not reflected in these results. For a constructive layout of the disk, further constraints regarding the geometry, for instance, have to be considered. With the solution procedure developed, it is possible to consider the special characteristics of ceramic materials in the design process. The structure of the optimization procedure SAPOP in combination with the FE-program ANSYS allows an optimal layout of further structures made of ceramics. In a preliminary study, however, it has to be decided whether the material can be used for the concrete example.

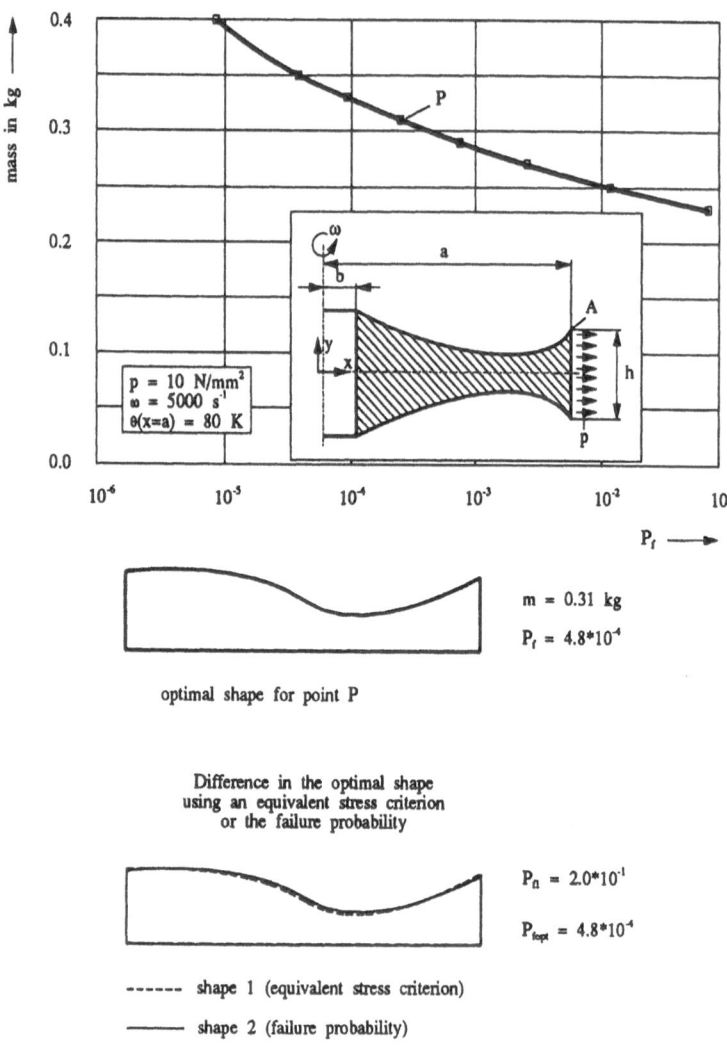

optimal shape for point P

m = 0.31 kg

$P_f = 4.8*10^{-4}$

Difference in the optimal shape
using an equivalent stress criterion
or the failure probability

$P_{fl} = 2.0*10^{-1}$

$P_{fopt} = 4.8*10^{-4}$

------ shape 1 (equivalent stress criterion)

———— shape 2 (failure probability)

Fig. 11. Functional-efficient boundaries as a solution of the optimization problem for the optimization model 2.

References

Batdorf, S.B.; Crose, J.G., 1974: A Statistical Theory for the Fracture of Brittle Structures Subjected to Nonuniform Polyaxial Stress. Journal of Applied Mechanics, 41 (1974), 459-465.

Eschenauer, H.A.; Koski, J.; Osyczka, A. (eds.), 1990: Multicriteria Design Optimization. Springer-Verlag: Berlin, Heidelberg, New York, 1990.

Eschenauer, H.A.; Vietor, T., 1992: Aspects in the Shape Optimization Using Ductile and Brittle Materials. In: Hoeltzel, D.A. (ed.): Advances in Design Automation 1992, Vol. 1, ASME, New York. 18th Design Automation Conference, Scottsdale, Arizona, Sept. 13-16, 1992.

Eschenauer, H.A.; Geilen, J.; Wahl, H.J., 1993: SAPOP - An Optimization Procedure for Multicriteria Structural Design. International Series of Numerical Mathematics, Vol. 110. Birkhäuser-Verlag: Basel, 1993, 207-227.

Evans, A.G., 1978: A General Approach for the Statistical Analysis of Multiaxial Fracture. Journal of The American Ceramic Society, 61 (1978) 7-8, 302-308.

Freudenthal, A.M., 1968: Statistical Approach to Brittle Fracture. In Liebowitz, H. (ed.): Fracture. An Advanced Treatise. Academic Press, New York, London, 1968, 592-618.

Koski, J., 1990: Multicriteria Design of Ceramic Components. In: Eschenauer, H.A.; Koski, J.; Osyczka, A. (eds.), 1990: 447-463.

NN, 1992: ANSYS, Engineering Analysis System, User's Manual. Swanson Analysis Systems, Inc., Houston, Pennsylvania.

Acknowledgement

The authors express their thanks to the German Research Community (DFG) for the support of the project "Stochastic Structural Optimization" (Es 53/4-1,2).

Tolerance Dynamics for a High Precision Balance

Eberhard Franz and Friedrich Pfeiffer

Lehrstuhl B für Mechanik, TU-München, 80290 MÜNCHEN, Germany

Abstract. High precision balances with resolutions up to 10^7 are extremely sensitive to tolerances of the design parameters. The influence of these tolerances with respect to the dynamical behavior of the balance must be known for lay-out purposes of the balance control unit. The balance is modeled within a range of 4 kHz which results in about 200 degrees of freedom including a large number of elastic modes of all components. The system is linear with respect to dynamics but parameter influences are nonlinearly contained in the amplitude-frequency-function. Of central importance for control design is the following problem.

Given a nominal set of design parameters and their relevant tolerances as foreseen by design and realized by manufacturing. Given furtheron the nominal amplitude-frequency-function for that set. Search for that combination of parameters and their tolerances which generate maximum deviations from the nominal amplitude-frequency-function at each frequency. This optimization problem is solved with a series of constraints applying sensitivity analysis for a pre-selection and then maximizing the deviations. Out of nearly 500 parameters of the balance only about 20 parameters have 95 % of the influence on the deviations.

Keywords. Elastic multibody systems, dispersion frequency response, optimization, high precision balance system.

1 Introduction

High precision balances with resolutions up to 10^7 are widely used in chemical and pharmaceutical industry where in many applications a fast weighing process is desired. Todays balances typically display a stable weight reading within 2–3 seconds. Reducing this transient time very quickly leads to a series of problems involving a more dynamical balance design, a more sophisticated control concept and, as an essential point, the influence of parameter scattering. The assembly of balances affords certain tolerances and, moreover, every production process is connected with inaccuracies which are unavoidable. The balance under consideration includes nearly 500 parameters with tolerances, where fortunately only a few influence significantly the balance dynamics.

As a measure for this influence we introduce the concept of a dispersion frequency response which quantifies the tolerance influences on dynamics. The basic idea is simple but new. A high precision balance can be represented by an elastic multi-body system which as an approximation is linearized and exhibits around 200 DOF. The amplitude frequency response follows in a standard and straightforward way using the reference parameter set. The deviations from this reference response are evaluated with the help of an optimization process: Given the reference and given the parameter scattering find for a given frequency that combination of parameters which generate the maximum deviations from the reference response. This problem results in a nonlinear optimization with unilateral constraints which can be solved by standard techniques [9,11,17]. The dispersion of the frequency response represents a basis for improved control designs.

History and tendencies of balances may be found in [12,8]. The papers [4,5,14] describe the development from pure mechanical designs to electro-mechanical force compensation which is widely used today. Up to now only a few models have been realized to describe balance dynamics. In [15] a FEM-model of high order is presented and frequency responses are evaluated. The more control-oriented papers [2,3,16] consider digital control concepts on the basis of very simple mechanical models.

General informations on rigid and elastic multibody systems and on FEM-techniques may be taken from the text-books [1,7,20,22]. Further research efforts at our institute concern an order (n)-algorithm [6], practical aspects of elastic multibody systems [13,23], dynamics and control of elastic manipulators [18,19] and optimization processes in dynamics [21].

2 Mechanical Models

Figure 1 illustrates the design and the operations of a typical high precision balance [10]. The lever (1) is fixed in a rotating way to the support (15) by the springs (2). The force is transmitted to the lever by the coupling belt (3). Typical lever ratios are around 100 which of course cannot be illustrated by Fig. 1. The longer lever end contains a coil (4) which works together with a permanent magnet (5) and with an opto-electronical sensor (6). Together they form a force compensation unit. On the other hand the belt (3) is connected to the couple (7) which itself is fixed to the pan carrier (8). The lower dish (9) and the upper balance pan (10) are partially decoupled by rubber elements (12). The asymmetric performance of the elements (3,7 to 12) is counter-balanced by the spring levers or shackles (13,14), which realize a parallel construction. These spring levers possess a rotational degree of freedom relative to the support (15) which itself is connected to the shell (16).

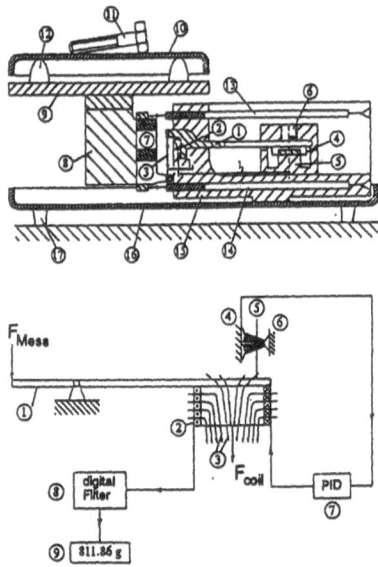

Fig. 1. High Precision Balance and its Principle

A weight on the balance moves its lever (1) and thus its opto-electronical system (4,5,6) (lower picture of Fig. 1). The signal enters a PID-control which generates a coil current and so a coil force which balances the weight. The current is a measure for the weight (8,9).

In establishing a mechanical model we must consider the high resolution performance which requires to consider a large frequency range from 0 Hz to 4 kHz. Therefore, most of the balance components have to be modelled as elastic bodies. Figure 2 illustrates the situation. Six components are assumed to be elastic and three components are rigid bodies. The spring symbols stand for a general force law which might be linear or non-linear. All numbers in Figure 2 correspond to those in Figure 1.

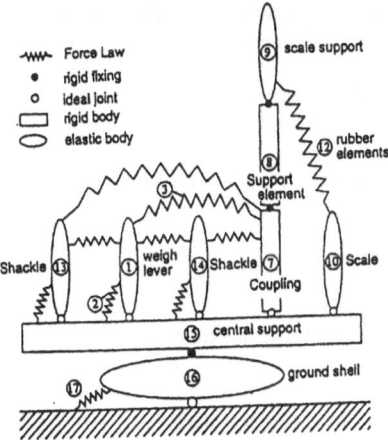

Fig. 2. Mechanical Model of the Balance System

We describe the system of Fig. 2 by a multibody approach starting with the principle of lost power (Jourdain, Lagrange, d'Alembert)

$$\sum_{\text{all bodies}} \int_{m_i} \left(\frac{\partial_R v}{\partial \dot{q}} \right)_i^T (_R a - _R f)_i \, dm_i = 0 \quad , \tag{1}$$

where $_R v_i \in \mathbb{R}^3$, $_R a_i \in \mathbb{R}^3$ are the absolute velocity, acceleration of body i in a body-fixed reference system, $_R f_i \in \mathbb{R}^3$ are all forces applied to body i and $\dot{q} \in \mathbb{R}^f$ are the generalized velocities. m_i is the mass of body i. The velocity and acceleration vectors must be evaluated in a body-fixed frame, too, which yields

$$
\begin{aligned}
_R v &= {}_R v_A + {}_R \tilde{\omega}_{IR} \left({}_R x_0 + {}_R \overline{r}_{el} \right) + {}_R \dot{\overline{r}}_{el} \quad , \\
_R a &= {}_R a_A + {}_R \tilde{\omega}_{IR} \, {}_R \tilde{\omega}_{IR} \, {}_R x_0 + {}_R \dot{\tilde{\omega}}_{IR} \, {}_R x_0 + {}_R \tilde{\omega}_{IR} \, {}_R \tilde{\omega}_{IR} \, {}_R \overline{r}_{el} \\
&\quad + {}_R \dot{\tilde{\omega}}_{IR} \, {}_R \overline{r}_{el} + 2 {}_R \tilde{\omega}_{IR} \, {}_R \dot{\overline{r}}_{el} + {}_R \ddot{\overline{r}}_{el} \quad .
\end{aligned}
\tag{2}
$$

The following abbreviations are used (index i always omitted):

$v = {}_R v_A$ absolute velocity, acceleration of the reference base R

$a = {}_R a_A$ expressed in the R-system,

$\omega = {}_R \omega_{IR}$ angular velocity of the reference system expressed in the R-sytem,

$x_0 = {}_R x_0$ vector from R to the mass element dm in the not deformed configuration,

$\bar{r} = {}_R \bar{r}_{el}$ deformation vector,

$J_T = \left(\dfrac{\partial {}_R v_A}{\partial \dot{q}} \right)$ Jacobian of translation, $J_T \in \mathbb{R}^{3,f}$,

$J_R = \left(\dfrac{\partial {}_R \omega_{IR}}{\partial \dot{q}} \right)$ Jacobian of rotation, $J_R \in \mathbb{R}^{3,f}$,

$\tilde{\omega} r = \omega \times r$ definition of serpent tensor.

Combining the equations (1), (2) and the above abbreviations we come out with the following set of equations of motion

$$\sum \int_{m_i} \left[J_T^T + J_R^T \left(\tilde{x}_0 + \tilde{\bar{r}} \right) + \left(\frac{\partial \dot{\bar{r}}}{\partial \dot{q}} \right)^T \right]$$
$$\left(a + \tilde{\omega}\tilde{\omega} x_0 + \dot{\tilde{\omega}} x_0 + \tilde{\omega}\tilde{\omega}\bar{r} + \dot{\tilde{\omega}}\bar{r} + 2\tilde{\omega}\dot{\bar{r}} + \ddot{\bar{r}} - f \right) dm = 0 \ , \tag{3}$$

where elastic influences are regarded as shown in Fig. 3. The radius vector r from an inertial frame (I) to a mass element dm_i of the deformed body B_i can be written as

$$r = r_{IR} + x_0 + \bar{r}_{el} \ , \tag{4}$$

where r_{IR} is the vector from (I) to (R), x_0 the vector to the mass element in the undeformed configuration and $\bar{r}_{el}$ the deformation vector. As in most cases of technical relevancy we assume that the elastic deformations of the balance components are very small compared to their geometric dimensions.

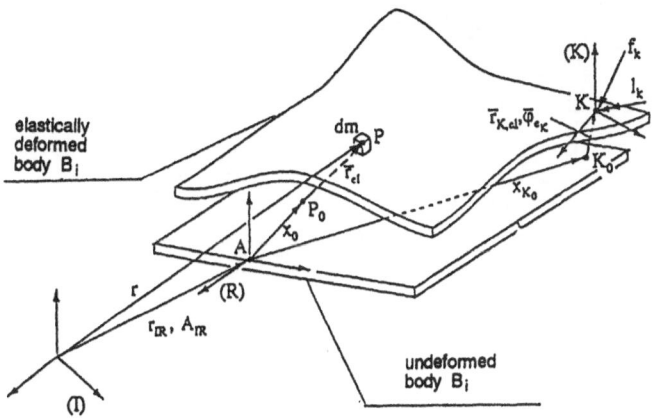

Fig. 3. Coordinates for a deformed Body B_i

This allows the introduction of a Ritz-approach for the elastic deformations [7,10,23]

$$\overline{r}_i(x_0,t) = W_i(x_0)q_{el,i}(t) \; ,$$

$$W_i = (w_1, \cdots, w_j, \cdots)_i = \begin{pmatrix} \overline{\overline{r}}_x^T \\ \overline{\overline{r}}_y^T \\ \overline{\overline{r}}_z^T \end{pmatrix}_i \; , \tag{5}$$

with $(q_{el} \in \mathbb{R}^{n_{el}})_i$ and $(W \in \mathbb{R}^{3,n_{el}})_i$. This well-known superposition of Ansatz- or shape-functions $w_{ji}(x_0)$ requires their completeness property [7]. For the elastic balance components (Fig. 2) we shall apply the eigenfunctions for appropriate boundary conditions as shape-functions. Figure 4 gives some typical examples, which have been determined by a FEM-code.

Combining equations (3,4,5) we must consider the dependencies $a(\ddot{q}), \dot{\omega}(\ddot{q}), \ddot{q}_{el,i}(\ddot{q})$ which says that all absolute and elastic accelerations depend on the generalized accelerations $\ddot{q}$ [10]. This property can be expressed by

$$\begin{aligned} a &= J_T\ddot{q} + \underline{a}(q,\dot{q},t) \; , \\ \dot{\omega} &= J_R\ddot{q} + \underline{\dot{\omega}}(q,\dot{q},t) \; . \end{aligned} \tag{6}$$

$$\ddot{q}_{el,i} = \frac{\partial q_{el,i}}{\partial q}\ddot{q} = \frac{\partial \dot{q}_{el,i}}{\partial \dot{q}}\ddot{q} = J_E\ddot{q} \; , \tag{7}$$

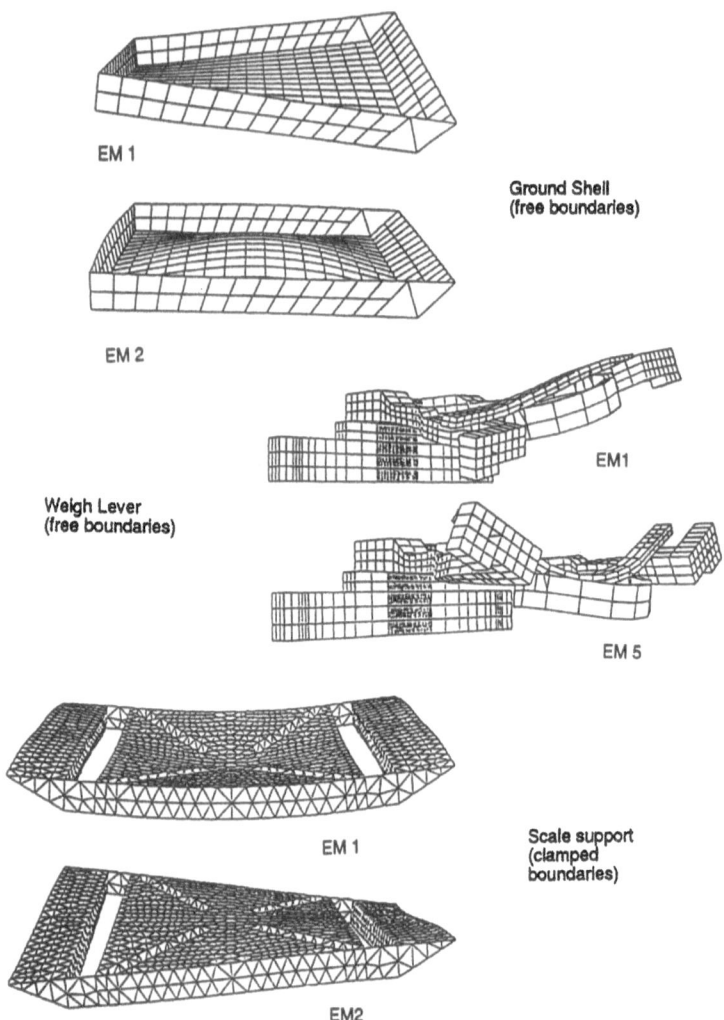

Fig. 4. Typical Eigenmodes applied as Shape Functions [10]

The equations (3) to (7) can now be put into the form

$$\sum_i \left\{ \int_{m_i} J_T^T \left[J_T - \left(\tilde{x}_0 + \tilde{\tilde{r}} \right) J_R + W J_E \right] dm + \right.$$
$$\int_{m_i} J_R^T \left(\tilde{x}_0 + \tilde{\tilde{r}} \right) \left[J_T - \left(\tilde{x}_0 + \tilde{\tilde{r}} \right) J_R + W J_E \right] dm +$$
$$\int_{m_i} J_E^T W^T \left[J_T - \left(\tilde{x}_0 + \tilde{\tilde{r}} \right) J_R + W J_E \right] dm \right\} \ddot{q} +$$
$$\int_{m_i} \left[J_T^T + J_R^T \left(\tilde{x}_0 + \tilde{\tilde{r}} \right) + J_E^T W^T \right]$$
$$\left[\underline{a} + \tilde{\omega} \tilde{\omega} \left(x_0 + \tilde{r} \right) + \underline{\dot{\omega}} \left(x_0 + \tilde{r} \right) + 2 \tilde{\omega} \dot{\tilde{r}} - f \right] dm = 0 \ . \tag{8}$$

The evaluation of these equations takes some further fifty pages (see [10]) which cannot be repeated here. The following steps are necessary:

- Determination of the applied forces f which are composed of gravity forces and elastic forces. The elastic forces follow Hooke's law and include the elastic deformations and thus $q_{el,i}$.

- Kinematical magnitudes are contained in the Jacobians, in the angular velocities ω_i, in the absolute velocities of the reference systems. They are calculated in a recursive way starting with the inertial system and proceeding to the last body of the tree-like structured system.

- Evaluation of the force elements as springs (12,13,14,17 in Fig. 1), bars (2,3), coil (4) and sensor (6). The last one is in reality a null-force-element, but is conveniently modeled as such. The most complicated and lengthy part of determining force elements consists in their relative kinematics.

- Determination of all integrals after splitting them up in spatial- and time-dependent parts.

After performing these steps and after introducing the control vector

$$u = \begin{pmatrix} f_c \\ l_c \end{pmatrix} \tag{9}$$

as electromagnetic forces and torques of the coil system we come out with an equation of motion

$$\ddot{q} = g(q, \dot{q}, u, p) \ , \tag{10}$$

where p is a set of construction parameters of the balance system. Equations (10) are now nonlinear, discretized ordinary differential equations which will be further simplified. For this purpose we first determine the equilibrium state (q_E, u_E) by setting $\dot{q} = 0, \ddot{q} = 0$ which results in

$$g(q_E, 0, u_E, p) = 0 \ . \tag{11}$$

From these nonlinear algebraic equations we get $q_E = q_E(p)$ and $u_E = u_E(p)$.

The motion of all balance parts is extremely small. Therefore we may linearize around (q_E, u_E) by assuming

$$q = q_E + \overline{q} \ , \qquad u = u_E + \overline{u} \ . \tag{12}$$

Equations (10) and (12) give together

$$M\left[q_E(p), p\right]\ddot{\overline{q}} + D\left[q_E(p), p\right]\dot{\overline{q}} + K\left[q_E(p), p\right]q = \overline{h}\left[u, q_E(p), p\right] \ , \tag{13}$$

which are the basic equations for all further considerations.

3 Control System

The control system is not a central topic of this paper, it is taken from existing works. On the other side a real balance simulation is not possible without including control. Therefore, we shall give a short illustration of the control concept in the following.

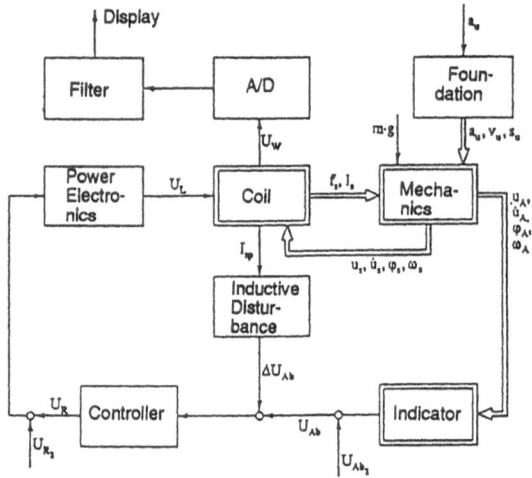

Fig. 5. Control for a High Precision Balance [10]

A detailed mechanical description has been given in chapter 2. The motion of the weigh lever ((1) in Fig. 1) has direct influence on the coil and the indicator (Fig. 5). The indicator generates the input signals for the controller, these signals additionally are disturbed by inductive coil effects. The controller controls the coil via its power electronics in such a way that the weight is exactly counterbalanced by the coil magnetic forces. The voltage applied for this balancing process is a measure for the weight. It passes a A/D-converter and a filter before being used in a display. The mechanical construction of the balance is additionally loaded by vibrations of the foundation which may cause considerable disturbances.

The interconnection of mechanics and control is, as always, twofold. Deviations from the weight-coil force balance are indicated by a diode system (Fig. 6) where a slit diaphragm fixed to the weigh lever distributes the light coming from the right diode to two photodiodes. If balance is perfect both photodiodes receive the same amount of light, as a result the voltage difference is zero. If the weigh lever moves a voltage difference signal will be generated which serves as control input (Fig. 5).

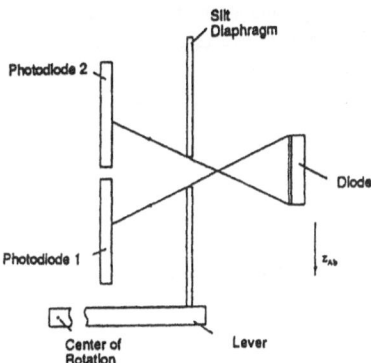

Fig. 6. Diode Sensor System

On the other hand the control output is converted into a force by the coil system which altogether behaves slightly nonlinear. The coil itself is fixed to the lever so that the magnetic forces act directly on the balance system. The local combination of sensor and actuator guarantees an efficient control of the balance.

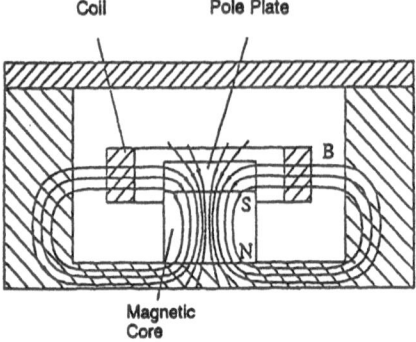

Fig. 7. Coil System as Magnetic Actuator

Combining the equations (13) of the mechanical system with the control concept of Fig. 5 allows a closed loop simulation of the balance system and a comparison with measurements. Figure 8 illustrates a case where a sinusoidal excitation at U_{R2} (Fig. 5) generates vibrations of the complete balance. The amplitudes A of Fig. 8 represent the relation (U_{ab}/U_w). Agreement even for higher frequencies is very satisfactory.

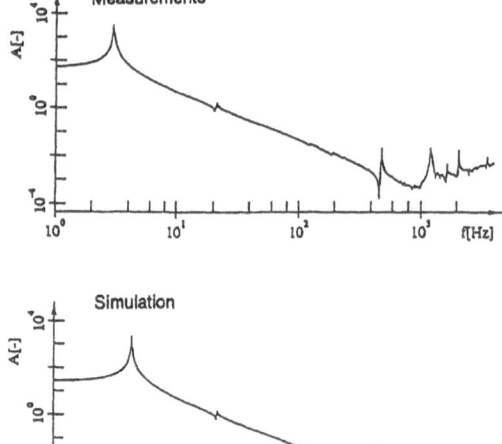

Fig. 8. Comparison Simulation-Measurement ($A = U_{ab}/U_w$)

4 Frequency Response Dispersion

4.1 Parameter Uncertainties and Optimization of Dispersion

The control problem of designing and operating high precision balances is illustrated by Fig. 9, which shows the frequency responses of two balances of exactly the same construction.

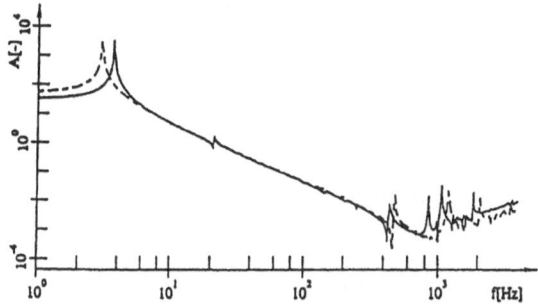

Fig. 9. Measurements for two Balances of the same Type (A = slit diaphragm position/coil force)

In spite of a very tight design with very small tolerances and in spite of a very careful production of these balances we recognize large deviations of the two response curves especially around resonance areas (note the logarithmic scale!). This influences the balance performance which can be avoided to a certain extent by an appropriate control design. In doing so the response dispersion must be known in advance which requires an adequate algorithm.

Dynamics and control of the balance depend on a parameter set $p \in \mathbb{R}^{n_p}$, each component p_i of which has tolerances and therefore a lower and an upper bound ($p_l \leq p \leq p_u$). Let us assume that p_0 is our parameter reference value. Then, transforming eqs. (13) into the frequency domain we are able to determine for each frequency value Ω_j a reference response $\hat{y}_{j0}(\Omega_j)$, where $\hat{y}_{j0}$ is the position of the slit of the diode sensor system (Fig. 6). Input for this response $\hat{y}_{j0}$ is the electromagnetic coil force $\hat{u}$ at the weigh lever. The output $\hat{y}$ is a function of p, and as ($p_l \leq p \leq p_u$), it is a function of the bounds $\hat{y} = \hat{y}(p, p_l, p_u)$. From this it must be possible to find that combination of $(p_{lk}, p_{uk})_{k \in n_p}$ which makes the deviations $\Delta \hat{y}_j$ from $\hat{y}_{j0}$ to an extremum. This will give us the maximum dispersal.

Evaluating these dispersal magnitudes for each frequency Ω_i we have to look at additional constraints, which refer to the static balance conditions of the weigh system: In equilibrium the coil current is a linear function of the weight. Dead load properties require that in equilibrium and for half the load only no current must be generated in the coil. For not centered loads on the weighing dish the display signal should remain more or less the same as for centered loads. These conditions lead to additional nonlinear constraints $n(p) \in \mathbb{R}^{n_c}$ with lower and upper bounds n_l, n_u.

Combining the above considerations we come out with the following optimization problem:

$$
\begin{array}{ll}
\underset{\substack{\Omega_l \le \Omega \le \Omega_u \\ p \in \mathbb{R}^{n_p}}}{\text{extremum}} & \{\zeta(p) : g(p) \le 0\}
\end{array}
\tag{14}
$$

$$
g(p) := \begin{pmatrix} p_l - p \\ p - p_u \\ n_l(p) - n(p) \\ n(p) - n_u(p) \end{pmatrix} \in \mathbb{R}^{2n_p + 2n_c}
\tag{15}
$$

$$
\zeta(p) := \left| \frac{\hat{y}(p)}{\hat{u}} \right| = \sqrt{\left[Re\left(\frac{\hat{y}(p)}{\hat{u}} \right) \right]^2 + \left[Im\left(\frac{\hat{y}(p)}{\hat{u}} \right) \right]^2} \ .
\tag{16}
$$

The optimization process requires a fast computation of $\zeta(p)$ and its derivatives with regard to p. This can be performed to a large extent analytically. Let us consider in a first step the output $\hat{y}$ as a function of $\hat{u}$.

We start with eqs. (13). A transformation into frequency domain yields

$$
\hat{q}(s,p) = (Ms^2 + Ds + K)^{-1}\hat{h}(u(s),p) \ ,
\tag{17}
$$

where all matrices depend on p and $\hat{h}$ is a Laplace transform of h. The only measurement is y, the only input the electromagnetic force u. Therefore, we may write

$$
y = c^T q \ , \qquad h = bu \ ,
\tag{18}
$$

which together with eqs. (17) results in

$$
\hat{y} = c^T (Ms^2 + Ds + K)^{-1} b\hat{u} \ ,
\tag{19}
$$

the magnitudes $(\hat{y}, \hat{u})$ being the Laplace transforms of (y, u). From eq. (19) we can evaluate eq. (16) immediately.

With regard to parameter uncertainties we must consider six groups of parameter properties

- tolerances in position and orientation of the force elements,

- assembly tolerances in position and orientation of all bodies,

- inertia (mass) tolerances of all bodies,

- uncertainties of the eigenfrequencies of the elastic bodies,

- tolerances in all measures being applied for balancing the weigh system with respect to static null-behaviour,

- tolerances of geometry and stiffness of all force elements.

Of course we cannot discuss here nearly 500 parameters which are typical for a balance (Fig. 1). But in place of all these parameters we shall discuss the case of assembly tolerances for a force element. Figure 10 shows a characteristic example. A force element realized by a plate-like bar shaped out from a cast of forged metal piece may be shifted in position by a (spatial) vector u and it may be rotated in orientation by a small angular vector φ. These uncertainties possess significant influence on the force characteristic of such an element and from there on the dynamic behaviour of the overall balance system.

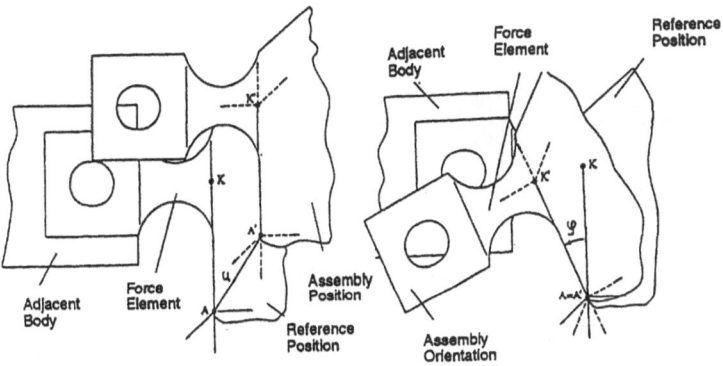

Fig. 10. Uncertainties for a Force Element due to Assembly

Tolerances of that kind are statistically known from measurements (quality assurance) in the production line so that a corresponding data base exists.

4.2   Reduction Processes

The severest problem in applying some kind of optimization to large mechanical systems consists in computing time. Therefore, a couple of measures have been taken to reduce computing time requirements. For the balance under consideration a frequency range 0 to 4 kHz has to be investigated which leads to a model with about 200 DOF, rigidly and elastically. The frequency range requires about 300 – 400 frequency points, for a complete simulation and optimization, which gives $4 \cdot 10^7$ sec computing time on a VAX 2540 VMS. The reduction measures to follow bring this down to $8 \cdot 10^3$ sec.

First of all we assume that the parameter tolerances $(\boldsymbol{p}_u - \boldsymbol{p}_l)$ (eq. 15) are small compared with the reference value $\boldsymbol{p}_0$:

$$\left(\frac{|\boldsymbol{p}_u - \boldsymbol{p}_l|}{|\boldsymbol{p}_0|} \right) \ll 1 \tag{20}$$

This allows a Taylor expansion of all parameter-dependent vectors and matrices:

$$
\begin{aligned}
\boldsymbol{M}(\boldsymbol{p}) &= \boldsymbol{M}(\boldsymbol{p}_0) + \sum_{i=1}^{n_p} \left(\frac{\partial \boldsymbol{M}}{\partial p_i} \right)_0 (p_i - p_{0i}) \ , \\
\boldsymbol{D}(\boldsymbol{p}) &= \boldsymbol{D}(\boldsymbol{p}_0) + \sum_{i=1}^{n_p} \left(\frac{\partial \boldsymbol{D}}{\partial p_i} \right)_0 (p_i - p_{0i}) \ , \\
\boldsymbol{K}(\boldsymbol{p}) &= \boldsymbol{K}(\boldsymbol{p}_0) + \sum_{i=1}^{n_p} \left(\frac{\partial \boldsymbol{K}}{\partial p_i} \right)_0 (p_i - p_{0i}) \ , \\
\boldsymbol{b}(\boldsymbol{p}) &= \boldsymbol{b}(\boldsymbol{p}_0) + \left(\frac{\partial \boldsymbol{b}}{\partial \boldsymbol{p}} \right)_0^T (\boldsymbol{p} - \boldsymbol{p}_0) \ , \\
\boldsymbol{c}(\boldsymbol{p}) &= \boldsymbol{c}(\boldsymbol{p}_0) + \left(\frac{\partial \boldsymbol{c}}{\partial \boldsymbol{p}} \right)_0^T (\boldsymbol{p} - \boldsymbol{p}_0) \ .
\end{aligned} \tag{21}
$$

Furtheron we set according to a frequently applied formula

$$D = \alpha M + \beta K \ , \tag{22}$$

and in addition we introduce a modal transform

$$q = V \xi \ . \tag{23}$$

The column-vectors of the modal matrix V are the eigenvectors v_i of the left-hand side of eqs. (13), and the modal coordinates ζ represent the eigenoscillations of the linear mechanical system. Combining eqs. (13) and (23) gives in the frequency domain

$$V^T \left(M s^2 + sD + K \right) V \hat{\xi} = V b \hat{u} \ , \tag{24}$$

which will be used to evaluate the derivatives of the eigenfrequencies ω_0 and of the eigenvectors v_i [10,21]. After some tedious but known operations we come out with

$$
\begin{aligned}
\frac{\partial \omega_0}{\partial p_k} &= \frac{v^T \left(-\omega_0^2 \dfrac{\partial M}{\partial p_k} + \dfrac{\partial K}{\partial p_k} \right) v}{2\omega_0} \\
\frac{\partial v_i}{\partial p_k} &= \sum_{j=1}^{n} c_{ijk} v_j \ .
\end{aligned}
\tag{25}
$$

$$
c_{ijk} = \begin{cases}
\dfrac{v_i^T \left[\left(\dfrac{\partial K}{\partial p_k} \right) - \omega_{0i}^2 \left(\dfrac{\partial M}{\partial p_k} \right) \right] v_i}{\omega_{0i}^2 - \omega_{0j}^2} \ , & i \neq j \ , \\[4ex]
-\frac{1}{2} v_i^T \left(\dfrac{\partial M}{\partial p_k} \right) v_i \ , & i = j \ ,
\end{cases}
\tag{26}
$$

We know now all derivatives to calculate the derivative of our penalty function $\zeta(p)$ (eq. 16). We get [10]

$$
\begin{aligned}
&\left\{ -\Omega^2 + j\Omega \left[\alpha + \beta \left(\omega_{0i} + \frac{\partial \omega_{0i}}{\partial p} p \right)^2 \right] + \left(\omega_{0i} + \frac{\partial \omega_{0i}}{\partial p} p \right)^2 \right\} \frac{\partial \hat{\xi}_i}{\partial p} + \\
&\left[2j\Omega\beta \left(\omega_{0i} + \frac{\partial \omega_{0i}}{\partial p} p \right) \frac{\partial \omega_{0i}}{\partial p} + 2 \left(\omega_{0i} + \frac{\partial \omega_{0i}}{\partial p} p \right) \frac{\partial \omega_{0i}}{\partial p} \right] \xi_i = \frac{\partial b_i^*}{\partial p} \ , \\[2ex]
&\frac{\partial \hat{y}}{\partial p} = \sum_{i=1}^{n} \frac{\partial c^{*T}}{\partial p} \hat{\xi} + c^{*T} \frac{\partial \hat{\xi}_i}{\partial p} \ , \\[2ex]
&\frac{\partial \zeta}{\partial p} = \frac{\mathrm{Re}(\hat{y})\mathrm{Re}(\partial \hat{y}/\partial p) + \mathrm{Im}(\hat{y})\mathrm{Im}(\partial \hat{y}/\partial p)}{\zeta} \ .
\end{aligned}
\tag{27}
$$

with

$$\frac{\partial b_i^*}{\partial p} = \frac{\partial v_i^T}{\partial p} b + v_i \frac{\partial b}{\partial p} ,$$
$$\frac{\partial c_i^*}{\partial p} = \frac{\partial c^T}{\partial p} v_i + c^T \frac{\partial v_i}{\partial p} . \tag{28}$$

The direct knowledge of $(\partial \zeta / \partial p)$ reduces the computing time considerably.

As a next step we reduce the number of parameters by applying a sensitivity analysis. We define two measures, a sensitivity measure $(\partial \zeta / \partial p_k)$ of the penalty function with respect to the k-th parameter and an importance function $[(\partial \zeta / \partial p_k)(p_{uk} - p_{lk})]$ which gives a better measure with regard to the possible modification band-width of p_k. It is (ν = excitation frequency)

$$_\nu S_k := \frac{\partial \zeta(\nu, p)}{\partial p_k} , \quad _\nu E_k := |_\nu S_k| (p_u - p_l)_k . \tag{29}$$

Summarizing all $_\nu E_k$ results in a measure for the total dispersion of the dynamics at the frequency ν. It is defined as

$$_\nu B := \sum_{k=1}^{n_p} {}_\nu E_k . \tag{30}$$

It makes sense to relate the importance function $_\nu E_k$ to the linear dispersion $_\nu B$ which gives us a dispersion number

$$_\nu s_k := \frac{_\nu E_k}{_\nu B} . \tag{31}$$

In the same way we define the sensitivity, the importance function and the dispersion number for the nonlinear constraints $n(p)$ (eq. 15):

$$S_k^{n_i} := \frac{\partial n_i}{\partial p_k} , \qquad E_k^{n_i} := |S_k^{n_i}| \cdot (p_u - p_l)_k \tag{32}$$

$$s_k^{n_i} := \frac{E_k^{n_i}}{\sum_{j=1}^{n_c} E_j^{n_i}} . \tag{33}$$

The total variation of the penalty function $\zeta(\boldsymbol{p})$ for a deviation $\Delta\boldsymbol{p}$ and a frequency ν is given by

$$\Delta\zeta(\nu,\boldsymbol{p}) = \sum_{k=1}^{n_p} {}_\nu S_k \Delta p_k \ . \tag{34}$$

As the optimization process is not only useful to determine the maximum dispersion magnitudes but also to find out the most important parameters generating dispersion we average the dispersion numbers over certain frequency ranges (ν_1,ν_2)

$${}_{\nu_1}^{\nu_2}\sigma_k := \frac{1}{\nu_2 - \nu_1} \int_{\nu_1}^{\nu_2} {}_\nu s_k d\nu \ . \tag{35}$$

The averaged dispersion number σ_k represents a measure for the influence of parameter p_k in the frequency range (ν_1,ν_2) under consideration.

5 Results

Figure 11 gives a typical result for a dispersion response for a high precision balance [10]. Dispersion is largest in the vicinity of resonance frequencies and comes out with factors around 100 which as a matter of fact significantly influences the balance dynamical performance.

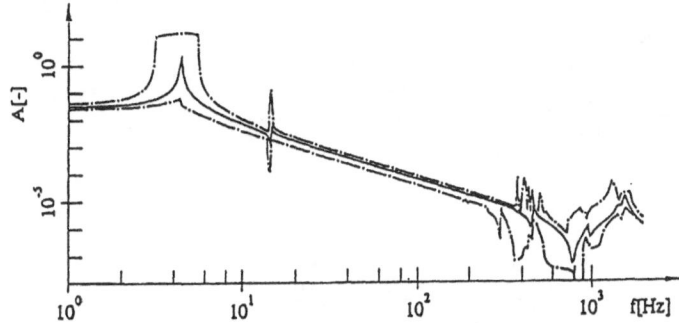

Fig. 11. Dispersive Response for a High Precision Balance (A = slit position/coil force)

The distribution of dispersion indicates possible improvements especially in connection with the most important parameters causing this dispersion. They are listed in three tables representing three different frequency ranges. A fourth table explains the symbols for the parameters.

Integral Dispersion	Parameter	Frequency 2 – 6 Hz
$\leq 30.0\%$	Weight 1 (Dead Load)	
$\leq 20.0\%$	Weight 2 (Dead Load)	
	d_3 connecting rod	
$\leq 10.0\%$	B lever	
	E-Modul connecting rod	
	d_2 connecting rod	
	d_4 connecting rod	
	E-Modul lever spring left	
	E-Modul lever spring right	
	d_4 lever spring left	
	d_4 lever spring right	
	r_2 connecting rod	
	r_3 connecting rod	
	w lever spring left (central support)	
	w lever spring right (central support)	
	d_5 lever spring left	
	d_5 lever spring right	
	d_3 lever spring right	
	d_3 lever spring left	
$\leq 1.0\%$	438 other Parameters $\sum = 9.4\%$	

Table 1. Most Important Parameters for Disperson for 2 - 6 Hz

Integral Dispersion	Parameter	Frequency 13 – 16 Hz
$\leq 40.0\%$	x Weight	
$\leq 30.0\%$	Weight 1 (Dead Load)	
$\leq 20.0\%$	Weight 2 (Dead Load)	
$\leq 10.0\%$	B lever	
	Eigenfrequency 2 (Scale)	
	Eigenfrequency 3 (Scale)	
	y Weight	
	z_s lever	
	w lever spring left (central support)	
	w lever spring right (central support)	
$\leq 1.0\%$	447 other Parameters $\sum = 12.4\%$	

Table 2. Most Important Parameters for Dispersion for 13 - 16 Hz

Integral Dispersion	Parameter	
$\leq 40.0\%$	Weight 1 (Dead Load)	
$\leq 30.0\%$	Weight 2 (Dead Load)	
$\leq 20.0\%$	no Parameter	
$\leq 10.0\%$	Eigenfrequency 5 (lever)	
	x_s lever	
	y Weight 1 (Dead Load)	
	y Weight 2 (Dead Load)	
	B lever	
	y_s lever	
	z_s lever	
	x Weight 2 (Dead Load)	
	x Weight 1 (Dead Load)	
	u Coil	
	Eigenfrequency 1 (lever)	
	Eigenfrequency 2 (lever)	
$\leq 1.0\%$	443 other Parameters $\sum = 4.8\%$	

Frequency 300 – 2000 Hz

Table 3. Most Important Parameters for Dispersion for 300 – 2000 Hz

d	=	thickness rod
B	=	moment of inertia
r	=	curvature rod
w	=	tolerance in position of force element
x, y	=	location of weight or of the dead loads
x_s, z_s	=	center of mass (lever)
u	=	tolerance in position of any force element

Table 4. Explanation of Symbols in Tables 1 – 3

The results show that the specific balance under consideration is extremely sensitve with regard to the weight position and with regard to the so-called dead loads. They are a mean to balance the weigh system for the static reference position which means that the mechanical parts must be in equilibrium without a coil force. The tolerances of these dead loads are large due to the fact that forged weigh levers themselves exhibit large tolerances.

Another important result consists in the fact that only 2 % – 4 % of all 457 parameters influence the dispersion behaviour. Beyond the dominating dead loads and the weight mainly lever properties give rise to dispersion. Indirectly the dead load influence is caused by the lever as well, therefore lever design is a central problem of this type of balance system (Fig. 1).

6 Summary

A high precision balance system is modeled by applying elastic multibody theory. Due to a large frequency range under consideration (0 – 4 kHz) altogether 200 mechanical degrees of freedom must be considered. The system possesses linear kinematics but nonlinear force laws. Nevertheless it can be linearized around a reference which represents static equilibrium.

Control is realized by one sensor, a slit diaphragm/diode system, and one actuator in form of a coil system generating electromagnetic forces. The control concept is taken from the existing balance design.

The central problem of such high precision balances consists in dynamic dispersion due to parameter tolerances. To evaluate the maximum dispersion amplitudes from a reference response a specific optimization problem is formulated taking into account the advantages of linear system dynamics. This optimization process not only yields the maximum deviations from the reference response but also that set of parameters influencing the most dispersion. Several measures were taken to reduce computing time.

As a result it turns out that the weigh lever system must be designed and manufactured with special care. Comparisons of theory and experiments are excellent.

References

[1] BATHE, K.-J. *Finite-Elemente-Methoden*, Springer-Verlag, Berlin Heidelberg New York usw., 1990

[2] BECKER, W.-J., SIEBERT, P., *Elektromechanische Kompensationswaage mit digitaler Regelung*, Wägen + Dosieren, 6, pp. 2-7, 1991

[3] BECKER, W.-J., SIEBERT, P., *Elektromechanische Kompensationswaage mit modellgestützter Messung des Wägegutes*, Technisches Messen 58 (1991) 5, pp. 202-208

[4] BEETZ, B., *Ein Beitrag zur Leistungssteigerung bei der dynamischen Wägung mit selbsttätigen Kontrollwaagen*, Dissertation, Universität Kaiserslautern 1984

[5] BEETZ, B., KOCH, B., *Integration elektronischer Waagen in industriellen Prozessen* Technisches Messen 58 (1991) 5, pp. 193-195

[6] BRANDL, H., JOHANNI, R., OTTER, M., *A Very Efficient Algorithm for the Simulation of Robots and Similar Multibody Systems without Inversion of the Mass Matrix*, IFAC,/IFIP/IMACS Symposium on Theory of Robots, Wien, 1986

[7] BREMER, H., PFEIFFER, F., *Elastische Mehrkörpersysteme*, Teubner Studienbücher, Stuttgart 1992

[8] CLEVERMANN, K., *Entwicklungstendenzen in der automatisierten Wägetechnik*, Technisches Messen 58 (1991) 4, pp. 184-187

[9] ESCHENAUER, H., KOSKI, J., OSYCZKA, A., *Multicriteria Design Optimization*, Springer Verlag, Berlin, 1990

[10] FRANZ, E., *Dynamik von elektromechanischen Präzisionswaagen*, Fortschrittberichte VDI, Reihe 18, Nr. 124, VDI-Verlag, Düsseldorf, 1993

[11] GOSDIN, M., *Analyse und Optimierung des dynamischen Verhaltens eines Pkw-Antriebsstranges*, Fortschrittberichte Reihe 11, Nr. 69, VDI-Verlag, Düsseldorf, 1985

[12] HAEBERLE, K.E., *10000 Jahre Waage – Aus der Entwicklungsgeschichte der Wägetechnik*, Bizerba-Werke (Hrsg.), Wilhelm Kraut KG, Balingen, 1967

[13] JOHANNI, R., *Automatisches Aufstellen der Bewegungsgleichungen von baumstrukturierten Mehrkörpersystemen mit elastischen Körpern*, Lehrstuhl B für Mechanik, TU München, Diplomarbeit, 1984

[14] JENEMANN, H.R., *Die mechanische Analysenwaage: Weiterentwicklung und Vollendung in der Zeit von 1960 bis 1975*, Wägen + Dosieren, 4/1988 – 1/1989

[15] MAAS, S., *Dynamische Strukturanalyse eines gekoppelten elektromechanischen Meßsystems*, Fortschrittberichte, VDI-Verlag, Düsseldorf, 1992

[16] MAIER, R., SCHMIDT, G., *Integrated Digital Control and Filtering for an Electrodynamically Compensated Weighing Cell*, IEEE Transactions on Instrumentation and Measurement, Vol. 38, No. 5, 1989, pp. 999-1003

[17] PAPAGEORGIOU, M., *Optimierung*, Oldenbourg-Verlag, München Wien, 1991

[18] PFEIFFER, F., BREMER, H., *Elastic Multibody Theory Applied to Elastic Manipulators*, IUTAM Symposium Moskau 1990, Dynamical Problems of Rigid-Elastic Systems and Structures, Springer Verlag, Berlin Heidelberg New York, 1991

[19] RICHTER, K., *Kraftregelung elastischer Roboter*, Fortschrittberichte, Reihe 8, Nr. 259, VDI-Verlag, Düsseldorf, 1991

[20] ROBERSON, R.E., SCHWERTASSEK, R., *Dynamics of Multibody Systems*, Springer Verlag, Berlin Heidelberg New York, 1988

[21] ROSS, CH., *Strukturoptimierung mit Nebenbedingungen aus der Dynamik*, Fortschrittberichte, Reihe 20, Nr. 38, VDI-Verlag, Düsseldorf, 1991

[22] SHABANA, A.A., *Dynamics of Multibody Systems*, John Wiley & Sons, New York, 1989

[23] SORGE, K., *Mehrkörpersysteme mit starr-elastischen Subsystemen*, Fortschrittberichte, Reihe 11, Nr. 184, VDI-Verlag, Düsseldorf, 1993

On Boundary — Initial Value Problem for Linear Hyperbolic Thermoelasticity Equations with Control of Temperature

Jerzy Gawinecki and Lucjan Kowalski

Department of Mathematics, Military Technical Academy 01-489 Warsaw, POLAND

Abstract. In this paper we consider the boundary-initial value problem for linear hyperbolic thermoelasticity equations with control of temperature.

The obstacles for temperature are formulated in the term of maximal monotone set. Existence and uniqueness of the solution of this problem are proved in Sobolev Spaces.

Keywords. obstacle for temperature, optimal control, maximal monotone operator, temperature controller, boundary — initial value problem, Hilbert Space, Sobolev Space, stochastic equations

1. Introduction

In this section we shall describe two systems of equations for coupled thermoelasticity theory in three-dimensional space: hyperbolic — parabolic and hyperbolic coupled system of thermoelasticity. We will present below the fundamental differential equations of the thermoelasticity theory, in which the deformation and temperature fields are coupled:

$$\nabla^2 u + (\lambda + \mu)\,\text{grad}\,\text{div}\,u + X = \partial_t^2 u + \gamma\,\text{grad}\,\Theta \tag{1.1}$$

$$\left(\nabla^2 - \frac{1}{\kappa}\,\partial_t\right)\Theta - \eta\,\text{div}\,\dot{u} = -\frac{Q}{\kappa} \tag{1.2}$$

In the equations of motion (1.1) there appears the temperature terms $(\text{grad}\,\Theta)$ whereas in the equations of thermal conductivity (1.2) — the deformation terms $(\text{div}\,\dot{u})$. The classical theory of thermoelasticity was investigated and developed by Nowacki [22]. The system of equations (1.1), (1.2) was studied also by Kupradze [17], Dafermos [6], Sneddon [26] and other authors [7],[8]. If we neglect in our considerations the influence of deformation on the change of temperature, i.e. $\text{div}\,\dot{u} = 0$, then equation (1.2) takes the form of the classical equation of thermal conductivity:

$$\left(\nabla^2 - \frac{1}{\kappa}\,\partial_t\right)\Theta = -\frac{Q}{\kappa} \tag{1.3}$$

We obtain equation (1.3) putting the Fourier law of the form:

$$q = -k\,\text{grad}\,\Theta \tag{1.4}$$

into the equation of the balance of the entropy. The equation of thermal conductivity (1.3) is a parabolic differential equation. Physically, this means that the thermal perturbations propagate with an infinite velocity.

At first sight this outcome of the theory seems to contradict the physical intuition. The aim then was to modify the Fourier law. Cattaneo [5] proposed to replace the Fourier law $q = -k \operatorname{grad} \Theta$ with the generalized Fourier law of the form:

$$q + \varepsilon \dot{q} = -k \operatorname{grad} \Theta \tag{1.5}$$

Substituting the equation (1.5) into the equation of the balance of the entropy, we obtain the hyperbolic differential equation of the form:

$$\left(\nabla^2 - \frac{1}{\kappa} \partial_t \right) \Theta - \frac{1}{\nu^2} \partial_t^2 \Theta = 0, \qquad \nu^2 = \frac{k}{\tau \varepsilon} \tag{1.6}$$

The term $\dot{q}$ occurring in the equation (1.5) is responsible for the finite velocity of propagation of thermal perturbations. Taking into account the coupling between deformation and temperature fields, we get the following equations of motion:

$$\nabla^2 u + (\lambda + \mu) \operatorname{grad} \operatorname{div} u + X = \partial_t^2 u + \gamma \operatorname{grad} \Theta \tag{1.7}$$

and the generalized equation of thermal conductivity:

$$\nabla^2 \Theta - \partial_t (1 + \tau \partial_t) \left(\frac{1}{\kappa} \Theta + \eta \operatorname{div} u \right) = 0 \tag{1.8}$$

Equations (1.1), (1.2) are called the equations of classical thermoelasticity theory and the equations (1.7), (1.8) are called the equations of generalized thermoelasticity theory with one relaxation time (cf. [19], [25]). The equation (1.7), (1.8) was investigated by Nowacki [22], Podstrigač [25] and others (cf. [8]).

Below, we present a new hyperbolic system of thermoelasticity theory which was derived according with the constitutive relations.

We observe (cf. (1.2), (1.1)) that the classical theory leads to a parabolic differential equation for temperature distribution in rigid heat conductors, implying that thermal perturbations are felt instantaneously in every part of the body. Although, at first sight, this outcome of the theory seems to be contradictory to physical intuition, it can be justified by resorting to the fact that molecular motion, which plays a crucial part in transport phenomena, is very rapid except at extremely low temperatures. Hence a finite velocity of propagation for thermal perturbations is usually nonobservable unless experiments are performed in some neighborhood of absolute zero, such in the case of liquid helium. In fact thermal waves, commonly known as *secound sound*, are detected in some metals cooled approximately down to 20 K. For a short survey the reader is referred to the work of Ackerman and Guyer [1], Taylor at al [28] and Jackson, Walker [14]. Below, we shall describe a theory of thermoelasticity [27], by considering the temperature rate dependence and assigning an appropriate constitutive function for the antropy flux. Such a theory leads to a hyperbolic differential equation for thermal perturbations other than equation (1.6).

One approach to remedy this apparent flaw (the fact that thermal perturbations are felt instantaneously in every part of the body) is to include the temperature rate among the constitutive variables, which leads to the presence of a second-order time derivative of the temperature field in the energy balance.

However, the Clausius-Dühem inequality, in the form we have employed up to now eliminates the temperature rate dependence from all the constitutive functions except for the constitutive function of heat flux. Hence, in order to obtain a properly posed theory for hyperbolic thermoelasticity theory, we have to resort to an entropy principle in its full generality presented in [20]. Such a theory of thermoelasticity was proposed by Müller in [20], who advocated rather special constitutive relations for the entropy supply in rigid conductors, which are simple generalizations of the conventional form Suhubi [27] extended his results to the thermoelasticity and obtained equations, which in the case of anisotropic and inhomogeneous medium and in the three-dimensional space have the following form [9,27]:

$$\frac{\partial^2 u_i}{\partial t^2} = f_i + \frac{\partial}{\partial x_j}\left(c_{ijhk}(x)\,\varepsilon_{hk}(u)\right) - \frac{\partial}{\partial x_j}\left(g_{ij}(x)\,\frac{\partial \Theta}{\partial t}\right) \qquad (1.9)$$

$$\frac{\partial^2 \Theta}{\partial t^2} + \frac{\partial \Theta}{\partial t} - \frac{\partial}{\partial x_j}\left(k_{ij}(x)\frac{\partial \Theta}{\partial x_i}\right) + g_{ij}\,\frac{\partial}{\partial t}\,\varepsilon_{ij}(u) +$$
$$+ (a_0^i - a^i)\left\{\frac{\partial}{\partial x_i}\left(\frac{\partial \Theta}{\partial t}\right)\right\} - \left(\frac{\partial}{\partial x_i}a^i\right)\frac{\partial \Theta}{\partial t} = q \qquad (1.10)$$

where:

$u = u(x,t) = \left(u_1(x,t), u_2(x,t), u_3(x,t)\right)'$ — denotes the displacement vector field,

$\Theta = \Theta(x,t)$ — the temperature of the medium,

$f = (f_1(x,t), f_2(x,t), f_3(x,t))'$ — denotes the body force vector,

$S = [c_{ijhk}]$, $i,j,k,h = 1,2,3$ — the elasticity tensor,

$e = [\varepsilon_{ij}]$, $i,j = 1,2,3$ — the strain tensor, which is related to displacement by

$$\varepsilon_{hk}(u) = \frac{1}{2}\left(\frac{\partial u_h}{\partial x_k} + \frac{\partial u_k}{\partial x_h}\right), \quad G = \{g_{ij}\}, i,j = 1,2,3$$

— the tensor of thermal expansion,

$K = [k_{ij}], i,j = 1,2,3$ — the tensor of thermal conductivity,

a_0^i, a^i — the physical coefficients (cf. [27]),

$q = q(x,t)$ — the intensity of heat sources, $'$ — denotes the transposition.

Our paper is organized as follows.

In the section 2 the statement of the problem is given. i.e. — we formulated the problem to the system (1.9), (1.10) with obstacle for the derivative with respect to t of the temperature and we prove a theorem of existence and uniqueness of the solution above problem.

Section 3 is devoted to the boundary — initial value problem with control of temperature. Finally, in the last section some concluding remarks are given.

Notation:

Besides other standard notation we use the symbol $W^{m,p}(\mathbb{R}^3)$ ($1 \le p \le \infty$, $m \in \mathbb{N} \cup \{0\}$) for the well-known Sobolev spaces (cf. [29], [2]).

2. Statement of the Problem. Existence and Uniqueness of the Weak Solution of the Problem

2.1. Statement of the Problem

Let Ω — be a bounded domain in $\mathbb{R}^3$ with smooth enough boundary $\Gamma = \partial\,\Omega$. Let t — be time variables, taking values from the interal $[0, T] \in \mathbb{R}$, and let $Q = \Omega \times (0, T)$. We consider the anisotropic and inhomogeneous thermoelastic medium occupied domain $\overline{\Omega}$. For $x \in \mathbb{R}$ and $t \in (0, T)$ by, $u = u(x, t)$ we denote the displacement vector field of the medium, and by $\Theta = \Theta(x, t)$ we denote the temperature of the medium. The pair (u, Θ) will be satisfying the equation (1.9) and instead of equation (1.10) some relation describing below (such relation will be describing the obstacle for the temperature and the obstacle for the derivative with respect to t of the temperature).

Now, we describe two values of the temperature $\psi_1(x)$, $\psi_2(x)$ ($x \in \Omega$, $\psi_1(x) \le \psi_2(x)$) and we need that the temperature of the medium taking the value from the interval $[\psi_1(x), \psi_2(x)]$ for any $t \in (0, T)$. In order to obtain this aim, we must to control the voluminal heat source satisfying the role of the temperature controller (in the algebraic sense) with intensity $\widehat{g}$.

Let the intensity of this heat source belongs to the interval $[g_1, g_2]$ (we assume that $0 \in [g_1, g_2]$, g_i may be also equal to the infinity). We control the intensity of the additional heat source as follows:

(i)   If $\Theta(x, t) \in [\psi_1(x), \psi_2(x)]$, then $\widehat{g} = 0$

(ii)   If $\Theta(x, t) \notin [\psi_1(x), \psi_2(x)]$, then we lead the heat which is directly proportional to the difference between the temperature $\Theta(x, t)$ and the interval $[\psi_1(x), \psi_2(x)]$. So, we have:

$$\Theta(x, t) > \psi_2(x) \Rightarrow \begin{cases} -\widehat{g} = k_2(\Theta - \psi_2) & \text{if } k_2(\Theta - \psi_2) \le g_2 \\ -\widehat{g} = g_2 & \text{if } k_2(\Theta - \psi_2) > g_2 \end{cases} \tag{2.1}$$

$$\Theta(x, t) < \psi_1(x) \Rightarrow \begin{cases} -\widehat{g} = k_1(\Theta - \psi_1) & \text{if } k_1(\Theta - \psi_1) \ge g_1 \\ -\widehat{g} = g_1 & \text{if } k_1(\Theta - \psi_1) < g_1 \end{cases} \tag{2.2}$$

This above two cases, we can write as follows:

$$-\widehat{g} = \beta(\Theta) \tag{2.3}$$

where β is the function defined by the formulae:

$$\beta(\Theta) = \begin{cases} g_1 & \text{if } \Theta \le \psi_1 + 1/k_1\, g_1 \\ k_1(\Theta - \psi_1) & \text{if } \psi_1 + 1/k_1\, g_1 \le \Theta \le \psi_1 \\ 0 & \text{if } \psi_1 \le \Theta \le \psi_2 \\ k_2(\Theta - \psi_2) & \text{if } \psi_2 \le \Theta \le \psi_2 + 1/k_2\, g_2 \\ g_2 & \text{if } \Theta \le \psi_2 + 1/k_2\, g_2 \end{cases} \tag{2.4}$$

We can consider the problem move generally putting

$$-\hat{g} \in \beta(\Theta) \tag{2.5}$$

where β is any maximal monotone operator in $\mathbb{R}$ (β may be also multivalued operator). Now, we shall describe some examples of the operator β:

Example 2.1.
$k_1, k_2 < +\infty$; $-\infty < g_1, g_2 < +\infty$ In this case, the intensity of the heat sources is bounded. We have so called: *proportional control.*

Example 2.2.
$k_1 = k_2 = +\infty$; $-\infty < g_1$; $g_2 < +\infty$ In this case the intensity of the heat sources is bounded. We have so called: *jumping control.*

$$\beta(\Theta) = \begin{cases} g_1 & \text{if } \Theta < \psi_1 \\ [g_1, 0] & \text{if } \Theta = \psi_1 \\ 0 & \text{if } \psi_1 \leq \Theta \leq \psi_2 \\ [0, g_2] & \text{if } \Theta = \psi_2 \\ g_2 & \text{if } \Theta > \psi_2 \end{cases} \tag{2.6}$$

Example 2.3.
$\psi_1 = -\infty$, $k_1 = k_2 = +\infty$; $g_1 = 0$; $g_2 = +\infty$ We have the upper restriction for the temperature.

$$\beta(\Theta) = \begin{cases} 0 & \text{if } \Theta < \psi_2 \\ [0, +\infty] & \text{if } \Theta = \psi_2 \\ \emptyset & \text{if } \Theta > \psi_2 \end{cases} \tag{2.7}$$

Example 2.4.
$\psi_1 = +\infty$, $k_1 = k_2 = +\infty$; $g_1 = -\infty$; $g_2 = 0$ Than we have the lower restriction the temperature

$$\beta(\Theta) = \begin{cases} \emptyset & \text{if } \Theta < \psi_1 \\ [-\infty, 0] & \text{if } \Theta = \psi_1 \\ 0 & \text{if } \Theta > \psi_1 \end{cases} \tag{2.8}$$

Example 2.5.
$k_1 = k_2 = +\infty$; $-g_1 = g_2 = +\infty$ In this case we have lower and upper restriction for the temperature

$$\beta(\Theta) = \begin{cases} \emptyset & \text{if } \Theta < \psi_1 \\ (-\infty, 0] & \text{if } \Theta = \psi_1 \\ 0 & \text{if } \psi_1 \leq \Theta \leq \psi_2 \\ \emptyset & \text{if } \Theta > \psi_2 \end{cases} \tag{2.9}$$

Remark 2.1.
We can also consider the control problem not for temperature but for the derivative with respect to t of the temperature $\Theta(x, t)$ i.e. for $\frac{\partial}{\partial t}\Theta(x, t)$ in the same

way as described above as well (putting in the above formulae $\frac{\partial \Theta}{\partial t}(x,t)$ instead of $\Theta(x,t)$). At first, we consider the first boundary - initial value problem for linear hyperbolic thermoelasticity theory with obstacle for the derivatives with respect to t of the temperature (i.e with obstacle for the velocity of the temperature). Let $\beta : \Omega \times \mathbb{R} \longrightarrow \mathbb{R}$, be multivalued. $\forall\ x \in \Omega$; $\ \psi \longrightarrow \beta(x, \psi)$ is maximal monotone operator (cf. [3]). (We assume that $D(\beta) = \mathbb{R}$). We will be seaking the displacement vector field u and the temperature Θ of the thermoelastic medium which satisfy the equations (1.9) i.e.

$$\frac{\partial^2 u_i}{\partial t^2} = f_i + \frac{\partial}{\partial x_i}\left(c_{ijhk}(x)\,\varepsilon_{hk}(u)\right) - \frac{\partial}{\partial x_j}\left[g_{ij}(x)\,\frac{\partial \Theta}{\partial t}\right] \tag{2.10}$$

relation (cf. (1.10):

$$\frac{\partial^2 \Theta}{\partial t^2} + \frac{\partial \Theta}{\partial t} - \frac{\partial}{\partial x_j}\left(k_{ij}(x)\frac{\partial \Theta}{\partial x_i}\right) + g_{ij}\frac{\partial}{\partial t}\,\varepsilon_{ij}(u) +$$
$$(a_0^i - a^i)\left(\frac{\partial}{\partial x_i}\left(\frac{\partial \Theta}{\partial t}\right)\right) - \left(\frac{\partial}{\partial x_i}a^i\right)\frac{\partial \Theta}{\partial t} + \beta\left(x, \frac{\partial \Theta}{\partial t}\right) \ni q \tag{2.11}$$

initial conditions:

$$\Theta(x, 0) = \Theta^0(x) \qquad \text{for } x \in \Omega \tag{2.12}$$

$$\frac{\partial \Theta}{\partial t}(x, 0) = \Theta^1(x) \qquad \text{for } x \in \Omega \tag{2.13}$$

$$u(x, 0) = u^0(x) \qquad \text{for } x \in \Omega \tag{2.14}$$

$$\frac{\partial u}{\partial t}(x, 0) = u^1(x) \qquad \text{for } x \in \Omega \tag{2.15}$$

and boundary conditions:

$$\Theta(x, t) = 0 \qquad \text{for } (x, t) \in \Gamma \times (0, T) \tag{2.16}$$

$$u(x, t) = 0 \qquad \text{for } (x, t) \in \Gamma \times (0, T) \tag{2.17}$$

for given β, f, q, Θ^0, Θ^1, u^0, u^1.

Let H be a Hilbert space:

$$H = [L^2(\Omega)]^{13} \tag{2.18}$$

and let $\tilde{\beta} \subset \left(L^2(\Omega) \times L^2(\Omega)\right)$ be the operator defined as follows:

$$D(\tilde{\beta}) = \left\{u \in L^2(\Omega): \ \exists\ v \in L^2(\Omega),\ v \in \beta(u(x)) \text{ a.e. } x \in \Omega\right\} \tag{2.19}$$

$$\tilde{\beta}(u) = \left\{v \in L^2(\Omega): \ v \in \beta\,(u(x)) \text{ a.e. } x \in \Omega\right\} \qquad u \in D(\tilde{\beta}) \tag{2.20}$$

Since Ω is bounded, then $\tilde{\beta}$ is also maximal monotone operator. Now, we define the operator B:

$$B: H \longrightarrow H \qquad B = \begin{bmatrix} 0 & 0 \\ 0 & \tilde{\beta} \end{bmatrix} \tag{2.21}$$

$$D(B) = [L^2(\Omega)]^{12} \times D(\tilde{\beta}) \tag{2.22}$$

B is maximal monotone operator (cf. [15]). Introducing the notation $\dot{U} = \frac{\partial u}{\partial t}$ we can write the equation (2.10) and relation (2.11) as follows:

$$\ddot{U} - D' \, SDU + D'\Gamma\dot{\Theta} = f \tag{2.23}$$

$$\ddot{\Theta} + \dot{\Theta} - \nabla' \, k\nabla\Theta + \Gamma'D\dot{U} + (A_0 - A)'\nabla\dot{\Theta} - \nabla'A\dot{\Theta} + \beta(\dot{\Theta}) \ni q \tag{2.24}$$

where

$$A = [\, a_1, \ a_2, \ a_3\,]' \ ; \quad A_0 = [a_0^1, \ a_0^2, \ a_0^3]' \tag{2.25}$$

$'$ — denotes the transposition:

$$U = \begin{bmatrix} u_1 \\ u_2 \\ u_3 \end{bmatrix} ; \quad \nabla = \begin{bmatrix} \partial_1 \\ \partial_2 \\ \partial_3 \end{bmatrix} ; \quad f = \begin{bmatrix} f_1 \\ f_2 \\ f_3 \end{bmatrix} \tag{2.26}$$

and

$$\Gamma = \begin{bmatrix} \gamma_1 \\ \gamma_2 \\ \gamma_3 \\ \gamma_4 \\ \gamma_5 \\ \gamma_6 \end{bmatrix} \qquad D = \begin{bmatrix} \partial_1 & 0 & 0 \\ 0 & \partial_2 & 0 \\ 0 & 0 & \partial_3 \\ 0 & \partial_3 & \partial_2 \\ \partial_3 & 0 & \partial_1 \\ \partial_2 & \partial_1 & 0 \end{bmatrix} \qquad \begin{aligned} S &= [c_{ijhk}], \quad i,j,h,k = 1,2,3 \\ K &= [k_{ij}], \quad i,j = 1,2,3 \end{aligned} \tag{2.27}$$

where:

$\gamma_1 = g_{11}, \ \gamma_2 = g_{22}, \ \gamma_3 = g_{33}, \ \gamma_4 = g_{23}, \ \gamma_5 = g_{31}, \ \gamma_6 = g_{12}$

Next, we convert above boundary — initial value problem (2.10) — (2.17) into evolution equation in Hilbert space H.

$$\begin{cases} \dot{V} + AV + BV \ni' F & t \in (0, T) \\ V(0) = V^0 \end{cases} \tag{2.28}$$

where: $A : H \longrightarrow H$ is the linear operator of the form:

$$A = \begin{bmatrix} 0 & SD & 0 & 0 \\ D' & 0 & 0 & -D'\Gamma \\ 0 & 0 & 0 & \nabla \\ 0 & -\Gamma'D & \nabla'K & \nabla'A - (A_0 - A)'\nabla - I \end{bmatrix} \tag{2.29}$$

The domain of the operator $D(A)$ is dense in H.

$$D(A) = \{V \in H : \ V_2 \in [H_1^0(\Omega)]^3; \ V_4 \in H_1^0(\Omega); \ AV \in H\} \tag{2.30}$$

and

$$
V = \begin{bmatrix} SDU \\ \dot{U} \\ \nabla\Theta \\ \dot{\Theta} \end{bmatrix}, \quad F = \begin{bmatrix} 0 \\ f \\ 0 \\ q \end{bmatrix}, \quad V^0 = \begin{bmatrix} SDU^0 \\ U^1 \\ \nabla\Theta^0 \\ \Theta^1 \end{bmatrix} \tag{2.31}
$$

The symbol $\in'$ has the following meaning:

$$
\begin{aligned}
(x_1, \ldots, x_{12}, x_{13}) &\in' A_1 \times \ldots A_{12} \times A_{13} \Leftrightarrow \\
&\Leftrightarrow A_i = [x_i] \quad i = 1, \ldots, 12, \quad x_{13} \in A_{13}
\end{aligned} \tag{2.32}
$$

2.2. Existence and Uniqueness of the Weak Solution of the Problem

In this section we proved the existance and uniqueness of the solution to the problem (2.28) using the semigroup theory of linear operators. At first, we investigate the properties of the operator A.

Now, we introduce in the space X new scalar product (with scales) $(\cdot, \cdot)_Q$.

$$
(u, v)_Q = (u, \ Qv) ; \quad u, v \in H \tag{2.33}
$$

where

$$
Q = \begin{bmatrix} S^{-1} & 0 & 0 & 0 \\ 0 & I & 0 & 0 \\ 0 & 0 & K & 0 \\ 0 & 0 & 0 & I \end{bmatrix} \tag{2.34}
$$

Let H_Q be the Hilbert space H with the scalar product $(\cdot, \cdot)_Q$. The operator A has the following properties:

Lemma 2.1.

(i) $\quad D(A^*) = D(A)$, where

$$
A^* = \begin{bmatrix} 0 & -SD & 0 & 0 \\ -D' & 0 & 0 & D'\Gamma \\ 0 & 0 & 0 & -\nabla \\ 0 & \Gamma'D & -\nabla'K & -\nabla'A + (A_0 - A)'\nabla - I \end{bmatrix} \tag{2.35}
$$

(ii) $\quad A$ — is closed operator

If we additional assume that $\|\nabla'A_0\| \leq 1$, then

(iii) $\quad A$ — is monotonical operator

(iv) $\quad (0, +\infty) \subset \rho(A)$ and $\forall \lambda > 0$, $\|(\lambda - A)^{-1}\| \leq \lambda^{-1}$.

Proof: The following inclusion $D(A) \subset D(A^*)$ is true if $U \in D(A)$ and $U = A^*U$, so we have:

$$
(U, AV)_Q = (h, V)_Q \qquad \forall V \in D(A)
$$

Now we prove that $D(A^*) \subset D(A)$.
Let $U \in D(A^*)$ i.e. $\exists h \in H$ such that:

$$
(U, AV)_Q = (h, V)_Q \qquad \forall V \in D(A)
$$

it means that:

$(U_1, DV_2) + (U_2, D'V_1 - D'\Gamma V_4) + (U_3, K\nabla V_4) - (u_3, \Gamma'DV_2 - \nabla'KV_3 + V_4 + \nabla'AV_4 + (A_0 - A)'\nabla V_4) = (U_1, S^{-1}V_1) + h_2, V_2) + h_3, KV_3) + (H_4, V_4)$.

Let $V_2 = 0$, $V_3 = 0$, $V_4 = 0$, V_1 — be arbitrary. Then from Korn's inequality it follows: $h_1 = -SDU_2$, $U_2 \in [H_0^1(\Omega)]^3$.

Similarly, putting $V_2 = 0$, $V_1 = 0$, $V_4 = 0$, V_3 — arbitrary, we get: $h_3 = -\nabla U_4$, $U_4 \in H_0^1(\Omega)$.

If $V_1 = 0$, $V_3 = 0$, $V_4 = 0$ and V_2 — arbitrary, we have: $h_2 = -D'U_1 + D'\Gamma U_4$.

Analogous, if $V_1 = 0$, $V_3 = 0$, $V_2 = 0$ and V_4 — arbitrary, we obtain: $h_4 = \Gamma'DU_2 - \nabla'KU_3 - U_4 - A'\nabla U_4 + \nabla'(A_0 - A)U_4$.

So, we get:

$$AU \in H \text{ and } U \in D(A) \qquad \text{Q.E.D.}$$

Ad (ii). We have: $\overline{D(A)} = \overline{D(A^*)} = H$.

So, (cf. [16]) A^* is closed operator. So, we have:

$$\forall U_n \in D(A^*), \text{ if } V_n \to 0 \text{ that}$$
$$- SDV_2^n \to 0,$$
$$- D'V_1^n + D'\Gamma V_4^n \to 0,$$
$$\nabla V_4^n \to 0,$$
$$\Gamma'DV_2^n + \nabla'KV_3^n - V_4^n - A'\nabla V_4^n + \nabla'(A_0 - A)V_4^n \to 0.$$

Now, we prove that A is also closed operator.

Let $\forall V_n \in D(A)$ and $V_n \to 0$ $AV_n \to V$. It is enought to prove that $V = 0$. From the fact $AV_n \to V$ it follows:

$$SDV_2^n \to V_1,$$
$$D'V_1^n - D'\Gamma V_4^n \to V_2,$$
$$\nabla V_4^n \to V_3,$$
$$\Gamma'DV_2^n + \nabla'KV_3^n - V_4^n - \nabla'AV_4^n + (A_0 - A)'\nabla V_4^n \to V_4.$$

From the last relation and the fact that $V_n \to 0$, we get $V = 0$.

Ad (iii). For $\forall V \in D(A)$, after applyng Green's theorem, we get:

$(AV, V)_Q = (DV_2, V_1) + (D'V_1, V_2) - (D'\Gamma V_4, V_2) - (\Gamma'DV_2, V_4) + (\nabla'KV_3, V_4) + (\nabla V_4, KV_3) - (V_4, V_4) + (\nabla'AV_4, V_4) - ((A_0 - A)'\nabla V_4, V_4) = (\nabla'AV_4, V_4) - \|V_4\|^2 \leq 0$.

Ad (iv). Let $\lambda > 0$ and $(\lambda - A)V = 0$. We have:

$$0 = ((\lambda - A)V, V)_Q = \lambda \|V\|^2 - (AV, V) \geq 0$$

From the last relation, we get:

$V = 0$ and also $\text{Ker}(\lambda - A) = 0$, $\text{Im}(\lambda - A)^\perp = \text{Ker}(\lambda - A^*)$.

Now, let $(\lambda - A^*)V = 0$, then acting as above we get: $\underline{V = 0}$.

So, $\text{Ker}(\lambda - A^*) = 0$. From this fact, we obtain that $\overline{\text{Im}(\lambda - A)} = H$, $\forall \lambda > 0$.

So, we have that the operator $(\lambda - A)^{-1}$ is continuous. Let $(\lambda - A)V = W$. So, from Hölder's inequality we get:

$$\lambda\|V\|^2 - (AV, V)_Q = (V, W)_Q \leq |(V, W)_Q| \leq \|V\|\,\|W\|.$$

So, we have:

$$\lambda\|V\|^2 \leq \|V\|\,\|W\| \text{ and } \|V\| \leq \|W\|\lambda^{-1}$$

So, $\|(\lambda - A)^{-1}\| = \sup_{\|W\|=1} \|(\lambda - A)^{-1}W\| = \sup_{\|V\|=1} \|V\| \leq \lambda^{-1}$ Q.E.D

So, in view of this fact the assumptions of the Hille-Yosidy Theorem are satisfied (cf. [4], [12], [24]). From this theorem it follows that A is the generator some continuous linear contraction semigroup $T(t)$, $t \geq 0$. Since the operators A and B are maximal monotone and $D(A) \cap \text{Int } D(B) \neq 0$, then from Rockafaller's-Moreau's Theorem it follows that $A + B$ is maximal monotone. So, (cf. [2], p. 101) we can prove the following theorem.

Theorem 2.1. If the coefficients of the equations (1.9) — (1.10) are continuously differentiable, $\|\nabla' A_0\| \leq 1$, $0 \in \text{Int } D(B)$ and:

$$f \in W^{1,1}\left((0, T); [L^2(\Omega)]^3\right), \quad q \in W^{1,1}\left((0, T); [L^2(\Omega)]\right) \tag{2.36}$$

$$U^1 \in \left[H_1^0(\Omega)\right]^3, \quad \Theta^1 \in H_1^0(\Omega), \quad (SDU^0, U^1, \nabla\Theta^0, \Theta^1) \in H \tag{2.37}$$

then there exists unique solution of the boundary — initial value problem (2.10) — (2.15) such that:

$$(SDU, U, \nabla\Theta, \Theta) \in W^{1,\infty}\left((0, T); H\right) \tag{2.38}$$

3. Boundary — Initial Value Problem with Obstacle for Temperature

Let $\beta : \Omega \times \mathbb{R} \longrightarrow \mathbb{R}$ be multivalued mapping such that: $\forall\, x \in \Omega$, $\Theta \longrightarrow \beta(x, \Theta)$ is maximal monotone operator. We assume that $D(\beta) = \mathbb{R}$. Now, we consider the first boundary - initial value problem for linear hyperbolic thermoelasticity theory with obstacle for temperature. We seek the displacement vector field u and the temperaturte Θ of the thermoelastic medium, which satisty the equations:

$$\frac{\partial^2 u_i}{\partial t^2} = f_i + \frac{\partial}{\partial x_j}\left(c_{ijhk}(x)\,\varepsilon_{hk}(u)\right) - \frac{\partial}{\partial x_j}\left(g_{ij}(x)\,\frac{\partial\theta}{\partial t}\right) \tag{3.1}$$

relation:

$$\frac{\partial^2 \Theta}{\partial t^2} + \frac{\partial\Theta}{\partial t} - \frac{\partial}{\partial x_j}\left(k_{ij}(x)\,\frac{\partial\Theta}{\partial x_i}\right) + g_{ij}(x)\,\frac{\partial}{\partial t}\,\varepsilon_{ij}(u) + $$
$$+ (a_0^i - a^i)\left(\frac{\partial}{\partial x_i}\left(\frac{\partial\Theta}{\partial t}\right)\right) - \left(\frac{\partial}{\partial x_i}a^i\right)\frac{\partial\Theta}{\partial t} + \beta(x, \Theta) \ni q \tag{3.2}$$

initial conditions:

$$\Theta(x,0) = \Theta^0(x) \qquad \text{for } x \in \Omega \qquad (3.3)$$

$$\frac{\partial \Theta}{\partial t}(x,0) = \Theta^1(x) \qquad \text{for } x \in \Omega \qquad (3.4)$$

$$u(x,0) = u^0(x) \qquad \text{for } x \in \Omega \qquad (3.5)$$

$$\frac{\partial u}{\partial t}(x,0) = u^1(x) \qquad \text{for } x \in \Omega \qquad (3.6)$$

and boundary conditions:

$$\Theta(x,t) = 0 \qquad \text{for } (x,t) \in \Gamma \times (0,T) \qquad (3.7)$$

$$u(x,t) = 0 \qquad \text{for } (x,t) \in \Gamma \times (0,T) \qquad (3.8)$$

for given f, q, Θ^0, Θ^1, u^0, u^1. Since (cf. section 2) the temperature Θ didn't apear in as an independent variable in the problem (2.28) we need to modify the evolution equation (2.28) (we would like to look after optimal control for temperature Θ). We consider Hilbert space:

$$H = \left[L^2(\Omega) \right]^{14} \qquad (3.9)$$

Let $\widetilde{\beta} \subset L^2(\Omega) \times L^2(\Omega)$) be the operator given by formulae (2.19). We define operator $B: \quad H \longrightarrow H$

$$B = \begin{bmatrix} 0 & 0 & 0 & 0 & 0 \\ 0 & 0 & 0 & 0 & 0 \\ 0 & 0 & 0 & 0 & 0 \\ 0 & 0 & 0 & 0 & \widetilde{\beta} \\ 0 & 0 & 0 & 0 & 0 \end{bmatrix} \qquad (3.10)$$

with

$$D(B) = \left[L^2(\Omega) \right]^{13} \times D(\widetilde{\beta}) \qquad (3.11)$$

B — is maximal monotone operator (cf. [16]). Now, we can convert boundary — initial value problem (3.1)–(3.8) into evolution equation in Hilbert space.

$$\begin{cases} \dot{V} + AV + BV \ni' F & t \in (0,T) \\ V(0) = V^0 \end{cases} \qquad (3.12)$$

where: $A: \quad H \longrightarrow H$ is linear operator of the form:

$$A = \begin{bmatrix} 0 & SD & 0 & 0 & 0 \\ D' & 0 & 0 & -D'\Gamma & 0 \\ 0 & 0 & 0 & \nabla & 0 \\ 0 & -\Gamma D & \nabla' K & \nabla' A - (A_0 - A)'\nabla - I & I \\ 0 & 0 & 0 & -I & 0 \end{bmatrix} \qquad (3.13)$$

with domain $D(A)$ dense in H.

$$D(A) = \left\{ v \in H : V_2 \in \left[H_1^0(\Omega) \right]^3, \ V_4 \in H_1^0(\Omega), \ V_5 \in H_1^0(\Omega) \ AV \in H \right\} \tag{3.14}$$

and

$$V = \begin{bmatrix} SDU \\ U \\ \nabla\Theta \\ \dot{\Theta} \\ \Theta \end{bmatrix}, \qquad F = \begin{bmatrix} 0 \\ f \\ 0 \\ q \\ 0 \end{bmatrix}, \qquad V^0 = \begin{bmatrix} SDU^0 \\ U^1 \\ \nabla\Theta^0 \\ \Theta^1 \\ \Theta^0 \end{bmatrix} \tag{3.15}$$

Using the semigroup methods of nonlinear operators we can prove the following theorem:

Theorem 3.1. If the coefficients of the equations (2.9)–(2.10) statisty the assumptions of the theorem 2.1., than the problem (3.1)–(3.8) has a unique solution

$$(SDU, \ V, \ \nabla\Theta, \ \dot{\Theta}, \ \Theta) \ \in \ W^{1,\infty}\left((0,T) \ ; \ H \right) \tag{3.16}$$

Sketch of proof:
We can write the problem (3.12) as follows:

$$\begin{cases} \dot{V} + A_1 V + BV + A_2 V \ \ni' \ F & t \in (0,T) \\ V(0) = V^0 \end{cases} \tag{3.17}$$

where

$$A_1 = \begin{bmatrix} 0 & SD & 0 & 0 & 0 \\ D' & 0 & 0 & -D'\Gamma & 0 \\ 0 & 0 & 0 & \nabla & 0 \\ 0 & -\Gamma'D & \nabla'K & \nabla'A - (A_0 - A)'\nabla - I & 0 \\ 0 & 0 & 0 & 0 & 0 \end{bmatrix} \tag{3.18}$$

and

$$A_2 = \begin{bmatrix} 0 & 0 & 0 & 0 & 0 \\ 0 & 0 & 0 & 0 & 0 \\ 0 & 0 & 0 & 0 & 0 \\ 0 & 0 & 0 & 0 & I \\ 0 & 0 & 0 & -I & 0 \end{bmatrix} \tag{3.19}$$

The operator A_1 has the same properties as operator A. So, the operator A_1 and B are maximal monotone operators, we have also $D(A_1) \cap \text{Int } D(B) \neq \emptyset$. Applying Rockafaller's-Moreau's Theorem we get that the operator $A_3 = A_1 + B$ is maximal monotone operator. In this case we can applied the theorem (cf. [2], p. 135) about "perturbation" equation, because our problem given by equation (3.12) we can write in the following form:

$$\begin{cases} \dot{V} + A_4 V - V \ \ni' \ F & t \in (0,T) \\ V(0) = V^0 \end{cases} \tag{3.20}$$

where: $A_4 = A_3 + A_2 + I$ and $A_3 + A_2 = A_4 - I$. The operator A_4 is maximal monotone operator (cf. [2], p. 82).

4. Concluding Remarks

Remark 4.1.
In order to obtain in the problems of optimal control of the temperature move regular solutions we can substitute the maximal monotone operator β by smooth function β_g. For example if we take into accunt the maximal monotone operator $\beta(\psi)$ of the form:

$$\beta(\psi) = \begin{cases} g_1 & \text{if } \psi < \psi_1 \\ [g_1, 0] & \text{if } \psi = \psi_1 \\ 0 & \text{if } \psi_1 \leq \psi \leq \psi_2 \\ [0, g_2] & \text{if } \psi = \psi_2 \\ g_2 & \text{if } \psi > \psi_2 \end{cases} \tag{4.1}$$

for fixed ψ_1, ψ_2 $(\psi_1 < \psi_2)$; g_1, g_2 $(0 \in g_1, g_2)$ we can accepted Fridrick's Regularization of the continuous function $(\beta_c(\psi))$:

$$\beta_c(\psi) = \begin{cases} g_1 & \text{if } \psi < \psi_1 - \varepsilon \\ \frac{g_1}{2\varepsilon}(\psi_1 + \varepsilon - \psi) & \text{if } \psi_1 - \varepsilon \leq \psi < \psi_1 + \varepsilon \\ 0 & \text{if } \psi_1 + \varepsilon \leq \psi < \psi_2 - \varepsilon \\ \frac{g_2}{2\varepsilon}(\psi + \varepsilon - \psi_2) & \text{if } \psi_2 - \varepsilon \leq \psi < \psi_2 + \varepsilon \\ g_2 & \text{if } \psi > \psi_2 + \varepsilon \end{cases}$$

for fixed, arbitrary small $\varepsilon > 0$. Then the initial conditions have very important influence for the regularity of the solution. Under some assumption (cf. [16]) we get the solution with required reqularity. For the function $\beta_c(\psi)$ given above we can also consider the weak solutions (cf. [16]).

Remark 4.2.
We can also external our results to the stochastic thermoelastic equations. Such equation contain in the right hand side of the equations describing thermoelastic medium, the term $\partial W / \partial t$ which is the weak derivative of a Wiener process. The distribution of the process W is given by the covariance operator Q. The initial condition U_0 is a random variable. Using the method of Hilbert space, we can obtain the behavior of probality distribution μ_t of $U(t, \cdot)$.

Acknowledgement

This research was carried out in the frame-work of the Polish GRANT No 211659101 supported during the years 1991–1994 by the State Committee for Scientific Research (KBN).

References

[1] Ackerman G.C., Guyer R.A., Temperature pulses in dielectric solids, Ann. Phys., 51, 128–185, 1968

[2] Barbu V., Nonlinear semigroups and differential equations in Banach spaces, Leyden, 1976

[3] Brezis H., Operatours maximaux monotones et semigroupes de contractions dans les espaces de Hilbert, Math.Studies, 5, North Holland, 1973

[4] Brezis H., Analyse Fonctionnelle, Paris, 1983

[5] Cattaneo G., Sulla conduzione del calone, Atti.del Semminatis Mat. Fis. Univ. Modena, 3, 83–101, 1948

[6] Dafermos C.D., On the existence and asymptotic stability of solution to the equations of linear thermoelasticity, Arch. Rat. Mech. Anal., 4, 241–271, 1972

[7] Duvaut G., Lions J.L., Les inéquations en mécanique et en physique, Dunod, Paris, 1978

[8] Gawinecki J., The Faedo-Galerkin method in thermal stresses theory, Comm. Math., 27, 83–101, 1987

[9] Gawinecki J., The Cauchy problem for the linear hyperbolic equations system of the theory of thermoelasticity of the temperature-rate-dependent solids, Bull. Acad. Polon. Sci. (35), 7–8, 421–433, 1987

[10] Gawinecki J., Global solution in non-linear hyperbolic thermoelasticity, Math. Meth. in the Appl. Sci., vol.15, 223–237, 1992

[11] Gawinecki J., Kowalski L., Ebihara Y., Initial-boundary value problem in nonlinear hyperbolic thermoelasticity, — in print

[12] Goldstein J.A., Semigroups of linear operators and applications, New York, 1985

[13] Green A.E., Lindsay K.E., Thermoelasticity, J.Elasticity, 2, 1–7, 1972

[14] Jackson H.E., Walker T.C., Thermal conductivity second-sollud and phonon--phonon interaction in NaF, Phys. Rev., B–3, 1428–1435, 1971

[15] Kowalski L., On the first boundary-initial value problem for linear thermal stress equations with obstacles for temperature, Bull. Acad. Sci., 36, No. 3–4, 13–20, 1988

[16] Kowalski L., Existance and uniqueness of the solution to the first boundary--initial value problem of the linear thermoelasticity with obstacle for temperature, (in Polish) Biul. WAT., XL, No 1, 1991

[17] Kupradze V.D., Three-dimensional problems of mathematical theory of elasticity and thermoelasticity (in Russian), Moskva 1976

[18] Leis R., Lectures on initial-boundary value problems in mathematical physics, Bonn, 1984

[19] Lord H.W., Shulman Y., Generalized dynamical theory of thermoelasticity, J. Mech. Phys. Solids, 15, 299–309, 1968

[20] Müller J., The coldness as universal function in thermoelastic bodies, Arch. Rat. Mech. Anal., 41, 319–332, 1971

[21] Naumann J., Einführung in die Theorie parabolischer Variationsungleichungen, Teubner-Texte zur Mathematik, No.64, Leipzig, 1984

[22] Nowacki W., Dynamical problems of thermoelasticity (in Polish), PWN, Warsaw, 1966

[23] Panagiotopoulos P., Inequality problems in mechanics and applications, Boston-Stuttgart, 1985

[24] Pazy A., Semigroups of linear operators and applications to partial differential equations, New York, 1983

[25] Postrigač J.S., Koliano J.B., Generalized thermomechanics (in Russian), Kiev, 1976

[26] Sneddon N.T., The propagation of thermal stresses in thin metallic rods, Proc. Roy. Soc. Edin., 65, 121–142, 1959

[27] Suhubi E.S., Thermoelastic solids, in A.C. Eringen, Continuum Physics, vol.2, New York, 1975

[28] Taylor B., Marris H.J., Elbaum C., Phononfocusing in soilds, Phys. Rev. Lett., 23, 416–419, 1969.

[29] Yosida K., Functional analisis, Springer-Verlag, Berlin–Göttingen Heidelberg, 1965

Optimal Trajectory Planning for Robots under the Consideration of Stochastic Parameters and Disturbances
– Computation of an Efficient Open-Loop Strategy –

K. Marti and S. Qu

Institute of Mathematics and Computer Science
Federal Armed Forces University Munich
85577 Neubiberg, Germany

Abstract. The stochastic variations of many mechanical parameters and certain mechanical disturbances may be very large. To reduce the high expenses for on-line sensors and on-line processing time, the available statistical information about the stochastic parameters of the robot and disturbances should be considered already for off-line programming of robots. This paper proposes some substitue problems considering the stochastic variations of the model parameter of robots. Numerical methods for solving the substitute problems are discussed, and the dependence of the solutions on the variances of stochastic parameters is analysed.

Keywords. Robotics, Trajectory Planning, Uncertainty, Penalty Function, Stochastic Optimization

1 Introduction

The increasing application of robots requires investments of considerable extent. Therefore, the design and the programming of robots and robotic cells must be performed from the beginning by considering safety and economic efficiency. In order to reduce the expensive shut-down time during the manufacturing process and at the assembly line, one is more interested in simulating robots and their enviroment by applying a 3D-CAD Simulation System. Since the operation procedure is determined especially during the assembly process, the most important factor is the determination of the time steps by optimizing the velocity profile of the robot along the given path. Of utmost importance in regards to quality and reliability of the results is the consideration of known statistical features of all possible disturbances already during the simulation phase. Corresponding requirements are increasingly transferred by the user of the simulation systems to the developer of off-line systems.

For a given collision-free path in the environment of a robot, the problem is to determine an optimal velocity profile along the path. When optimizing the velocity profile, one generally tends to minimize the total performing time (or the consumption of energy) [2,5,8,12], under the constraints required by the underlying mechanical system and its production environment, and by fully utilizing the reserves of the robot, however, without exceeding the available torque limits and/or allowable velocity limits in manipulator joints.

The stochastic variations of many mechanical parameters (e.g. payloads, center of mass, inertia tensor) and certain mechanical disturbances (e.g. manufacturing errors, friction, clearance in joint links, thermical expansion, assembly error of robotic coordinates and elastic deformations) can be very large. Hence, since on-line corrections can be very expensive due to high expenses for on-line sensors and on-line processing time, the available statistical informations (especially probability distributions) about the stochastic parameters of the robot and disturbances should be, therefore, already considered for off-line programming of robots, so that - with the help of an efficient open-loop strategy [1,14] - only a *simple* on-line feed-back control is necessary for the corrections.

2 Trajectory planning under uncertainty

In order to perform a given task the robot must move along a specified path in work space. From the kinematic equation:

$$T(q, p) = x, \tag{2.1}$$

where

x is a n-dimensional vector describing the position und orientation of the end-effector of the robot,

q is a n-dimensional vector describing the configuration coordinates of the robot,

$T: R^n \rightarrow R^n$ is the kinematic operator, and

p is a vector of model parameters (e.g. mass, length, manufacturing errors, joint space),

one can determine the configuration coordinates q in accordance with x . Let $x = x(s)$, $0 \le s \le s_e$, be a function of a path parameter s (e.g. the length of the path to a certain position), then from (2.1) the configuration variable q can be also described as a function of s, i.e. $q = q(s), 0 \le s \le s_e$.

A robot is driven by the torques and forces generated by the motors of the robot. The torques and forces have to be calculated in a way that the end-effector of the robot runs through the specified path in work space. For this we need the dynamical equation:

$$\sum_{j=1}^{n} J_{ij}\ddot{q}_j + \sum_{j,k=1}^{n} D_{ijk}\dot{q}_j\dot{q}_k + G_i = \tau_i \ , \ i = 1, 2, ...n, \tag{2.2}$$

where

$J_{ij} = J_{ij}(q, p)$ are the elements of the inertia matrix of the robot,

$D_{ijk} = D_{ijk}(q, p)$ are the coefficients of the centrifugal and Coriolis forces,

$G_i = G_i(q, p)$ are the gravity forces,

q_i are the elements of q,

$\dot{q}_i, \ddot{q}_i$ are the first and second derivates of q_i with respect to time t,

τ_i are the torques and forces generated by the motors.

Describing the path parameter s as a function of time t, $s = s(t)$, yields

$$\dot{q}_i(s) = q_i'(s)\dot{s} \tag{2.3}$$

$$\ddot{q}_i(s) = q_i'(s)\ddot{s} + q_i''(s)\dot{s}^2 \tag{2.4}$$

$$\ddot{s} = \frac{1}{2}\frac{d}{ds}(\dot{s}^2), \tag{2.5}$$

where

$$(\quad)' = \frac{d}{ds} \quad and \quad (\dot{\ }) = \frac{d}{dt} \ .$$

With (2.3)-(2.5), equation (2.2) can be rewritten as follows:

$$a_i(s, \mathbf{q}, \mathbf{q}', \mathbf{p})v' + b_i(s, \mathbf{q}, \mathbf{q}', \mathbf{q}'', \mathbf{p})v + c_i(s, \mathbf{q}, \mathbf{p}) = \tau_i, \quad i = 1, 2, ...n, \tag{2.6}$$

where

$$v = \dot{s}^2 \tag{2.7}$$

$$a_i(s, \mathbf{q}, \mathbf{q}', \mathbf{p}) = \frac{1}{2}\sum_{j=1}^{n} J_{ij}(\mathbf{q}, \mathbf{p})q_j' \quad, i = 1, 2, ...n \tag{2.8}$$

$$b_i(s, \mathbf{q}, \mathbf{q}', \mathbf{q}'', \mathbf{p}) = \sum_{j=1}^{n} J_{ij}(\mathbf{q}, \mathbf{p})q_j'' + \sum_{j,k=1}^{n} D_{ijk}(\mathbf{q}, \mathbf{p})q_j'q_k', \quad i = 1, 2, ...n \tag{2.9}$$

$$c_i(s, \mathbf{q}, \mathbf{p}) = G_i(\mathbf{q}, \mathbf{p}), \quad i = 1, 2, ...n. \tag{2.10}$$

If the vector of model parameters p is *known exactly*, then for a given path $\mathbf{x} = \mathbf{x}(s), 0 \leq s \leq s_e$, in work space of the robot, we can find the path $\mathbf{q} = \mathbf{q}(s, \mathbf{p})$ in the configuration space and then $\mathbf{q}'$ and $\mathbf{q}''$ by solving (2.1). If $v = v(s)$ is also known, the torques and forces can be calculated from (2.6) .

The problem of the trajectory planning for a given path is to find a function $v = v(s)$, so that the torques and forces calculated from (2.6) do not exceed the given bounds $\tau_{min,i} \leq \tau_i \leq \tau_{max,i}$, some conditions for v, e.g. $v(s) \geq 0$, $v(0) = 0, v(s_e) = 0$, are satisfied and an objective function, e.g.

- for time optimal problems

$$t_e = \int_0^{s_e} \frac{1}{\sqrt{v}}ds$$

- for energy optimal problems

$$E = \int_0^{s_e} \sum_{i=1}^{n} \kappa_i \tau_i^2 \frac{ds}{\sqrt{v}}$$

with motor-dependent constants κ_i,

is optimized.

Sometimes the velocities of the end-effector and those of the joint angles (v and $\dot{q}_i$) should also be constrained. For a given path, this is equivalent to the condition $v \leq v_{max}(s, \mathbf{p})$.

The problem of optimal trajectory planning can then be defined mathematically as follows:

$$\min_{v, q} \int_0^{s_e} f_0(s, q, q', q'', v, v') ds \tag{2.11}$$

subject to

$$v(0) = 0 \quad , \quad v(s_e) = 0 \tag{2.12}$$

$$\tau_{min,i} \leq a_i v' + b_i v + c_i \leq \tau_{max,i}, \quad i = 1, 2, \dots n, \tag{2.13}$$

$$0 \leq v \leq v_{max}(s, \mathbf{p}) \tag{2.14}$$

and

$$\mathbf{T}(\mathbf{q}(s), \mathbf{p}) = \mathbf{x}(s). \tag{2.15}$$

This problem was solved in [2, 5, 8, 12]. Especially, for the time optimal problem there is a so-called *Phase-Plan-Method* [2, 5, 8, 12].

Remark on (2.13), (2.14) and (2.15):

a_i, b_i, c_i in (2.13) and v_{max} in (2.14) depend on the model parameter $\mathbf{p}$. The solution $\mathbf{q} = \mathbf{q}(s)$ of the inverse kinematic problem (2.15) is also dependent on $\mathbf{p}$. Due to the very often existing uncertainty about the model parameters $\mathbf{p}$, e.g.

- uncertainty of physical nature (e.g. mechanical, thermical, and environmental effects upon the robot)
- uncertainty of statistical nature (e.g. measurement errors of the parameters)
- uncertainty of the mathematical models,

we must reformulate the constraints (2.13), (2.14) and (2.15). In the following some substitute problems are proposed [6], so that the stochastic variation of the parameter $\mathbf{p}$ can be taken into account *within* the optimal trajctory planning process.

2.1 The first substitute problem

For the sake of simplicity, the parameter $\mathbf{p}$ in (2.15) is, at first, replaced by its *expected value* $\bar{\mathbf{p}}$. Thus we can determine the path $\mathbf{q} = \mathbf{q}(s, \bar{\mathbf{p}})$ in the configuration space for the given path $\mathbf{x} = \mathbf{x}(s)$ in the work space by solving the inverse kinematic problem (2.15).

The violations of the constraints (2.13) and (2.14) lead, by means of some penalty functions, to penalty costs, which are set into the objective function. This yields a modified *variation problem*:

$$\min_{v} \int_0^{s_e} f(s, q, q', q'', v, v') ds \qquad (2.16)$$

subject to

$$v(0) = 0 \quad , \quad v(s_e) = 0 \qquad (2.16.1)$$

where

$$f = f_0(s, q, q', q'', v, v') + \sum_{i=1}^{n} \gamma_i u_i(\tau_i, \tau_{min,i}, \tau_{max,i}) + \gamma_0 u_0(v, v_{max}), \qquad (2.17)$$

$u_i(\tau_i, \tau_{max,i}, \tau_{min,i})$ and $u_0(v, v_{max})$ are penalty functions, and γ_i and γ_0 are weight coefficients.

Because some parameters in (2.13) and (2.14) are stochastic variables, we consider now the *expected penalty cost*:

$$\bar{f} := f_0(s, q, q', q'', v, v') + \sum_{i=1}^{n} \gamma_i E_{\mathbf{p}} u_i(\tau_i, \tau_{min,i}, \tau_{max,i}) + \gamma_0 E_{\mathbf{p}} u_0(v, v_{max}).$$
$$\qquad (2.18)$$

We obtain then the first substitute problem:

$$\min_{v} \int_0^{s_e} \bar{f}(s, q, q', q'', v, v') ds. \qquad (2.19)$$

subject to

$$v(0) = 0 \quad , \quad v(s_e) = 0 \qquad (2.19.1)$$

2.2 The second substitute problem

As in the first substitute problem, the parameter $\mathbf{p}$ in (2.15) is replaced by its expected value $\bar{\mathbf{p}}$. Hence, we can solve the kinematic equation (2.15) for q and get the path in the configuration space $q = q(s, \bar{\mathbf{p}})$, and then q' and q'' too.

The violations of the constraints (2.13) and (2.14) are evaluated numerically, as above, with the help of some chosen penalty functions. It is demanded then, that some prescribed upper bounds $\delta_{u,i}$ $i = 1, 2, ...n$, and $\delta_{u,0}$ should not be exceeded. This yields the second substitute problem:

$$\min_{v} \int_0^{s_e} f_0(s, q, q', q'', v, v') ds \qquad (2.20)$$

subject to

$$v(0) = 0, \quad v(s_e) = 0 \tag{2.21}$$

$$E_{\mathbf{p}} u_i(\tau_i, \tau_{min,i}, \tau_{max,i}) \le \delta_{u,i}, \quad i = 1, 2, \dots n \tag{2.22}$$

$$E_{\mathbf{p}} u_0(v, v_{max}) \le \delta_{u0}. \tag{2.23}$$

3 Solving the substitute problems

In Section 2 some substitute problems for trajectory planning considering uncertainty of the robotic model are proposed. Because a robot itself is already a very complex system, we can solve these problems usually only numerically. In this section some numerical methods will be discussed.

3.1 Numerical calculation of $E_p u_i$, $E_p u_0$ and $E_p u_T$

In both substitute problems we have to calculate $E_p u_i$ and $E_p u_0$. The numerical calculation of the expected values $E_p u_i$ and $E_p u_0$ is usually very difficult because $u_i(\tau_i, \tau_{min,i}, \tau_{max,i})$ and $u_0(v, v_{max})$ could be some complicated functions of p. If the derivatives of $u_i = u_i(s, p)$ with respect to p up to the $(K+1)$-th order exist, then by means of Taylor expansion of u_i at the expectation $\bar{p}$ of p we get

$$
\begin{aligned}
u_i(s, p) &= u_i(s, \bar{p}) + \sum_{k=1}^{K} \frac{1}{k!} \frac{\partial^k u_i}{\partial p^k}(s, \bar{p}) \cdot (p - \bar{p})^k \\
&\quad + \frac{1}{(K+1)!} \frac{\partial^{(K+1)} u_i}{\partial p^{(K+1)}}(s, \tilde{p}_i) \cdot (p - \bar{p})^{(K+1)} \\
&\approx u_i(s, \bar{p}) + \sum_{k=1}^{K} \frac{1}{k!} \frac{\partial^k u_i}{\partial p^k}(s, \bar{p}) \cdot (p - \bar{p})^k,
\end{aligned} \tag{3.1}
$$

where $\tilde{p}_i$ is a point between p and $\bar{p}$. Let μ_k be the system of k-th central moments of p. If all moments of the stochastic variable p up to the K-th order exist, then $E_p u_i(s, p)$ can be described approximatively by

$$E_p u_i(s, p) \approx u_i(s, \bar{p}) + \sum_{k=2}^{K} \frac{1}{k!} \frac{\partial^k u_i}{\partial p^k}(s, \bar{p}) \cdot \mu_k. \tag{3.2}$$

For $E_p u_0$ there is an analogous expression.

3.2 The solution of the first substitute problem

The first substitute problem is a simple variation problem with given boundary conditions. As a *necessary condition* for the optimal velocity profile $v = v(s)$, we have the *Euler differential equation*

$$\frac{\partial \bar{f}}{\partial v} - \frac{d}{ds} \frac{\partial \bar{f}}{\partial v'} = 0, \tag{3.3}$$

with the boundary conditions

$$v(0) = 0, \qquad v(s_e) = 0. \tag{3.4}$$

Let $\bar{f}$ be a quadratic function of v and v', and suppose that $\dfrac{\partial^2 \bar{f}}{\partial v'^2} > 0$. Then (3.3) is a linear differential equation:

$$v'' + a_1(s)v' + a_2(s)v + a_3(s) = 0. \tag{3.5}$$

Let $v_1 = v_1(s)$ be the solution of (3.5) with the initial values

$$v_1(0) = 0, \qquad v_1'(0) = 0 \tag{3.6}$$

and $v_2 = v_2(s)$ be the solution of

$$v'' + a_1(s)v' + a_2(s)v = 0 \tag{3.7}$$

with the initial values

$$v_2(0) = 0, \qquad v_2'(0) = 1. \tag{3.8}$$

If

$$v_2(s_e) \neq 0 \tag{3.9}$$

holds, then the boundary value problem (3.3) and (3.4) has a unique solution which can be expressed by

$$v(s) = v_1(s) - \frac{v_1(s_e)}{v_2(s_e)} v_2(s). \tag{3.10}$$

If

$$v_2(s) \neq 0, \quad \text{for all } s, \ 0 \leq s \leq s_e, \tag{3.11}$$

then $v(s)$ as defined in (3.10) is the optimal solution of the first substitute problem.

3.3 The solution of the second substitute problem

The second substitute problem is a variation problem with inequality constraints. If we express v as a linear combination

$$v(s) = \sum_{j=1}^{J} \beta_j B_j(s), \tag{3.12}$$

where

β_j, j=1,2,...J, are unknown coefficients, and

$B_j(s)$, j=1,2,...J, are given functions,

we can approximate the second substitute problem as follows:

$$\min_{\beta} \ F(\beta) \qquad (3.13)$$

subject to

$$v(0) = 0, \qquad v(s_e) = 0 \qquad (3.14)$$

and

$$U_i(\beta) \leq 0, \qquad i = 1, 2, ...n, \qquad (3.15)$$

$$U_0(\beta) \leq 0, \qquad (3.16)$$

where

$$F(\beta) := \int_0^{s_e} f_0(s, q, q', q'', \sum_{j=1}^{J} \beta_j B_j(s), \sum_{j=1}^{J} \beta_j B_j'(s)) ds \qquad (3.17)$$

and

$$U_i(\beta) := \max_s \ Eu_i(\tau_i(s), \tau_{min,i}, \tau_{max,i}) - \delta_{ui}, \quad i = 1, 2, ...n, \qquad (3.18)$$

$$U_0(\beta) := \max_s \ Eu_0(\sum_{j=1}^{J} \beta_j B_j(s), v_{max}(s)) - \delta_{u0}. \qquad (3.19)$$

Hence, the second substitute problem (2.20) -(2.23) is reduced to a nonlinear finite-dimensional parameter optimization problem. There are many numerical methods available to solve this problem, e.g. *mathematical programming* and some *deterministic* and *stochastic search methods*.

The problem (3.13)-(3.19) means, in fact, to find an optimal solution for the second substitute problem in the function space spanned by the basis functions $B_j(s)$. This approximative problem depends naturally on the choice of the basis $B_j(s)$. The *Spline functions*, which are used very often in practice, are recommendable.

4 Analytical and numerical solutions

In this section we are discussing some concrete applications to demonstrate the methods proposed above.

4.1 One-arm robot : Analytical solutions

The dynamic equation for a rotary massless arm with a point payload in its hand (see **Figure 4.1**) can be written in the form:

$$mL^2\ddot{q} = \tau, \qquad (4.1)$$

where m is the mass of the point payload, and L is the length of the arm.

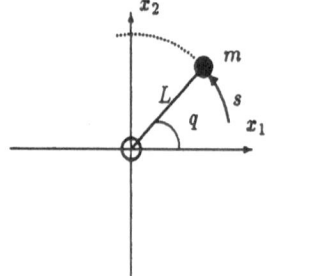

Figure 4.1

Suppose that the mass m and the length L are independent *random variables* with

$$E(m) = \bar{m} \quad , \quad Var(m) = \sigma_m^2, \tag{4.2}$$

and

$$E(L) = \bar{L} \ , \ Var(L) = \sigma_L^2 \ , \ E(L - \bar{L})^3 = 0 \ , \ E(L - \bar{L})^4 = \sigma_{LL}^4, \tag{4.3}$$

where E and Var denote the *Expectation* and the *Variance*, respectively. While $\bar{m}$ and $\bar{L}$ can be interpreted as the nominal values of the parameters m and L, σ_m^2, σ_L^2 and σ_{LL}^4 indicate the dispersions of the parameters about their nominal values. Obviously, increasing uncertainty is described by increasing values of σ_m^2, σ_L^2 and σ_{LL}^4.

From Figure 4.1 we have (by $p \to \bar{p}$ in equation (2.15))

$$q = \frac{s}{L} + q_0, \tag{4.4}$$

where s is the length of the path from the starting point, and q_0 is the initial angle.

Differentiating (4.4) yields

$$q' = \frac{1}{L} \tag{4.5}$$

and

$$q'' = 0. \tag{4.6}$$

From (4.1) (see (2.6)) we obtain

$$\frac{1}{2L} m L^2 v' = \tau. \tag{4.7}$$

Thus

$$E(\tau^2) = \frac{1}{4\bar{L}^2}(\bar{m}^2 + \sigma_m^2)(\bar{L}^4 + 6\bar{L}^2\sigma_L^2 + \sigma_{LL}^4)v'^2. \tag{4.8}$$

For the time optimal problem, the velocity v should be as high as possible, therefore, we define:

$$f_0 := -v. \tag{4.9}$$

Moreover, we choose $u(\tau) = \tau^2$ as the penalty function. The function $\bar{f}$ defined in (2.18) takes then the form:

$$\bar{f} = -v + \gamma \frac{1}{4\bar{L}^2}(\bar{m}^2 + \sigma_m^2)(\bar{L}^4 + 6\bar{L}^2\sigma_L^2 + \sigma_{LL}^4)v'^2 \tag{4.10}$$

where γ is a parameter relating $-v$ and $E(\tau^2)$. The Euler differential equation reads

$$\gamma \frac{1}{2\bar{L}^2}(\bar{m}^2 + \sigma_m^2)(\bar{L}^4 + 6\bar{L}^2\sigma_L^2 + \sigma_{LL}^4)v'' = -1. \tag{4.11}$$

For the sake of simplicity, we take $\bar{m} = 1$, $\bar{L} = 1$ and $s_e = 1$. With the boundary conditions $v(0) = 0$ and $v(1) = 0$, we obtain the fowllowing analytical solution of the first substitute problem:

$$v = \frac{1}{\gamma(1 + \sigma_m^2)(1 + 6\sigma_L^2 + \sigma_{LL}^4)} s(1 - s). \tag{4.12}$$

From the optimal solution (4.12) we find

$$\int_0^1 (-v)ds = -\frac{1}{6} \frac{1}{\gamma(1 + \sigma_m^2)(1 + 6\sigma_L^2 + \sigma_{LL}^4)} \tag{4.13}$$

and

$$t_e = \int_0^1 \frac{ds}{\sqrt{v}} = \sqrt{\gamma(1 + \sigma_m^2)(1 + 6\sigma_L^2 + \sigma_{LL}^4)} \; \pi; \tag{4.14}$$

replacing m and L in (4.5) by their expected values, we get

$$\bar{\tau} := \frac{1 - 2s}{2\gamma} \frac{1}{(1 + \sigma_m^2)(1 + 6\sigma_L^2 + \sigma_{LL}^4)}. \tag{4.15}$$

Moreover, the expected value and variance of the torque are given by

$$E(\tau) = \frac{1 - 2s}{2\gamma} \frac{1 + \sigma_L^2}{(1 + \sigma_m^2)(1 + 6\sigma_L^2 + \sigma_{LL}^4)} \tag{4.16}$$

and

$$Var(\tau) = \frac{(1 - 2s)^2}{4\gamma^2} \frac{4\sigma_L^2 + \sigma_{LL}^4 - \sigma_L^4 + \sigma_m^2(1 + 6\sigma_L^2 + \sigma_{LL}^4)}{(1 + \sigma_m^2)^2(1 + 6\sigma_L^2 + \sigma_{LL}^4)^2} \tag{4.17}$$

If σ_m^2, σ_L^2 and σ_{LL}^4 are very small, we can approximate the variance of τ by

$$Var(\tau) \simeq \frac{(1 - 2s)^2}{4\gamma^2}(4\sigma_L^2 + \sigma_{LL}^4 - \sigma_L^4 + \sigma_m^2) \tag{4.18}$$

Having solved the first substitute problem for the one-arm robot we consider now the second one.

Situation 1

Suppose first that v is a quadratic polynomial. To satisfy the conditions $v(0) = 0$ and $v(1) = 0$, v must take the form:

$$v = \beta_1 s(1 - s). \tag{4.19}$$

The objective function (performing time) reads

$$F(\beta_1) = \int_0^1 \frac{ds}{\sqrt{v}} = \frac{1}{\sqrt{\beta_1}} \pi. \qquad (4.20)$$

The penalty function u is chosen as before. From (4.8) and (3.15) we have

$$\beta_1^2 \le \frac{4\delta_{u1}}{(1+\sigma_m^2)(1+6\sigma_L^2+\sigma_{LL}^4)}. \qquad (4.21)$$

Hence,

$$\beta_1 \le \frac{2\sqrt{\delta_{u1}}}{\sqrt{(1+\sigma_m^2)(1+6\sigma_L^2+\sigma_{LL}^4)}}. \qquad (4.22)$$

According to (4.20) and (4.22), the solution of (3.13) - (3.16) is

$$\beta_1^* = \frac{2\sqrt{\delta_{u1}}}{\sqrt{(1+\sigma_m^2)(1+6\sigma_L^2+\sigma_{LL}^4)}}. \qquad (4.23)$$

Inserting (4.23) into (4.19) and (4.20), for the optimal velocity profile we have

$$v = \frac{2\sqrt{\delta_{u1}}}{\sqrt{(1+\sigma_m^2)(1+6\sigma_L^2+\sigma_{LL}^4)}} \, s \, (1-s), \qquad (4.24)$$

and the performing time is given by

$$t_e^{(1)} := \frac{\sqrt[4]{(1+\sigma_m^2)(1+6\sigma_L^2+\sigma_{LL}^4)}}{\sqrt[4]{4\delta_{u1}}} \pi. \qquad (4.25)$$

If γ in (4.10) is determined such that the solution of (4.11) also satisfies the condition (3.15), from (4.8) and (3.15), we then get

$$\gamma^2 \ge \frac{1}{4(1+\sigma_m^2)(1+6\sigma_L^2+\sigma_{LL}^4)\delta_{u1}}. \qquad (4.26)$$

If

$$\gamma = \frac{1}{2\sqrt{(1+\sigma_m^2)(1+6\sigma_L^2+\sigma_{LL}^4)\delta_{u1}}}, \qquad (4.27)$$

the solutions of the first and second substitute problems ((4.12) and (4.24)) agree with each other!

Situation 2

Let v now be a piecewise quadratic polynomial:

$$v(s) = \begin{cases} p_1(s), & 0 \le s < \frac{1}{2} - \alpha \\ p_2(s), & \frac{1}{2} - \alpha \le s < \frac{1}{2} + \alpha \\ p_3(s), & \frac{1}{2} + \alpha \le s \le 1 \end{cases} \qquad (4.28)$$

where

$\alpha \in (0, \frac{1}{2})$ is a constant,
p_i is a quadratic polynomial, and
v has a continuous derivative on $(0, 1)$.

Then v takes the form :

$$v = \beta_1 + \beta_2 s + \beta_3 s^2 + \beta_4 (s - \frac{1}{2} + \alpha)_+^2 + \beta_5 (s - \frac{1}{2} - \alpha)_+^2 \qquad (4.29)$$

We choose $u = \tau^2$ as penalty function once more. According to (4.8) and (3.15),

$$(v')^2 \leq \frac{4\delta_{u1}}{(1 + \sigma_m^2)(1 + 6\sigma_L^2 + \sigma_{LL}^4)}. \qquad (4.30)$$

For the time optimal problem, v should be as large as possible. Thus, considering (4.30) and $v(0) = 0$ and $v(1) = 0$, for $p_1(s)$, $p_3(s)$ we obtain

$$p_1 = \frac{2\sqrt{\delta_{u1}}}{\sqrt{(1 + \sigma_m^2)(1 + 6\sigma_L^2 + \sigma_{LL}^4)}} \, s, \qquad (4.31)$$

and

$$p_3 = \frac{2\sqrt{\delta_{u1}}}{\sqrt{(1 + \sigma_m^2)(1 + 6\sigma_L^2 + \sigma_{LL}^4)}} \, (1 - s). \qquad (4.32)$$

Hence,

$$\beta_1 = 0 \, , \quad \beta_2 = \frac{2\sqrt{\delta_{u1}}}{\sqrt{(1 + \sigma_m^2)(1 + 6\sigma_L^2 + \sigma_{LL}^4)}} \, , \quad \beta_3 = 0,$$

and

$$\beta_4 = -\beta_5 = -\frac{1}{2\alpha} \frac{2\sqrt{\delta_{u1}}}{\sqrt{(1 + \sigma_m^2)(1 + 6\sigma_L^2 + \sigma_{LL}^4)}},$$

and therefore

$$v(s) = \frac{2\sqrt{\delta_{u1}}}{\sqrt{(1 + \sigma_m^2)(1 + 6\sigma_L^2 + \sigma_{LL}^4)}} \left(s - \frac{1}{2\alpha} ((s - \frac{1}{2} + \alpha)_+^2 - (s - \frac{1}{2} - \alpha)_+^2) \right). \qquad (4.33)$$

The performing time is then given by

$$t_e^{(2)}(\alpha) := \frac{\sqrt[4]{(1 + \sigma_m^2)(1 + 6\sigma_L^2 + \sigma_{LL}^4)}}{\sqrt[4]{4\delta_{u1}}} I_0(\alpha), \qquad (4.34)$$

where

$$I_0(\alpha) := \sqrt{(\frac{1}{2} - \alpha) + 2\sqrt{2}\,\alpha \sin^{-1} \sqrt{\frac{\alpha}{(1 - \alpha)}}}. \qquad (4.35)$$

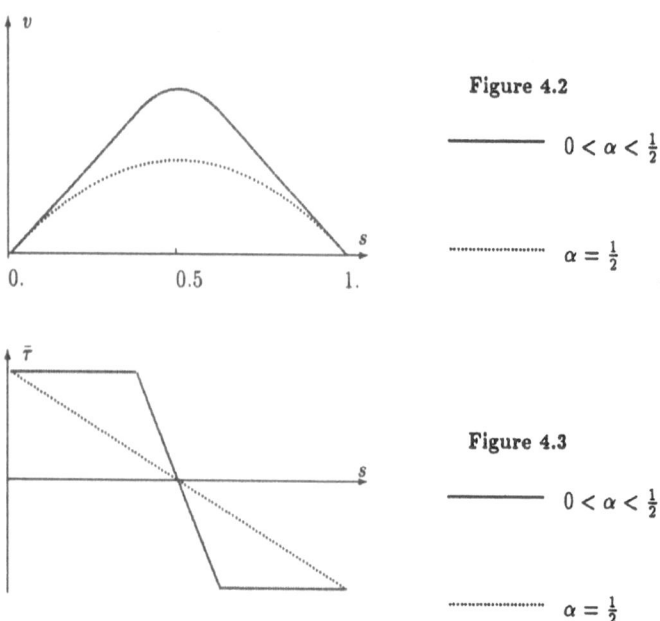

Figure 4.2

——————— $0 < \alpha < \frac{1}{2}$

··················· $\alpha = \frac{1}{2}$

Figure 4.3

——————— $0 < \alpha < \frac{1}{2}$

··················· $\alpha = \frac{1}{2}$

The shapes of v and $\bar{\tau}$, see (4.15), can be seen from Figure 4.2 and Figure 4.3 for some values of α. The solutions depend on α. If $\alpha = \frac{1}{2}$, (4.33) and (4.13) agree with each other once more! If $\alpha \to 0$, τ approximates a Bang-Bang-Control which is the absolute optimal solution of the second substitute problem.

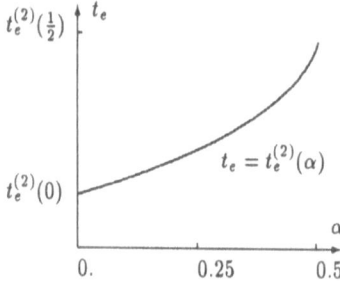

Figure 4.4 The dependence of the performing time on α

Figure 4.4 shows us how the performing time t_e depends on α, where

$$t_e^{(2)}(\frac{1}{2}) = \frac{\sqrt[4]{(1+\sigma_m^2)(1+6\sigma_L^2+\sigma_{LL}^4)}}{\sqrt[4]{4\delta_{u1}}}\pi = t_e^{(1)} \tag{4.36}$$

and

$$t_e^{(2)}(0) = \frac{\sqrt[4]{(1+\sigma_m^2)(1+6\sigma_L^2+\sigma_{LL}^4)}}{2\sqrt[4]{\delta_{u1}}}. \tag{4.37}$$

From (4.34) (4.36) and (4.37) one can see that *the performing time is an increasing function of* σ_m^2, σ_L^2 *and* σ_{LL}^4 representing the uncertainty about the model parameters m and L. This dependence of the performing time on σ_m^2, σ_L^2 and σ_{LL}^4 means that *an increasing uncertainty demands a slower motion!* It will be shown that this is also true for other robotic manipulators.

4.2 A two-arm robot: Numerical solution

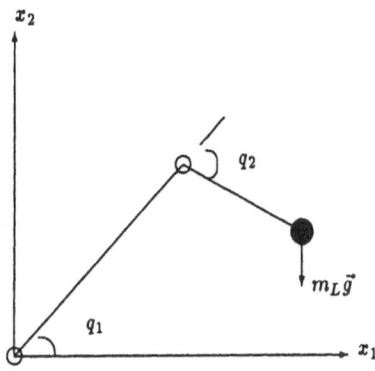

Figure 4.5 a two-arm robot

Figure 4.5 shows a two-arm robot which moves on a vertical plane and carries a payload m_L. For this robot, the kinematic and the dynamic equations can be written as follows [1, 7, 9]:

$$x_1 = T_1(q, p) = L_1 \cos q_1 + L_2 \cos(q_1 + q_2), \tag{4.38}$$

$$x_2 = T_2(q, p) = L_1 \sin q_1 + L_2 \sin(q_1 + q_2), \tag{4.39}$$

and

$$H_1 \ddot{q}_1 + H_{12} \ddot{q}_2 - 2h\dot{q}_1\dot{q}_2 - h\dot{q}_2^2 + G_1 = \tau_1, \tag{4.40}$$

$$H_{12} \ddot{q}_1 + H_2 \ddot{q}_2 + h\dot{q}_1^2 + G_2 = \tau_2, \tag{4.41}$$

where

$$H_1 = I_1 + I_2 + I_L + L_{c1}^2 m_1 + (L_1^2 + L_{c2}^2 + 2L_1 L_{c2} \cos q_2)m_2$$
$$+ (L_1^2 + L_2^2 + 2L_1 L_2 \cos q_2)m_L, \tag{4.42}$$

$$H_{12} = I_2 + I_L + (L_{c2}^2 + L_1 L_{c2} \cos q_2)m_2 + (L_2^2 + L_1 L_2 \cos q_2)m_L, \tag{4.43}$$

$$H_2 = I_2 + I_L + L_{c2}^2 m_2 + L_2^2 m_L, \tag{4.44}$$

$$h = L_1 L_{c2} \sin q_2 m_2 + L_1 L_2 \sin q_2 m_L, \tag{4.45}$$

$$G_1 = L_{c1} \cos q_1 g m_1 + (L_1 \cos q_1 + L_{c2} \cos (q_1 + q_2)) g m_2$$
$$+ (L_1 \cos q_1 + L_2 \cos (q_1 + q_2)) g m_L, \tag{4.46}$$

$$G_2 = L_{c2} \cos (q_1 + q_2) g m_2 + L_2 \cos (q_1 + q_2) g m_L, \tag{4.47}$$

and

I_i : moment of inertia of the i-th arm,
m_i : mass of the i-th arm,
I_L : moment of inertia of the payload,
m_L : mass of the payload,
L_i : length of the i-th arm,
L_{ci} : distance from the center of mass of i-th arm to its rotary axis.

Inserting $v = v(s)$ into (4.21) and (4.22), we have

$$\frac{1}{2}(H_1 q_1' + H_{12} q_2')v' + (H_1 q_1'' + H_{12} q_2'' - 2h q_1' q_2' - h q_2'^2)v + G_1 = \tau_1, \tag{4.48}$$

$$\frac{1}{2}(H_{12} q_1' + H_2 q_2')v' + (H_{12} q_1'' + H_2 q_2'' + h q_1'^2)v + G_2 = \tau_2. \tag{4.49}$$

Supposing that

$$\tau_{min,i} = -\tau_{max,i}, \tag{4.50}$$

we choose

$$u_i := \tau_i^2 \tag{4.51}$$

as our penalty functions $u_i = u_i(\tau_i, \tau_{min,i}, \tau_{max,i})$. Assuming that the payload is stochastic, we may represent I_L and m_L by

$$I_L = \bar{I}_p + \sum_{j=1}^{J} C_I^{(j)} \frac{w^j}{j!} \tag{4.52}$$

and

$$m_L = \bar{m}_L + \sum_{j=1}^{J} C_m^{(j)} \frac{w^j}{j!}, \tag{4.53}$$

where w is a random variable with

$$E(w) = 0, \qquad Var(w) = \sigma^2. \tag{4.54}$$

and $\bar{I}_L$, $\bar{m}_L$, $C_I^{(j)}$, and $C_m^{(j)}$ are certain constants. If the constraints about the velocity and angle velocities can be neglected, we can approximate function (2.18) as follows:

$$\bar{f} = -v + \gamma_1(\bar{\tau}_1^2 + \sigma^2((\frac{\partial \bar{\tau}_1}{\partial w})^2 + \bar{\tau}_1 \frac{\partial^2 \bar{\tau}_1}{\partial w^2}))$$
$$+ \gamma_2(\bar{\tau}_2^2 + \sigma^2((\frac{\partial \bar{\tau}_2}{\partial w})^2 + \bar{\tau}_2 \frac{\partial^2 \bar{\tau}_2}{\partial w^2})), \tag{4.55}$$

where

$$\bar{\tau}_i := \tau_i|_{w=0}, \tag{4.56}$$

$$\frac{\partial \bar{\tau}_i}{\partial w} := \frac{\partial \tau_i}{\partial w}\Big|_{w=0}, \tag{4.57}$$

and

$$\frac{\partial^2 \bar{\tau}_i}{\partial w^2} := \frac{\partial^2 \tau_i}{\partial w^2}\Big|_{w=0}. \tag{4.58}$$

Now we consider the following task for the robot: move the payload m_L from point (0.4, -0.65) to point (0.4, 0.65). For this, we choose the straight line from point (0.4, -0.65) to point (0.4, 0.65) as path in work space (see Figure 4.6 left). The results for this path are given in Figure 4.7, in comparison with the results of the phase-plane-method (dotted lines).

The other task is to move the payload from (0.6, -0.6) to (0.6, 0.6), while an obstacle stands at point (0.6, 0). We choose, therefore, a piece of arc as the path in work space (see Figure 4.6 right). The results of this task are given by Figure 4.8.

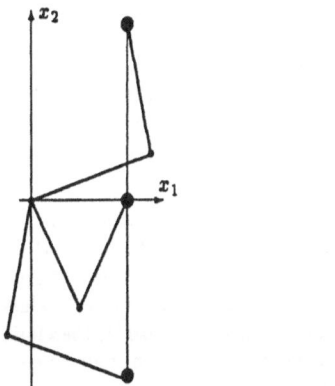 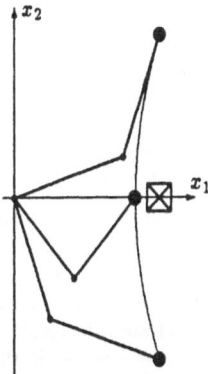

Figure 4.6 The paths in the work space

Table 4.1 contains the chosen data for the robot model.

L_1	L_2	m_1	m_2	$\bar{m}_L$	I_L	$\tau_{max,1}$	$\tau_{max,2}$
0.5	0.5	4	4	1	0.1	50	40

Table 4.1

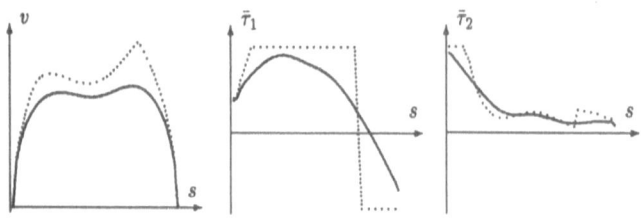

Figure 4.7 The solutions of the first substitute problem for a straight line path

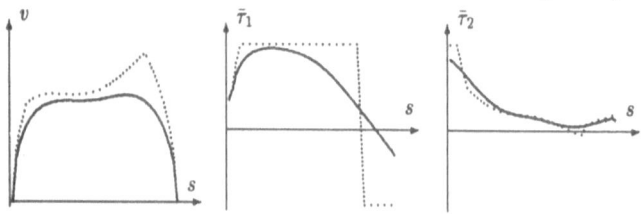

Figure 4.8 The solutions of the first substitute problem for an arc path

The dependence of the results of the first substitute problem on parameters γ_i and σ

In Section 2 the first substitute problem for trajectory planning under stochastic uncertainty is proposed. The violations of the constaints (2.13) and (2.14) for the torques τ_i are considered by inserting the expected penalty costs into the objective function, for which the weight factors γ_i are needed. The expected values $E_p u_i(\tau_i, \tau_{max,i}, \tau_{min,i})$ are calculated by means of Taylor expansion, therefore, σ_i^2, the variances of p appear also in the objective function. In the following we study the influence of γ_i and σ_i on the results.

For the *one-arm robot* it is clear, see (4.14) and (4.15), that larger γ, σ_m and σ_L lead to a larger performing time and smaller torques. A larger value of γ implies a sharper restriction on torque τ. A smaller torque leads then to a longer performing time. If there is greater uncertainty at m or L (larger values of σ_m or σ_L), the open loop strategy τ calculated from (4.5) takes smaller values. This results also in a longer performing time.

For the *two-arm robot* we can not get the explicit solution. But from the following graph we can also see the influence of the parameters γ_i and σ on the solutions. Figure 4.9 - 4.11 show how the paramerters γ_i and the random uncertainty σ effect the performing time and the maximum absolute values of τ_i. For fixed σ, time t is monotonous increasing, and the maximum absolute values of τ_i are monotonous decreasing functions of γ_i . Larger γ_i mean stricter constraints on τ_i. This causes smaller torques τ_i and naturally longer performing time. If σ_i increases, the maximal absolute values of τ_i become smaller and time

t longer. That means that smaller open loop strategies τ_i should be applied in case of greater parameter uncertainty. So we have a longer time again.

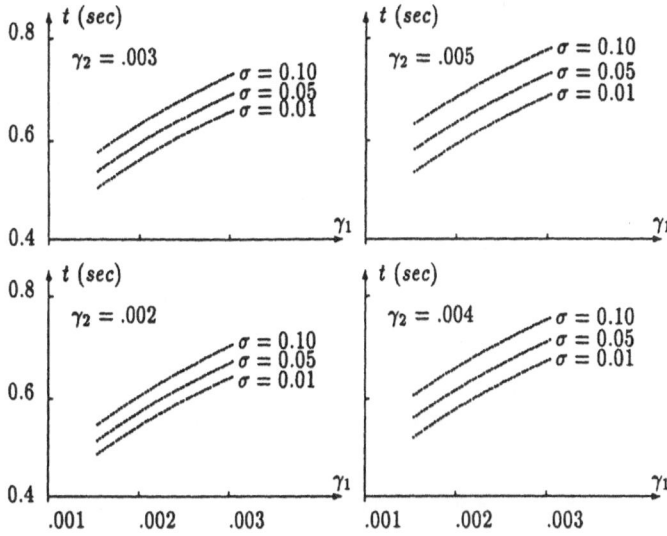

Figure 4.9 The dependence of time on γ_i and σ

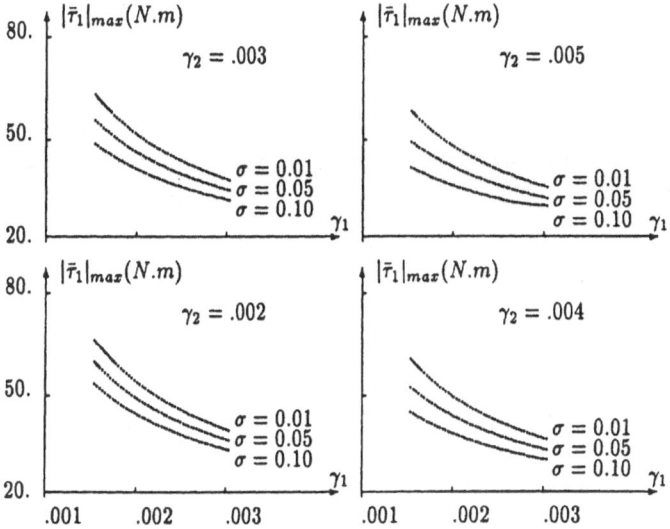

Figure 4.10 The dependence of the maximum absolute values of $\bar{\tau}_1$ on γ_i and σ

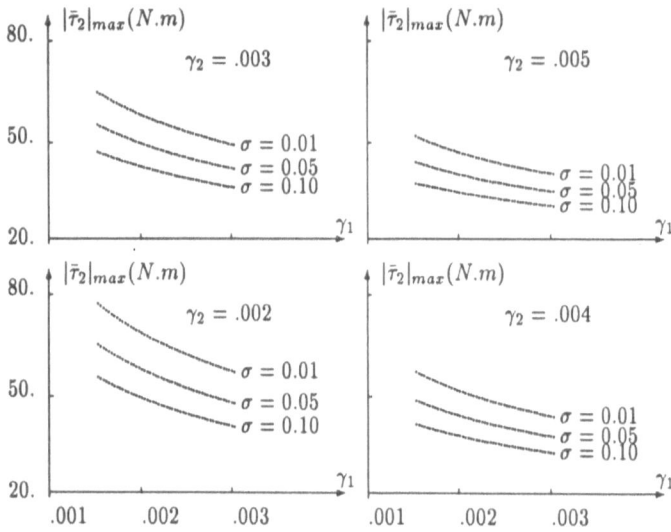

Figure 4.11 The dependence of the maximum absolute values of $\bar{\tau}_2$ on γ_i and σ

Larger σ_i leads to a longer performing time. However, the control is safer due to the consideration of the random parameter variation in calculating the optimal velocity profile $v = v(s)$.

Numerical solution of the second substitute problem

Having chosen B-spline functions as basis functions in (3.12), we solve the nonlinear finite-dimensional parameter optimization problem (3.13)-(3.15) and obtain the solutions of the second substitute problem for both tasks, as drawn in Figure 4.12 and 4.13:

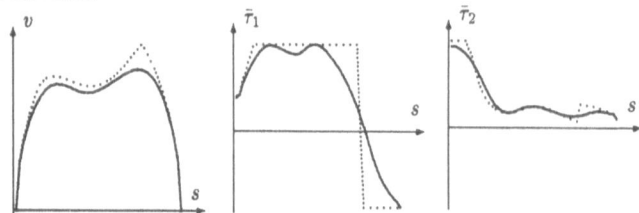

Figure 4.12 The solutions of the second substitute problem for a straight line path

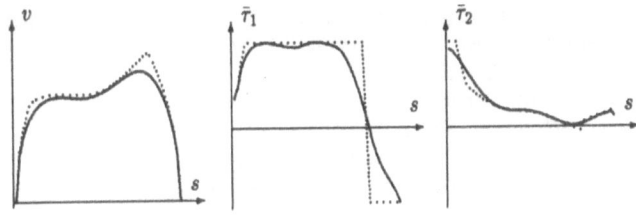

Figure 4.13 The solutions of the second substitute problem for an arc path

5 Further substitute problems

If the stochastic variations of parameter p is large, we may not simply replace p in (2.15) by its expectation. Analogous to the first and the second substitute problem, the violation of the constraint (2.15) can be evaluated numerically with the help of a further penalty function $u_T(\cdot)$, and then inserted into the objective function or constraint functions. This way we can construct further substitute problems.

References

1. Asada, H. and Slotine, J. J. E.: *Robot Analysis and Control*, John Wiley, New York, 1986
2. Bobrow, J. E., Dubowsky, S. and Gibson, J. S.: *Time-optimal Control of Robotic Manipulaters* , Int. J. Robot. Res. 4(3), p.3-17,1985
3. Gill, P.E. and Murray, W.: *Numerical Methods for Constrained Optimization*, Academic Press, London, 1974
4. Goldenberg, A., Benhabib, B. and Fenton, R.: *A Complete Generalized Solution to the Inverse Kinematics of Robots*, IEEE J. Robot. Automat. RA-1(1), p.14-20, 1985
5. Johanni, R. and Pfeiffer, F.: *Optimale Bahnplanung für Industrieroboter*, Robotersysteme 3, p.29-36, 1987
6. Marti, K.: *Seminar über Stochastische Optimierung*, Universität der Bundeswehr München, 1991
7. Paul, R.P.: *Robot Manipulators: Mathematics, Programming and Control*, MIT Press, Cambridge, 1981
8. Pfeiffer, F. and Johanni, R.: *A Concept for Manipulator Trajectory Planning*, IEEE J. Robot. Automat. RA-3(3), p.115-123, 1987
9. Pfeiffer, F. and Reithmeier, E.: *Roboterdynamik* , B.G.Teubner, Stuttgart, 1987
10. Rappl, G.: *Konvergenzraten von Random-Search-Verfahren zur Globalen Optimierung*, Dissertation, Universität der Bundeswehr München, 1984
11. Schiller, Z. and Dubowsky, S.: *Robot Path Planning with Obstacles, Actuator, Gripper and Payload Constraints*, Int. J. Robot. Res. 8(6), p.3-18, 1989

12. Shin, K.G. and McKay, N.D.: *Minimum-time Control of Robotic Manipulator with Geometric Path Constraints*, IEEE Trans. Automat. Control AC-30(6), p.370-375,1985
13. Tucker, M. and Perreira, N.D.: *Generalized Inverses For Robotic Manipulators*, Mech. Mach. Theory Vol. 22, No.6, p.507-514,1987
14. Zuehlke, D.: *Offline-Programming numerisch gesteuerter Industrieroboter*, Fortschrittberichte VDI, Reihe 2, Nr.54, VDI-Verlag, Düsseldorf, 1983

Performance Characteristics of OC Methods with Applications in Topology Design

G.I.N. Rozvany and M. Zhou, FB 10, Essen University, D-45117 Essen, Germany

Abstract. One of the major difficulties in structural optimization at present is the computational discrepancy between analysis capability and optimization capability. This discrepancy can be overcome, and even reversed, by employing *new optimality criteria (OC) methods* which achieve the correct optimal solution (i.e. Kuhn-Tucker point) while treating active stress constraints at the element level. This means that their optimization capability is only limited by the *number of global* (e.g. displacement or natural frequency) *constraints.* Since the latter is usually small for typical structural systems, these new techniques increase our optimization capability by several orders of magnitude.

The power of the new optimality criteria methods is demonstrated by considering a variety of problems in sizing and topology optimization. The proposed methods become highly economical for *sizing optimization* if the considered system is very large, with many thousands of variables and active stress constraints. Moreover, the considered methods are found to be almost unavoidable in *topological optimization,* because in the so-called ground structure approach the number of potential members must be very high for a reasonable accuracy. Finally, optimal topologies obtained by discretized methods are verified by comparing them with recently derived closed form analytical solutions.

1 Introduction: Advantages and Disadvantages of Optimality Criteria Methods

Methods of structural optimization fall into one of three broad categories, namely

- optimality criteria methods;
- mathematical programming methods using gradients (sensitivities); and
- random search methods.

In *optimality criteria methods*, all necessary (and sometimes sufficient) conditions of optimality are generated for a given problem. The corresponding set of equations are then solved, partly explicitly (if possible) and partly iteratively.

In *gradient-based mathematical programming methods*, the gradients of objective (cost) function and constraint gradients are calculated for a given design and then the next design is generated in a systematic way, by trying to maximize the reduction in cost in the feasible domain (or to minimize the computational effort to find an optimum). The above procedure is repeated until the change in cost or design parameters becomes small.

Random search techniques (e.g. genetic algorithms) are based on some randomized procedure which generates systematically a large number of designs, out of which the best one is selected.

A method is termed *robust*, if it can be used "blindly" or as a "black box" for a broad range of problems, without having to understand at depth the particular mathematical structure of the optimal solutions involved. Clearly, search methods are, in general, the most robust and optimality criteria methods the least robust.

A method is called *efficient* if it has a higher optimization capability in terms of number of variables or number of active constraints, or alternatively, if it uses less computer time and/or storage space than other methods. The relative efficiency of optimality criteria methods is usually the highest and that of search methods is, in general, the lowest. However, search methods may be unavoidable if the number of local minima is very high.

Gradient-based *mathematical programming* methods may also be divided into two subclasses, i.e.

- *primal methods*, in which the variable space contains the design variables; and

- *dual methods*, in which the variables are the Lagrange multipliers of the constraints.

Clearly, primal methods are uneconomical if the number of variables is large and dual methods are inefficient if the number of active constraints is too high. This means, however, that for problems with many variables and many active constraints, both primal and dual methods are inadvisable. This situation may appear hopeless; fortunately, however, most active behavioural constraints for structural systems are *local* (stress or side) *constraints* and these can often be eliminated in modern optimality criteria methods (COC/DCOC) through explicit calculations at the element level. This means that, in effect, only *global* constraints contribute to the size of the dual space, and hence the above methods become several orders of magnitude more efficient for very large structural systems than traditional techniques.

Another way of avoiding stress constraints in the dual space is to use a *fully stressed design* (FSD) for stress controlled elements and to treat other constraints properly in the dual space. This method, however, is known to result, in general, in non-optimal solutions.

We may add that the above classifications are somewhat oversimplified. It can be shown that certain dual programming methods and optimality criteria methods are, in fact, identical. Moreover, optimality criteria can be expressed through sensitivities and hence the distinction between programming and optimality criteria methods is far from being sharp.

The newest optimality criteria method (DCOC) can also be regarded as a modified version of the older discretized optimality criteria (DOC) method combined with fully stressed design (DOC-FSD). The difference is that DCOC is based on rigorous fulfilment of the Kuhn-Tucker conditions, whereas DOC-FSD is an intuitive method. Whilst both methods use the same redesign formulae, in DCOC the virtual load systems, corresponding to displacement constraints, are modified through the application of prestrains, in order to fulfil the above conditions. The details of this operation are explained in Section 3.

2 Historical Roots of Modern Optimality Criteria Methods

The earliest optimality criteria methods were somewhat heuristic. These *intuitive optimality* criteria methods included the already mentioned *fully stressed design* (FSD), as well as *uniform energy dissipation* (UED), and *uniform mutual energy dissipation* (UMED). All the above methods give the correct optimal solution for a limited class of problems: FSD for statically determinate structures, UED for a compliance (total work) constraint and UMED for a displacement constraint, but all three are restricted to a single load condition.

Rigorous optimality criteria were derived for *discretized systems* by Berke and others (e.g. Berke 1970; Venkayya, Khot and Berke 1973). These optimality conditions were then modified, on a somewhat intuitive basis, into very efficient redesign formulae. This group of methods will be termed here DOC (discretized optimality criteria) methods.

Surprisingly, entirely independently from the above development but almost at the same time, a research school around Prager derived optimality criteria for continua (e.g. Prager and Shield 1967; Prager and Taylor 1968; for eigenvalue problems, see e.g. Olhoff 1981). The above conditions were derived usually for a single criterion and often for plastic design. However, they were gradually extended to general elastic systems with simultaneous stress and displacement constraints (e.g. Rozvany 1989).

One of the obvious differences between the above developments was the fact that one (DOC) used a discretized formulation in a finite dimensional vector space and the other one (COC) a continuum formulation in an infinite dimensional design space. Correspondingly, the branch of mathematics involved is differential calculus in one case (DOC) and calculus of variations in the other (COC). A recently found, more intrinsic difference is that, whereas in DOC the only variables are the cross-sectional dimensions (design variables), in COC an extended design space includes also the real and virtual internal forces. Another improvement is the introduction of an "adjoint structure" in COC, which is used for giving the optimality conditions some physical meaning. The external loads on the adjoint structure depend on active displacement constraints and the fictitious prestrains in members on active stress constraints (see Section 3).

The COC algorithm was applied iteratively to large discretized systems (Rozvany and Zhou *et al.* 1989, 1990; Rozvany and Zhou 1991b; Zhou and Rozvany 1993a), by first deriving the optimality criteria for continua and then discretizing the results for FE methods. This transformation introduced extra work and the results were only valid for a continuous variation of the design variables along the members or elements involved. For these reasons, it represented a considerable improvement when Zhou reformulated the new optimality criterion method directly in terms of matrix methods of the finite element approach (Zhou and Rozvany 1992, 1993b). The new *discretized continuum type optimality criteria method* has been termed DCOC for historical reasons. As will be seen in Section 3, DCOC was first derived using the flexibility formulation of structural analysis and the fundamental features of COC, employing a stiffness method in the implementation only. This has had the drawback that it was difficult to understand by researchers familiar with traditional optimality criteria (DOC) methods. Moreover, the implementation was cumbersome for more advanced elements whose flexibility relationships are not readily available in the modern FE technology. These difficulties have been removed by recent work of Zhou and Haftka (1994) who showed that DCOC can, in fact, be derived from the traditional DOC formulation. In the above paper a new, "derivative-based" DCOC formulation is also presented, which avoids the flexibility formulation completely, as well as the concept of adjoint structure, replacing the latter with sensitivity analysis of certain functions. This way, implementation of DCOC becomes very convenient.

In the next section, however, COC/DCOC will be outlined on the basis of their original formulation, because this will have some conceptual advantages in discussing topology optimization in later sections.

3 Basic Features of COC/DCOC

3.1 What is an "Adjoint Structure"?

As mentioned in Section 2, a fundamental feature of the new optimality criteria methods is the adjoint structure, which is a fictitious system having the same analysis equations as the real structure but different loads (the virtual loads for displacement constraints), prestrains (for active stress constraints) and possibly support settlements. The analysis of the adjoint structure replaces the usual sensitivity analysis of other methods, although it can be shown in some cases to reduce to conventional sensitivity analysis by the adjoint variable method. The implementation of the adjoint method is rather efficient, because the stiffness matrix is the same for the real and adjoint structures. As pointed out earlier, this concept can also be avoided to suit existing computational technology.

In this paper, we shall not include the derivation or the general formulae of COC and DCOC, because they are summarized in a comprehensive form elsewhere (Zhou and Rozvany 1992, 1993a; for a detailed summary see Rozvany and Zhou 1994). *For didactic reasons, we are only considering herein truss structures*, and illustrations through very simple examples. The DCOC method is, however, available for most static and some dynamic problems, including a number of refinements (see Rozvany and Zhou 1994 or Section 6.2 herein).

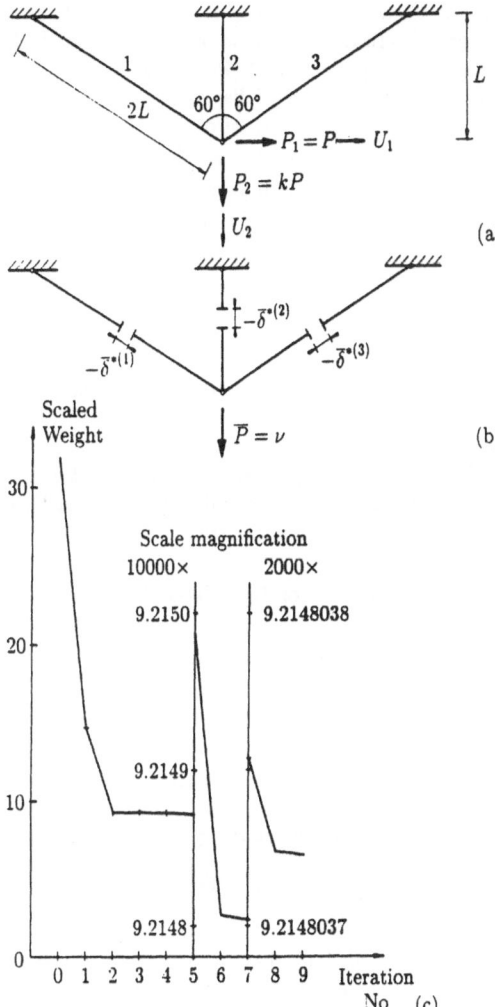

Fig. 1. Elementary example.

To explain the concept of the adjoint structure, consider the elementary problem in Fig. 1a, in which a three-bar truss is subject to the horizontal force P_1 and the vertical force $P_2 = kP$ and the weight of the truss is minimized subject to constraints on the stresses in compression and tension (σ_C and σ_T), and on the vertical deflection ($U_2 \leq t$). The horizontal displacement (U_1) is unconstrained. One of the shortcomings of this problem is the fact that a vertical displacement constraint is equivalent to a stress constraint for the vertical bar "2". This means that, in general, depending on the limiting values σ_T and t in the above problem, only one of the above constraints can be active. However, this restricting feature does not disturb the illustration of the DCOC method.

The *adjoint structure* for the above problem is shown in Fig. 1b. The loading on the adjoint structure is the Lagrange multiplier ν for the displacement constraint, acting at the location and in the direction of that constraint. The initial displacements (caused by fictitious prestrains) for the adjoint structure are $\overline{\delta}^{*(1)}$, $\overline{\delta}^{*(2)}$ and $\overline{\delta}^{*(3)}$, which are non-zero only if a stress constraint for a bar is active. For trusses, the above initial displacement is given by (Rozvany and Zhou 1994)

$$\overline{\delta}^{*e} = \frac{L^e}{\sigma^e} \left[\rho^e - \frac{f^e \overline{f}^e}{(x^e)^2 E^e} \right] \operatorname{sgn}(f^e), \tag{1}$$

where $L^e, \sigma^e, \rho^e, E^e$ and x^e are the bar length, permissible stress, specific weight, Young's modulus and cross-sectional area for the truss element e, whilst f^e and $\overline{f}^e$ are the corresponding bar forces in the real and adjoint trusses.

It can be seen that the adjoint structure is equivalent to the usual virtual load system with a scaling factor ν (in DOC or adjoint variable method of sensitivity analysis), and prestrains added for active stress constraints.

3.2 Iterative Computational Procedure for COC/DCOC

The basic steps in the new optimality criteria methods are as follows.

- Analysis of the real and adjoint structures.
- Updating cross-sectional parameters.
- Updating Lagrange multiplier(s) for global constraint(s).

Details of the latter two operations are given below.

3.3 Updating Cross-Sections in COC/DCOC

The redesign formula used depends on the type of active constraints for a given element. In numerical procedures, it is usually necessary to impose at least a lower limit on the cross-sectional parameter, e.g.

$$x^e \geq x^e_{\downarrow}, \tag{2}$$

where $x^e_{\downarrow}$ is a given minimum cross-sectional area for a bar e. Such restrictions are termed *side constraints*.

For trusses with stress and side constraints and a displacement constraint (e.g. Fig. 1), we have the following redesign formulae:

$$(x^e)^2 = \max \left\{ (x^e_d)^2, (x^e_\sigma)^2, (x^e_{\downarrow})^2 \right\}, \tag{3}$$

with

$$(x_d^e)^2 = \frac{f^e \overline{f}^e}{E^e \rho^e} , \qquad (4)$$

$$(x_\sigma^e) = \frac{|f^e|}{\sigma^e} . \qquad (5)$$

Clearly, (5) determines the cross-sections controlled by the stress constraint. The redesign relation (4) refers to truss members controlled by the displacement constraint. In general, elements of a truss fall into one of the three element sets, namely [see (3) above]

R_d : elements controlled by the displacement constraints,

R_σ : elements controlled by a stress constraint,

R_s : elements controlled by a side constraint.

3.4 Updating Lagrange Multipliers in COC/DCOC

Since $\overline{f}^e$ in (4) depends on the Lagrange multiplier ν (see Fig. 1b), the updated value of x_d^e also depends on it, whereas x_i^e and x_s^e are independent of ν. This means that by any change of ν only the elements in R_d are affected. The new value of ν can therefore be calculated by satisfying the *displacement constraint* as an equality using a work equation, substituting the cross-sectional areas from (4) and expressing ν from the above relation:

$$\sqrt{\nu} = \frac{\sum\limits_{e \in R_d} L^e \sqrt{\rho^e f^e \hat{f}^e / E^e}}{t - \sum\limits_{e \notin R_d} (L^e f^e \hat{f}^e / E^e x^e)} , \qquad (6)$$

in which $\hat{f}^e = \overline{f}^e / \nu$, with the value of ν taken from the prior iteration.

Using the above procedure for the elementary problem in Fig. 1a, with $L^e = E^e = \rho^e = 1$ and $\sigma_t^e = \sigma_C^e = 1.3$ for *all* elements as well as $P = P_1 = 1$, $k = P_2 = 8$ and $t = 1$, the iteration history is shown in Fig. 1c, in which an almost uniform convergence is achieved in spite of scale magnifications by 10000 and 2000. The optimal weight w_{opt} after 9 iterations was found to be 9.2148037239 and the cross-sectional areas $x^{(1)} = 0.882077$, $x^{(2)} = 7.430649$ and $x^{(3)} = 0.01$. In iteration 9, the weight value changed only from 9.2148037242 to the above value, satisfying the very stringent convergence criterion

$$\frac{|W_{new} - W_{old}|}{W_{new}} \leq T , \qquad (7)$$

with $T = 10^{-10}$. For practical applications such high accuracy is not necessary. The above iteration history (Fig. 1c) shows that after two iterations the (scaled) weight differs by only one per cent from the optimal weight.

3.5 Some Philosophical Remarks about Conceptual Clarity vs. Computational Expediency

The above dilemma will be illustrated through the formula for calculating the adjoint initial displacement in (1). Since for stress-controlled truss elements we have $f^e/x^e = \sigma^e \text{sgn} (f^e)$, the relation in (1) reduces to

$$\bar{\delta}^{*e} = L^e \left[\frac{\rho^e \text{sgn} (f^e)}{\sigma^e} - \frac{\bar{f}^e}{x^e E^e} \right] . \tag{8}$$

However, the external load on the adjoint truss causes an elastic member elongation of $L^e \bar{f}^e / x^e E^e$. Adding the prestrain to the strain due to external load, we obtain the total adjoint strain

$$\bar{\varepsilon}^e = \frac{\rho^e \text{sgn} (f^e)}{\sigma^e} , \tag{9}$$

which is simply *the ratio of the specific weight and the permissible stress*, with the sign of the real force in the considered truss element. Whilst the result in (9) is conceptually very valuable due to its simplicity, *in an iterative procedure* various quantities in (8) originate from different computational steps. This means that *for intermediate iterative values* of $\bar{f}^e$ (involving ν) and x^e, $\bar{\varepsilon}^e$ does not reduce to (9). Moreover, it is difficult to model the requirement that the total adjoint strain in stress-controlled members is independent of the adjoint force in the same member. Surprisingly, relation (1) is therefore more efficient computationally than (9).

Naturally, in all practical computations the most expedient formulae should be used, even when the latter completely obscures conceptually fundamental features of an optimal solution. In order to be able to check the correctness of discretized numerical solutions, or with a view to obtaining exact, explicit solutions (for example, in topology optimization), researchers should also be thoroughly familiar with the original, exact relations leading to a computational procedure.

4 Approximations Involved in Iterative COC/DCOC Methods and Their Effect on Computational Stability

Update formulae in the above iterative procedures involve a certain approximation, because *the effect of the cross-sectional areas on the internal forces is temporarily ignored*. The same applies to more complicated DCOC procedures for several load conditions and multiple displacement constraints, where the Lagrange multipliers for displacement constraints are calculated in a subiteration (Zhou and Rozvany 1993b). All the above formulae would be exact if the above dependence did not exist, that is, in the case of *statically determinate structures*. It follows from the above approximation concept that DCOC converges very satisfactorily if *the distribution of the internal forces does not depend significantly on the cross-sectional areas.* Convergence problems arise, and further refinements become necessary, if the problem is ill-conditioned in the sense that *the internal forces are highly dependent on the cross-sectional areas.* To avoid computational instability in the latter case, a procedure for automatic adjustments of the move limits can be adopted. A better solution of this problem is to introduce an improved approximation for the updating procedure.

It is further to be remarked that the redesign formula (4) is based on first order sensitivities but the relation in (5) is not related to derivatives. The latter, therefore, results in a slower convergence. In *DOC and dual methods*, stress constraints are usually replaced by equivalent displacement constraints, and hence first order redesign formulae are involved. This way the number of iterations is reduced, but the total computer time is increased because the calculation of the corresponding Lagrangians, which are coupled at the system level, becomes much more expensive.

5 A Review of Test Examples Demonstrating the Power and Efficiency of DCOC

A number of test examples based on DCOC were presented in recent publications (Zhou and Rozvany 1992, 1993b; Rozvany and Zhou 1994). Since for trusses and beams the COC and DCOC procedures are somewhat similar, earlier test examples using an iterative COC procedure are also relevant (Rozvany, Zhou *et al.* 1989, 1990). The above examples included

- a clamped beam with stress constraints and a displacement constraint;
- ten-bar truss examples with (i) stress constraints and one displacement constraint, (ii) stress constraints and two displacement constraints, (iii) only stress constraints with different permissible stress values, and (iv) two displacement constraints and a variable point load;
- a ten-storey, three-bay frame with stress constraints and a horizontal displacement constraint;
- the standard 25-bar truss;
- the standard 72-bar truss;
- the standard 200-bar truss problem and a modified version of it.

Out of the above test problems, the first one and the last one will be reviewed here in detail. This is because in those two problems, the *number of active stress constraints was relatively large,* which is not the case with most benchmark problems in structural optimization. The reason for this situation is probably the circumstance that authors advocating a certain method would often be having difficulties in applying it to a problem with a large number of elements and a large number of active stress constraints.

In addition, extensions of the beam problem by means of a so-called FE simulator will be briefly discussed (Section 5.2).

5.1 Clamped Beam Example with 100 and 1000 Elements Subject to Bending and Shear Stress Constraints and a Displacement Constraint

This first and often publicized example consists of a uniformly loaded (q) clamped beam of span $2a$, having a depth d and variable width, Young's modulus E, specific weight ρ and an admissible displacement t at mid-span. The bending and shear stress constraints take the form $x^e \geq k_1 |M^e|_{\max}$, $x^e \geq k_2 |V^e|_{\max}$ where $|M^e|_{\max}$ and $|V^e|_{\max}$ are the maximum absolute value of the moment or shear force on element e, whilst k_1 and k_2 are given constants. Mostly normalized parameters were used, $\rho = d = q = t = a = 1$ with $E = 12$, $k_1 = 0.23$ and $k_2 = 0.03$. The tolerance value T in (7) was 10^{-8} for the 100 element problem and 10^{-6} for the 1000 element problem, in which the element number refers to one half of the beam (due to symmetry). The above problems were solved by the DCOC, Dual and DOC-FSD methods, and the results are reproduced from another paper (Zhou and Rozvany 1992) in Table 1. The optimal width distributions are shown for the 100 and 1000 element beams in Fig. 2, in which R_d, R_σ and R_τ denote regions governed by displacement, axial and shear stress constraints, respectively. The following conclusions can be drawn from the above results:

Table 1. Results for the clamped beam example

100 Element Model			
	DCOC	DOC-FSD	Dual
Optimal Weight	0.064202389	0.064203018	0.064202389
Number of Analyses	29	28	9
CPU Optimization	7.84	3.63	535.28
Times Analysis	116.95	112.70	117.41
(sec.) Total	124.79	116.33	706.69
1000 Element Model			
	DCOC	DOC-FSD	
Optimal Weight	0.063996543	0.063999993	
Number of Analyses	18	20	
CPU Optimization	57.35	34.98	
Times Analysis	928.59	1126.41	
(sec.) Total	985.84	1161.39	

- The DOC-FSD method is about as economical as DCOC but results in a nonoptimal solution. In the current example, the weight difference between DOC-FSD and DCOC is negligible, but this is not so in some other

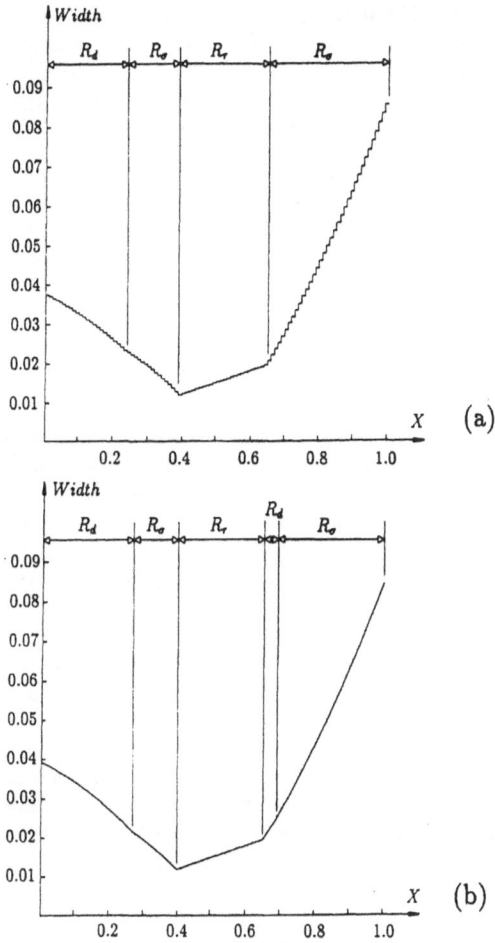

Fig. 2. Beam example with (a) 100 elements and (b) 1000 elements.

examples (Zhou and Rozvany 1993b, Section 5.1, ten-bar truss, weight by DCOC: 1497.6, weight by DOC-FSD 1725.2). The same can be seen in the case of the 200-bar truss herein (Section 5.3).

- The total CPU time for Dual is almost six times as much as for DCOC for the 100 element beam. The CPU times for Dual would have been prohibitively high for the 1000 element beam.

- The CPU time for "analysis", which includes that for the analysis of the adjoint system (DCOC), analysis of the virtual load system (DOC-FSD) and sensitivity analysis (Dual), is about the same for all three methods in the case of the 100 element beam.

- However, for the 100 element beam the optimization time for Dual is about 70 times higher than for DCOC, owing to 76 active stress constraints, which are coupled at the system level in Dual. Handling of about 760 active stress constraints for the 1000 element beam by the same method was entirely beyond the capability of Dual on a small computer.

- Due to using first order sensitivities in Dual (instead of zero order ones in DCOC) for stress constraints, the number of iterations was only 9 for the 100 element beam for Dual (against 29 for DCOC). This, however, did not help to make Dual more economical, owing to the high CPU requirement for optimization.

5.2 Beam Solutions with up to One Million Elements

Beam examples similar to the one in Fig. 2 were computed also for much larger element numbers. Since the analysis phase would have used too much CPU time if standard FE analysers had been used (Fig. 3a), the analysis was carried out in these examples by an *FE simulator*. These programs use *analytically based* computer calculations for a very large number of prismatic beam elements, having relatively simple support conditions for the beam (Fig. 3c). *Since here we are testing an optimizer*, and the FE simulator has been shown to give the same results as a standard FE program (but much faster), *the above numerical experiments are fully justified*. Results using 100000 elements were documented and compared with analytical solutions (Rozvany and Zhou *et al.* 1989, 1990; for a review see Zhou and Rozvany 1993a). Moreover, isolated beam examples with 1000000 elements were completed. The beam examples with 100000 elements required about 5 minutes CPU time on a relatively slow personal computer (in 1989). The above experiments, based on iterative COC, demonstrate the extremely high capability of the new OC methods. Admittedly, the beam examples considered are very well-conditioned for the approximation used (see comments in Section 4).

5.3 Modified 200-Bar Truss Problem

The standard 200-bar truss is shown in Fig. 4. The structure is subject to five load conditions as follows.

(1) A load of 1000 lbs in the positive X direction applied at nodes 1, 6, 15, 20, 29, 34, 43, 48, 57, 62, 71.

(2) A load of 1000 lbs in the negative X direction applied at nodes 5, 14, 19, 28, 33, 42, 47, 56, 61, 70, 75.

(3) A load of 10,000 lbs in the negative Y direction at nodes 1, 2, 3, 4, 5, 6, 8, 10, 12, 14, 15, 16, ..., 73, 74, 75.

(4) A combination of loading conditions 1 and 3.

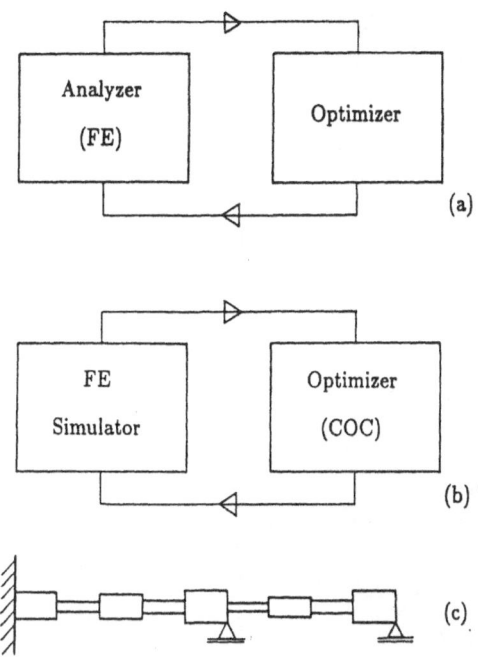

Fig. 3. The role of FE simulators in testing the capability of an optimizer.

(5) A combination of loading conditions 2 and 3.

Young's modulus for all members is $E = 3.0 \cdot 10^7$ psi, and the specific weight $\rho = 0.283$ lbs/in^3. The allowable displacements in this modified problem are for all nodes increased to ± 1.0 in in both X and Y directions (instead of the original ± 0.5 in). For elements of the two columns between nodes 2 and 76, as well as between nodes 4 and 77 the permissible stress is increased to $\sigma_a = 20000$ psi, for the other elements it remains $\sigma_a = 10000$ psi. The aims of the modifications of design conditions were as follows.

- Due to the higher value of allowable displacements, a much larger number of elements are stress-controlled than in the original problem.

- The non-uniform permissible stresses resulted in a significantly higher weight (by 13%) in the DOC-FSD solution than in the DCOC and Dual solutions.

Due to the large number of a active stress constraints, the total CPU time requirement for Dual (1948 sec) was found much higher than that for DCOC (110 sec). The distribution of cross-sectional areas given by DCOC and DOC-FSD is shown in Figs. 5a and b, respectively. Significant differences in the column cross-sections can easily be observed.

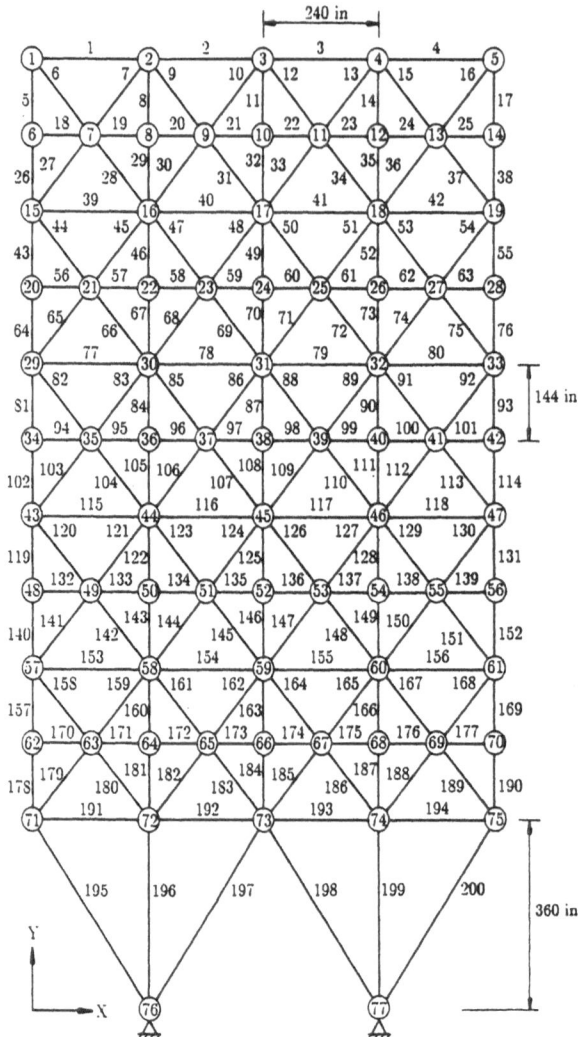

Fig. 4. The 200-bar truss.

6 Some Common Misconceptions about COC/DCOC and Comments by the Authors

In this section we mention certain criticisms of the proposed methods, together with arguments showing that they are not justified.

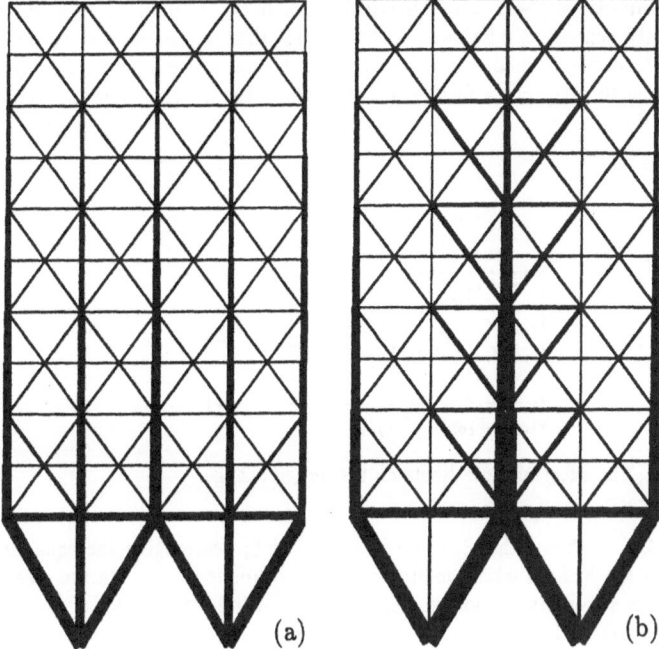

(a)

(b)

Fig. 5. Optimal solutions for the 200-bar truss: (a) by DCOC and (b) by DOC-FSD.

6.1 The New OC Methods are "Intuitive" and "Not Rigorous"

All optimality criteria used by DCOC were derived using the standard Kuhn-Tucker conditions for a local optimum in finite dimensional design space. Although the above conditions can break down in some special cases, this is *of no practical significance* in structural optimization.

The *update formulae* of DCOC are based on the Kuhn-Tucker conditions directly, whilst in other OC methods (DOC) the latter are modified to provide a control over the step size (see details below). As mentioned in Section 4, the update formulae take the values of the variables (real and adjoint forces, Lagrange multipliers, adjoint relative initial displacements, cross-sectional areas) from various past computational steps. The changes in values between two iterations depend on how strongly changes in the cross-sectional area affect the values of the internal forces. If the above dependence is very strong, the problem can become ill-conditioned and the procedure unstable, unless additional safeguards are built into the algorthim. One such method is to use a modified *step size* by multiplying the change in design $\{\Delta x\} = \{x_{new}\} - \{x_{old}\}$ by a given constant k with $k < 1$. The new modified design $\{x_{new\,mod}\}$ then becomes $\{x_{new\,mod}\} = \{x_{old}\} + k\{\Delta x\}$. The value of k is automatically determined if *move limits* are imposed on the design changes. If, for example, the move limits allow only a change of 0.5 times the old x-values, then the admissible new values are shown in Fig. 6 (shaded area) and $x_{\text{new mod}}$ lies on the boundary of the latter. It is usual to carry out a further optimization of the design *within the shaded area* in Fig. 6.

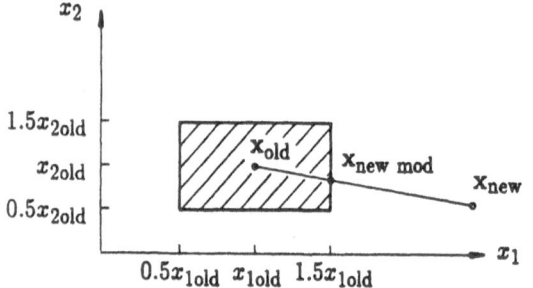

Fig. 6. The effect of move limits on the design change.

In DOC methods, the step size is controlled by rearranging the equations given by the Kuhn-Tucker conditions in some heuristic fashion. In the case of displacement constraints, for example, the original Kuhn-Tucker equation in (4) can be rearranged as

$$\frac{f^e \overline{f}^e}{E^e \rho^e (x^e)^2} = 1 \,, \tag{10}$$

and then we can raise both sides to the $(1/r)$-th power

$$\left[\frac{f^e \overline{f}^e}{E^e \rho^e (x^e)^2} \right]^{1/r} = 1^{1/r} = 1 \,. \tag{11}$$

Finally, both sides are multiplied by x^e, to get

$$x^e = x^e \left[\frac{f^e \overline{f}^e}{E^e \rho^e (x^e)^2} \right]^{1/r} \,. \tag{12}$$

This then provides the usual DOC redesign formula (e.g. Berke 1970)

$$x_{\text{new}}^e = \left\{ x^e \left[\frac{f^e \overline{f}^e}{E^e \rho^e (x^e)^2} \right]^{1/r} \right\}_{\text{new}} \,. \tag{13}$$

The parameter r, termed "step size parameter", will control the change in design. Note that the changes carried out in (11) and (12) were quite arbitrary, based on some intuitive insight. In DCOC, methods are currently developed to determine the optimal step size *systematically*. One possibility for doing this is a line search with a quadratic approximation.

6.2 "The New OC Methods Can Only Be Used for Some Simple, Idealized Problems and Structures"

Current extensions of DCOC include (e.g. Rozvany and Zhou 1994)

- stress constraints,
- several load conditions,
- several displacement constraints per load condition,
- elastic supports,
- given support settlements,
- contact problems,
- allowance for the cost of reactions,
- variable loads,
- variable prestrains,
- variable support settlements,
- allowance for the cost of variable loads,
- allowance for the cost of variable prestrains,
- allowance for the cost of variable support settlements,
- allowance for selfweight,
- allowance for member buckling,
- system stability constraints,
- natural frequency constraints,
- material nonlinearity.

Extensions to geometrical nonlinearity and to probabilistic design (which is subject of this meeting) are being developed currently.

As regards the type of structures used in test examples, DCOC has been applied to trusses, beams, frames, grillages, disks and plates. Extensions to shells and other structures are straightforward. In view of the above, the claim in the above subtitle cannot be sustained.

In fact, in some recent research papers by other authors, one of the topics listed above is discussed, without realizing that with the help of DCOC the treatment of the considered problem would nowadays become rather a student exercise than a piece of original research.

6.3 "A Test Example with One Million Elements is Absurd"

It was explained in Section 5.2 that the aim of the above example was to test the upper limits of the capability of DCOC as an optimizer. It is completely immaterial, whether we use some general purpose FE software or an FE simulator as analyser. Admittedly the beam examples used in the considered studies are well-conditioned from the point of view of DCOC approximations (see Section 4). However, a suitable control of step sizes should assure convergence even in ill-conditioned problems.

6.4 "DCOC is Just Another Fully Stressed Design Method"

Although the FSD updating formula is used for the design variables of stress-controlled cross-sections, the *exact Kuhn-Tucker conditions are enforced* in COC/DCOC by applying relative displacements (fictitious prestrains) to those elements of the adjoint structure, for which stress constraints are active.

7 Aims, Significance and Fields of Application of Topology Optimization

Topology of a structural system means the spatial sequence or configuration of members and joints or internal boundaries. The two main fields of application of topology optimization are layout optimization of grid-like structures and generalized shape optimization of continua or composites.

A *grid-type structure* has the basic feature that it consists of a system of intersecting members, the cross-sectional dimensions of which are small in comparison to their length, and hence the members can be idealized as one-dimensional continua. Consequences of this feature are that

- the influence of member intersections on strength, stiffness and structural weight can be neglected, and

- the specific cost (e.g. structural weight per unit area or volume) can be expressed as the sum of the costs (weights) of the members running in various directions. Examples of grid-type structures are trusses, grillages (beam systems), shell-grids and cable nets.

Layout optimization of grid-type structures consists of three simultaneous operations, namely

- *topological optimization* involving the spatial sequence of members and joints,

- *geometrical optimization* involving the coordinates of joints, and

- *sizing*, i.e. optimization of cross-sectional dimensions.

The above concepts are explained on an example in Fig. 7, in which all three trusses have the same topology, whilst the trusses in Figs. 7b and c have the same geometry but different cross-sectional dimensions.

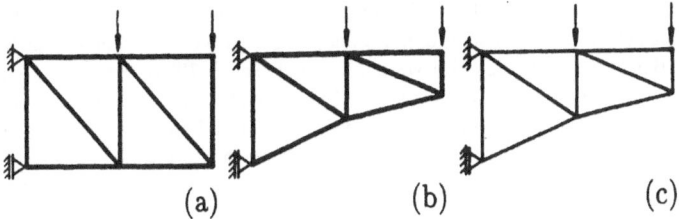

(a)   (b)   (c)

Fig. 7. Example illustrating topological, geometrical and cross-sectional properties of a grid-type structure.

Prager (Prager and Rozvany 1977b) regarded layout optimization as the *most challenging* class of problems in structural design because there exists an infinite number of possible topologies which are difficult to classify and quantify; moreover, at each point of the available space potential members may run in an infinite number of directions. At the same time, layout optimization is of *considerable practical importance*, because it results in much greater material savings than pure cross-section (sizing) optimization.

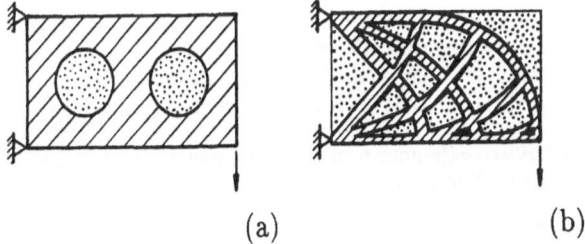

(a)   (b)

Fig. 8. Example illustrating generalized shape optimization.

The other important field of application of topology optimization is *generalized shape optimization*, in which a simultaneous optimization of both *topology and shape of boundaries* is required for continua and *of interfaces* between materials for composites. Fig. 8a, for example, shows the initial boundary shape and topology and Fig. 8b a hypothetical optimal shape and topology for a composite plate in plane stress, where the dotted regions denote the less stiff, weaker and lighter material. For a cellular structure (here: perforated plate), the dotted regions denote cavities (or "holes").

In the next two sections, a general description of a general theory of optimal layouts is given and its application to the derivation of exact analytical solutions is presented.

8 Basic Features of Optimal Layout Theory

The theory of optimal structural layouts has been covered extensively in the past, in particular in principal lectures at NATO meetings [Iowa 1980 (Rozvany 1981); Troia 1986 (Rozvany and Ong 1987); Edinburgh 1989 (Rozvany, Gollub and Zhou 1992); Berchtesgarden 1991 (Rozvany, Zhou and Gollub 1993); Sesimbra 1992 (Rozvany 1993)], and also in greater detail at a CISM course in Udine in 1990 (Rozvany 1992b) as well as in books (Rozvany 1976, 1989) and in book chapters (e.g. Rozvany 1984). Although the above exposure in the technical literature may appear excessive, the development in this field has also been both rapid and extensive, even during the last few months. For this reason, we can report a number of new developments in this paper.

Optimal layout theory, developed in the seventies by Prager and the first author (e.g. Prager and Rozvany 1977a and b) as a generalization of Michell's (1904) theory for trusses, was originally formulated for grid-like structures on the basis of the simplifying assumptions listed in Section 1. These simplifications were removed in a more advanced version of layout theory (e.g. Rozvany, Olhoff, Cheng and Taylor 1982; Rozvany, Olhoff, Bendsøe et al. 1987), which was applied to structures in which a high proportion of the available space was occupied by material. One could also say that the original, "classical" layout theory is concerned with structures having a low "volume fraction" (i.e. material volume/available volume ratio), whilst advanced layout theory deals with structures having a high volume fraction. In the latter, first the microstructure is optimized for a given ratio of the stiffnesses or forces in the principal directions and in a second operation, the optimal macroscopic distribution of microstructures is determined using methods of the layout theory.

The classical layout theory is based on *two underlying concepts*, namely

- the *structural universe* (in numerical methods: ground structure), which is the union of all potential members or elements, and

- *continuum-type optimality criteria* (COC), expressed in terms of a fictitious system termed *adjoint structure*, which were discussed in detail in Sections 1 to 6).

Since the above optimality criteria also provide adjoint strains for vanishing members, their fulfilment for the entire structural universe represents a necessary and sufficient condition of layout optimality if the problem is convex and certain additional requirements (e.g. existence) are satisfied. The above condition of convexity is fulfilled for certain so-called "self-adjoint" problems (e.g. Bendsøe and Haber 1993), the analytical treatment of which will be discussed in Section 9.4.

On the basis of optimal layout theory, two basic approaches have been developed using, respectively,

- *analytical methods* for deriving *closed form continuum-type solutions* representing the *exact optimal layout*, and

- *numerical, discretized iterative methods* for deriving *approximate* (but usually highly accurate) *optimal layouts*.

Fundamental differences between these two methods are listed in Table 2.

Table 2. A comparison of analytical and numerical methods based on layout theory

Computational Method	Analytical	Numerical
Structural Model	Continuum	Discretized (Finite Elements)
Procedure	Simultaneous Solution of All Equations	Iterative Solution
Structural Universe	Infinite Number of Members	Finite but Large Number of Members (Several Thousands)
Prescribed Minimum Cross-Sectional Area	Zero	Non-Zero but Small (10^{-8} to 10^{-12})

Layout optimization methods used by the numerical school (e.g. Kirsch 1989; Kirsch and Rozvany 1993) are usually based on the following two-stage procedure.

- First the topology is optimized for a given geometry (i.e. given coordinates of the joints), and then
- for this selected topology the geometry is optimized.

A drawback of this procedure is, of course, that for the new, optimized geometry the old topology may not be optimal any more. Until the introduction of COC/DCOC methods, however, the two-stage procedure was necessary because of the limited optimization capability of other methods, particularly for realistic problems with active stress constraints for a very large number of members in the structural universe.

The new optimality criteria methods (COC/DCOC), which are discussed in Sections 1–6, enable us to carry out a *simultaneous optimization of topology and geometry*, because the number of elements in the structural universe is either infinite (analytical methods) or very large (numerical COC/DCOC methods) and hence topological optimization achieves, in effect, also geometrical optimization.

In the next section, methods for deriving *exact analytical solutions* for structural layouts are presented, which are useful in checking the validity, accuracy and convergence of numerical methods.

9 Exact Analytical Solutions for Structural Layouts

9.1 Formulation and a Review of Proofs

A new unified formulation of the structural layout theory is presented in this section. It was mentioned in Table 2 that for layout problems we consider a structural universe with an infinite number of members, that is, with members running in all possible directions at all points of the available space. Moreover, any cross-sectional area can take on a zero value. The difference between this class of problems and cross-section optimization by DCOC is that in the latter case the number of elements is finite and a lower limit is imposed on cross-sectional areas.

The *proof* of the optimality conditions listed in this section can be based on *energy theorems* or *calculus of variations* but can also be derived from the *standard optimality conditions of DCOC*. In doing so, the following two limiting processes are employed:

- the lower limit on cross-sectional parameters tends to zero: $x_\downarrow^e \to 0$, and

- the number of members passing through a point of the available space tends to infinity.

It follows from the first of the above operations that for a zero cross-section we have an inequality in the optimality criterion (as in the case of $x^e = x_\downarrow^e$ in DCOC).

The second step above implies, due to the kinematic admissibility requirement of the adjoint strains in DCOC, that we can replace separate adjoint strains in bars running in all possible directions at a point with a single strain field (for plane trusses, for example, with a single plane strain field). Some mathematicians prefer avoiding the above transitions. For example, in a mathematically outstanding contribution, Strang and Kohn (1983) describe the "Michell truss" problem as minimizing the integral $\int \int (|\sigma_1| + |\sigma_2|)\, dx dy$ for a plane stress field. Michell's original problem would reduce to the Strang-Kohn problem after

- making the second transition above; and

- showing that the optimal member directions must coincide with the principal stress (and strain) directions.

The above steps are actually part of the standard optimal layout theory (e.g. Prager and Rozvany 1977a and b).

For all layout problems considered by the authors, the fundamental layout optimality criterion can be stated in terms of a *"criterion function"* ϕ^e for a given element e in the following form (Rozvany 1992a)

$$\phi^e = 1 \ \ (\text{for } A^e \neq 0), \quad \phi^e \leq 1 \ \ (\text{for } A^e = 0), \tag{14}$$

where A^e is the cross-sectional area of an element e. The element e as well as the criterion function f^e (and A^e) are location and direction dependent, that is,

$\phi^e = \phi^e(x, y, \theta)$ in the plane and $\phi^e = \phi^e(x, y, z, \theta, \eta)$ in 3D space where (x, y, z) are spatial coordinates and (θ, η) define the orientation of a line element.

9.2 Criterion Functions for Various Classes of Problems

For illustration purposes, we consider here *grid-type* structures with one design variable x^e and one nodal force f^e per element. Moreover, we assume that the element cost (weight) can be expressed as $c^e L^e x^e$, where c^e is a given constant and L^e is the length of the element, and the element stiffness takes the form $r^e x^e / L^e$, where r^e is a given constant. For *trusses*, for example, x^e is the cross-sectional area $(x^e := A^e)$, c^e is the specific weight $(c^e := \rho^e)$ and r^e is Young's modulus $(r^e := E^e)$. For *grillages* with beams of given depth d^e but variable width x^e, we have $A^e := x^e d^e$, $c^e := d^e \rho^e$, $r^e := E^e (d^e)^3 / 12$.

The *generalized strains* ε^e used below mean axial strain (usually ε^e) for trusses and curvatures $(\varepsilon^e := \kappa^e)$ for grillages. Moreover, the *generalized forces* f^e refer to member forces (previously f^e) in the case of trusses and to bending moments $(f^e := M^e)$ in the case of grillages.

Stress constraints can be formulated as $(|f^e|/a^e - x^e) \leq 0$, in which for trusses $a^e := \sigma^e$ and for grillages of given depth $a^e := \sigma^e (d^e)^2 / 6$, where σ^e is the permissible stress. For *plastic design*, $a^e = \sigma^e (d^e)^2 / 4$ and σ^e is the yield stress.

For various classes of *design conditions* the criterion functions are given below.
(a) *Structures with several load conditions* $(k = 1, \ldots, K)$ *and multiple displacement constraints* $(j = 1, \ldots, J_k)$. For this class of problems, the criterion function becomes (Rozvany 1992a)

$$\phi^e = (r^e/c^e) \sum_{k=1}^{K} (\varepsilon_k^e \bar{\varepsilon}_k^e), \tag{15}$$

with

$$\varepsilon_k^e = \frac{f_k^e}{r^e x^e}, \quad \bar{\varepsilon}_k^e = \frac{\sum_{j=1}^{J_k} \nu_{kj} \hat{f}_{kj}^e}{r^e x^e} = \frac{\bar{f}_k^e}{r^e x^e}, \tag{16}$$

where ε_k^e and $\bar{\varepsilon}_k^e$ are the real and adjoint strains for the element e and load condition k, f_k^e the corresponding generalized forces, $\hat{f}_{kj}^e$ is the virtual generalized force for load k and displacement j, $\bar{f}_k^e$ is the adjoint generalized force for the load and ν_{kj} the Lagrange multiplier for the load k and displacement constraint j. The value of ν_{kj} is nonzero only if the j-th displacement condition for the k-th load is active, that is, it is satisfied as an equality. For the optimal design variable x^e, we have from COC/DCOC (e.g. Rozvany 1992a; Rozvany and Zhou 1994):

$$x^e = \sqrt{\frac{\sum_k f_k^e \sum_j \nu_{kj} \hat{f}_{kj}^e}{r^e c^e}} = \sqrt{\frac{\sum_k f_k^e \bar{f}_k^e}{r^e c^e}}. \tag{17}$$

(b) *Structures with proportional displacement constraints.* These constraints restrict a weighted combination of displacements at the location and in the direction of point loads, with weighting factors proportional to the magnitude of the loads. For a single point load, this means constraints on a single displacement at and in the direction of that load. Proportional displacement constraints can be transformed into a *compliance constraint*, restricting the *total external work*.

Considering *structures with several load conditions* $(k = 1, \ldots, K)$ and a *proportional displacement constraint or compliance constraint* for each load condition, we have $f_k^e = \hat{f}_k^e$, $\overline{f}_k^e = \nu f_k^e$, $\overline{\varepsilon}_k^e = \nu_k \varepsilon_k^e$ and hence (15)–(17) reduce to

$$\phi^e = (r^e/c^e) \sum_{k=1}^{K} \nu_k (\varepsilon_k^e)^2 \,, \tag{18}$$

$$\varepsilon_k^e = \frac{f_k^e}{r^e x^e} \,, \quad \overline{\varepsilon}_k^e = \frac{\nu f_k^e}{r^e x^e} \,, \tag{19}$$

$$x^e = \sqrt{\frac{\sum_k \nu_k (f_k^e)^2}{r^e c^e}} \,. \tag{20}$$

(c) *Structures with one single proportional displacement constraint or compliance constraint (one load condition).* In this case, (18) and (20) reduce to

$$\phi^e = (r^e/c^e)\nu(\varepsilon^e)^2 \,, \tag{21}$$

$$x^e = \sqrt{\frac{\nu}{r^e c^e}} |f^e| \,. \tag{22}$$

(d) *Structures with several load conditions and stress constraints.* For this problem, we have

$$\phi^e = (a^e/c^e) \sum_{k=1}^{K} (\overline{\varepsilon}_k^e \; \mathrm{sgn} \; f_k^e) \,, \tag{23}$$

where the meaning of sgn f_k^e is sgn $f_k^e = 1$ for $f_k^e > 0$, sgn $f_k^e = -1$ for $f_k^e < 0$ and $(-1) \leq$ sgn $f_k^e \leq 1$ for $f_k^e = 0$. Since for $A^e = 0$ we have $f_k^e = 0$ (for all k) and (23) must be fulfilled for any values of the "sgn" functions, for $A^e = 0$ (14) and (23) reduce to

$$\phi^e = (a^e/c^e) \sum_{k=1}^{K} |\overline{\varepsilon}_k^e| \leq 1 \quad (\text{for } A^e = 0) \,. \tag{24}$$

A further optimality condition for the case of $A^e > 0$ is

$$\overline{\varepsilon}_k^e \neq 0 \quad \text{only if} \quad |f_k^e|/a^e = x^e \,, \tag{25}$$

or in other words, the adjoint strain is nonzero only if the stress constraint is active for the element (e) and load condition (k) involved.

Important Note. In the derivation of DCOC, only statical admissibility is assumed (Zhou and Rozvany 1992) and kinematic admissibility becomes an optimality criterion *if a displacement constraint is active.* Since in the considered class of problems no displacement constraint is involved, the optimality conditions in (14) with (23) represent an optimal solution only *if the strains ε_k^e in nonvanishing members are kinematically admissible.* This condition is always satisfied for statically determinate solutions, which is the case for a single load condition (ignoring non-unique optima) and sometimes for several load conditions. However, in *optimal plastic design* no kinematical admissibility of the real structure is required and hence a solution satisfying (14) and (23) always represents (in convex problems) the optimum.

(e) *Structures with one load condition and stress constraints.* In this case the criterion function becomes

$$\phi^e = (a^e/c^e)\bar{\varepsilon}^e \operatorname{sgn} f^e, \tag{26}$$

and then (14) and (26) imply

$$\bar{\varepsilon}^e = (c^e/a^e) \operatorname{sgn} f^e \ (\text{for } A \neq 0), \quad |\bar{\varepsilon}^e| \leq c^e/a^e \ (\text{for } A = 0). \tag{27}$$

For trusses with $c^e = \text{const.}$, $a^e := \sigma^e$ (27) reduces to the optimality conditions of Michell (1904). For grillages, (27) reduces to optimality conditions of Rozvany (1972; for a review see Prager and Rozvany 1977a).

9.3 General Procedure for Exact Layout Optimization

(a) Adopt continuum type strain fields ε_k^e and $\bar{\varepsilon}_k^e$ covering the entire available space and satisfying kinematic boundary and continuity conditions, such that
(b) the criterion function ϕ^e takes on a directionally maximum value of $\phi^e = 1$ at each point.
(c) Place members along lines with $\phi^e = 1$, calculate the member forces and dimension the members using optimality criteria for the cross-sectional parameters [e.g. for x^e in (17)].
(d) Check that the strains and adjoint strains in nonvanishing members (with $A^e \neq 0$) equal those given by the strain fields under (a).

Naturally, the above operations must be carried out simultaneously because all the conditions above must be fulfilled by the solution.

9.4. Self-Adjoint Problems

(a) One loading Condition. If we square both sides of (26) and adopt $(\phi^e)^2$ as a modified criterion function, we can see that structures of the type considered here (e.g. plane trusses, and grillages of given depth) have the *same optimal layout for one load condition and either a proportional displacement (or compliance constraint) or a stress constraint.*

Moreover, it follows from (27) that for one loading condition, *the optimal member directions must coincide with the principal directions of the adjoint strain field.*

This property was pointed out, in the context of trusses, already by Michell (1904) and was used by Hemp (1973) and Prager (1974) for deriving some *least-weight plane truss layouts for one load condition.* A more comprehensive treatment of the above class of problems was given by Rozvany and Gollub (1990), Lewiński, Zhou and Rozvany (1993, 1994) and Rozvany, Lewiński *et al.* (1993). For numerical examples, see Sections 11.1 and 11.2.

The *optimal layout of grillages for one load condition* has been discussed extensively in the literature (e.g. Rozvany 1972; Prager 1974; Prager and Rozvany 1977a; Rozvany 1976, 1981, 1984, 1989). The latest extension treated allowance for the cost of supports (Rozvany 1994).

(b) Several Loading Conditions. In the case of proportional displacement constraints or compliance constraints, optimal layouts have been determined for plane trusses with several load conditions (Rozvany, Zhou and Birker 1993; Rozvany, Birker and Lewiński 1994). It has been found that, even in the case of nonproportional displacement constraints, at any point of the optimal layout for plane trusses, *members may run in at most two directions,* which are *in general non-orthogonal.* A theory for *non-orthogonal, generalized Hencky-nets* for the layout of plane trusses with several load conditions has also been developed (Rozvany and Birker 1994). For an example with several load conditions see Section 11.3.

9.5 Non-Self-Adjoint Problems

The above result have been extended to problems with nonproportional displacement constraint, which are in general non-self-adjoint and non-convex (e.g. Rozvany, Sigmund, Lewiński *et al.* 1993). It was found that

- due to nonuniqueness, several solutions satisfying the optimality critria may need to be investigated;

- the global optimal solution may be nonstationary if too few displacement constraints are prescribed; and

- the solution given by the optimality criteria may represent a local maximum (and not a local minimum).

10 Generalized Shape Optimization

10.1 Historical Background and Problem Classification

As mentioned in the Introduction, the aim of generalized shape optimization is the simultaneous optimization of both *topology and shape* of the boundaries of two- or three-dimensional continua or of the interfaces between different materials in composites.

It was established by Kohn and Strang (1983) in the context of *plastic design* for torsion of a cross-section within a square area (Fig. 9) that generalized shape optimization may yield three types of regions in the solution, namely

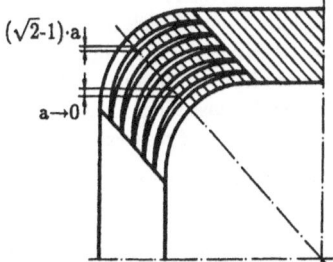

Fig. 9. One of the earliest solutions in generalized shape optimization (after Kohn and Strang 1983).

- solid regions (filled with material);
- empty regions (without material); and
- porous regions (some material, with cavities of infinitesimal size).

Considering elastic perforated plates in bending or in plane stress, it was found by various mathematicians (e.g. Lurie and Cherkaev 1984; Murat and Tartar 1985; Kohn and Strang 1986; Avellaneda 1987) that one optimal microstructure for a compliance constraint consists of *rank-2 laminates*, i.e. ribs of first and second order infinitesimal width in the two principal directions. An alternative, and more practical, microstructure was derived by Vigdergauz (1992).

10.2 Solid-Empty-Porous (SEP) Solutions

Analytical solutions based on rank-2 laminates for axisymmetric perforated plates, derived by Rozvany, Olhoff, Bendsøe *et al.* (1987) as well as Ong, Rozvany and Szeto (1988), have shown that

- a high proportion of the available space in exact optimal solutions consists of *porous regions*; and
- for low volume fractions (ratio of material volume to available volume) the solution tends to that for *least-weight trusses* (Michell 1904) for plane stress and *least-weight grillages* (e.g. Prager and Rozvany 1977a) for bending. This conclusion was also confirmed by Allaire and Kohn (1993).

Following the above developments, Bendsøe (e.g. 1989) opened up new avenues of research in topology optimization by investigating discretized optimal solutions derived by mathematical programming methods, using either the correct microstructure (rank-2 laminates) or a "sub-optimal" microstructure (square or rectangular holes). The above investigations have also made use of some important analytical relations regarding geometrical properties of optimal topologies (e.g. Pedersen 1989, 1990). Extensions of this work were reported by several investigators (e.g. Bendsøe and Kikuchi 1988; Suzuki and Kikuchi 1991; Díaz and Bendsøe 1992; Olhoff, Bendsøe and Rasmussen 1991; Allaire and Kohn 1993).

Numerical solutions using rank-2 laminates have a *considerable theoretical importance* because they represent (within a discretization error) an absolute limit on the structural weight for a class of compliance problems. However, other lines of research on topology optimization are justified, because solutions with rank-2 microstructures are somewhat *unpractical* for the following reasons:

- even an approximate version of rank-2 microstructures with finite rib widths would require *very high manufacturing costs*;

- rank-2 laminates in perforated structures (as distinct from composite structures) have a zero shear stiffness in one direction, which makes these designs completely *unstable* if the load direction is changed;

- solutions are only available for a *single compliance* or *natural frequency constraint* which are not realistic design problems. This was also demonstrated recently by Sankaranarayanan, Haftka *et al.* (1992).

Moreover, due to nonconvexity, the above solution could represent a *local optimum*.

Solutions with *sub-optimal microstructures* often turn out to be more practical, because they penalize and thereby suppress porous regions.

The solutions discussed in this subsection often appear in the literature under the term *homogenization*, which means that an inhomogeneous structural element, containing an infinite number of discontinuities in material or geometrical properties, is replaced by a homogeneous but anisotropic element, whose stiffness is direction but not location dependent within an element. The same idea was used in a very simple form much earlier by others (e.g. Michell 1904; Prager and Rozvany 1977b), introducing terms like "truss-like continua" or "grillage-like continua" (Prager 1974). To engineers this is a rather obvious idealization, but its rigorous mathematical treatment is much more involved (e.g. Bensousson *et al.* 1978). It would be incorrect, however, to make the term "generalized shape optimization" synonymous with "shape optimization by homogenization".

10.3 Solid-Empty (SE) Solutions and Solid, Isotropic Microstructure with Penalty (SIMP) for Intermediate Densities

The method described in this subsection has been used extensively by the authors, but it was also tried out earlier by Bendsøe (e.g. 1989), as an extension of a technique employed by Rossow and Taylor (1973).

From an engineering point of view, it is more useful to aim at solutions, in which porous regions are largely suppressed and then a second stage design procedure can produce a "practical" solution consisting of solid and empty regions only, with smooth boundaries to avoid stress concentrations. This procedure was lucidly demonstrated by Olhoff, Bendsøe and Rasmussen (1991) on an example involving a simply supported beam with a central point load.

It was suggested by the first author at a meeting in Karlsruhe in 1990 (Rozvany and Zhou 1991a; Zhou and Rozvany 1991) that porous regions could be suppressed by adding to the material costs the *cost of manufacturing* of holes, thereby penalizing porous regions. Once we decide that we want only solid and emply regions in the solution, *any* range of microstructures that includes the above two as limiting cases can be assumed in the solution process. We can therefore postulate that the specific material cost (e.g. weight) ϕ is proportional to the specific stiffness s of perforated regions (Fig. 10a). This would also be the case if we used a plate of variable thickness (Rossow and Taylor 1973), in which case we would have to penalize intermediate thicknesses. If the hypothetical microstructure contains some sort of holes, however, then the extra manufacturing cost would increase with the number or size of such holes, if we consider a casting or drilling process. However, for an empty region $(s = 0)$, the manufacturing costs also become zero. The corresponding specific fabrication cost is shown in Fig. 10b and the specific total cost in Fig. 10c, together with an approximation of the type

$$\phi = s^{1/n}, \tag{28}$$

where n is a given constant. The above relation is identical with that suggested by Bendsøe (1989, Eq. 7), although the first author was unaware of this when he proposed (28) above.

In selecting a hypothetical microstructure for obtaining a solid-empty (SE) solution, the following objectives should be considered:

- simplicity of analysis and optimization;

- selective suppression of porous regions by adjustable penalty; and

- capability of handling a variety of design conditions (e.g. combinations of stress, displacement, natural frequency and system stability constraints for several loading conditions).

It has been found that a *solid isotropic microstructure with* penalty (SIMP) for intermediate densities, as explained above (Fig. 10), largely fulfills these objectives (Rozvany, Zhou and Birker 1992; Rozvany, Zhou, Birker and Sigmund 1993; Rozvany, Zhou and Sigmund 1994).

In Fig. 11, three types of specific cost functions are compared for a plate in plane stress or bending. The straight line represents the normalized weight per unit area of a plate of variable thickness in plane stress or bending. The next curve shows the weight of a constant thickness perforated plate with rank-2 laminates having an equal stiffness in two directions (Rozvany, Olhoff, Bendsøe *et al.* 1987). Finally, the top curve represents the power type cost function of the SIMP formulation.

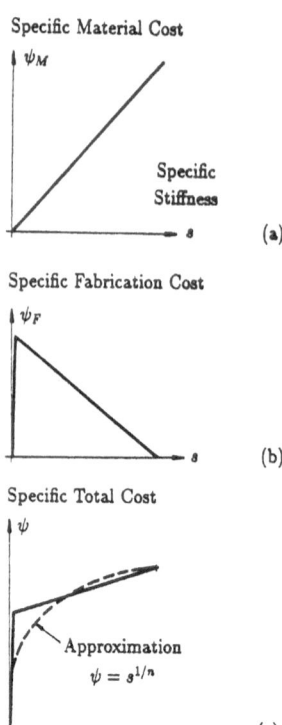

Specific Material Cost

ψ_M

Specific
Stiffness

s

(a)

Specific Fabrication Cost

ψ_F

s

(b)

Specific Total Cost

ψ

Approximation
$\psi = s^{1/n}$

s

(c)

Fig. 10. Suppression of porous regions in generalized shape optimization by taking fabrication costs into consideration.

et al. 1987). Finally, the top curve represents the power type cost function of the SIMP formulation.

When Bendsøe (1989) originally considered this power-type cost function, he used the term "direct approach" or "0-1 discrete optimization method with a suitable differentiable approximation" using an "artificial material". At the same time, Bendsøe expressed preference for optimal or sub-optimal anisotropic microstructures, reasoning that the solutions by the direct approach (now: SIMP) are (i) highly mesh-dependent and (ii) impossible to interpret physically. The authors are more optimistic in these respects, as can be seen from the following.

The SIMP model can be interpreted even for three-dimensional continua with stress and displacement constraints, assuming that *both Young's modulus and permissible stress* of a fictitious material (or a range of materials) *are proportional to* its (their) *density*, which varies between a given maximum value and a very small minimum value. It is then perfectly legitimate for the designer to penalize and thereby suppress intermediate densities, if he chooses to do so.

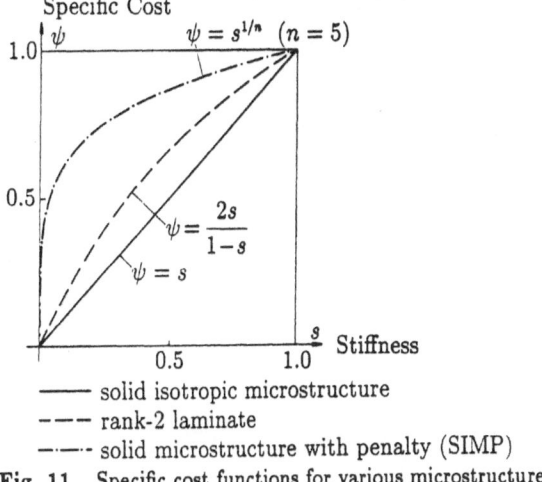

Fig. 11. Specific cost functions for various microstructures.

11 Comparisons of Exact Analytical Solutions for Optimal Truss Layouts with Numerical (DCOC) Solutions for Truss Layouts and with Numerical (DCOC/SIMP) Solutions for Perforated Plates under Plane Stress

A comparison of analytical and numerical solutions for optimal truss layouts is obviously justified. Moreover, it was shown through different methods by Rozvany, Olhoff, Bendsøe *et al.* (1987), Allaire and Kohn (1993) as well as Bendsøe and Haber (1993) that at low volume fractions the optimal topology of perforated plates under plane stress tends to that for least-weight trusses (Michell frames). For this reasons, solutions for the latter two types of problems should provide, through entirely different type of structures and solution procedure, an independent mutual confirmation of the methods presented.

11.1 Shallow Cantilever with a Point Load

Figure 12 shows the optimal solutions for the above problem using (a) a numerical (DCOC/SIMP) solution for a perforated plate; (b) an exact, closed form analytical solution for a truss; and (c) a numerical (DCOC) solution for a truss (Lewiński, Zhou and Rozvany 1994). The solution in 12a was based on 5400 constant-strain triangular elements. The analytical solution in Fig. 12b was obtained by using modified Bessel and Lommel functions. Finally, the numerical truss solution in Fig. 12c was based on a structural universe with 7204 truss elements. The close similarity of the three solutions is obvious. Moreover, the structural weight given by the solutions in Figs. 12b and c differed by less than 1 per cent.

320

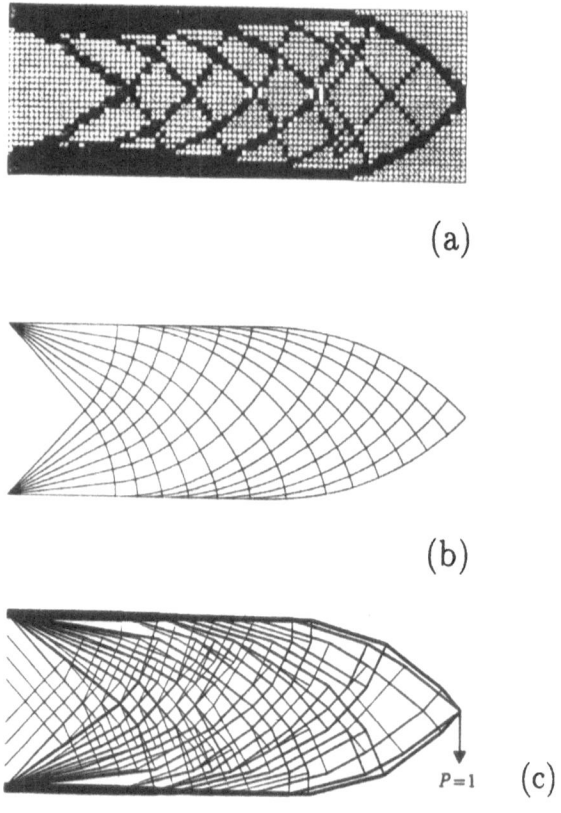

Fig. 12. Solutions for a shallow cantilever: (a) perforated plate (DCOC/SIMP); (b) analytical truss solution; (c) numerical truss solution (DCOC).

11.2 Rectangular Domain with Rotational Restraint at Both Ends and a Point Load

For this problem, the analytical truss solution, perforated plate solution and numerical truss solution are shown, respectively, in Figs. 13a, b and c (Rozvany, Lewiński, Gerdes and Birker 1993). Moreover, the solution in Fig. 13b was used for a second stage optimization, in which a proportional displacement constraint and stress constraints were imposed. The initial design and optimal design for the latter optimization process are shown in Fig. 14a and b. For a comparison, solutions with two circular holes of variable radius are given in Figs 14c (initial design) and Fig. 14d (optimal design). It can be seen that the weight of the

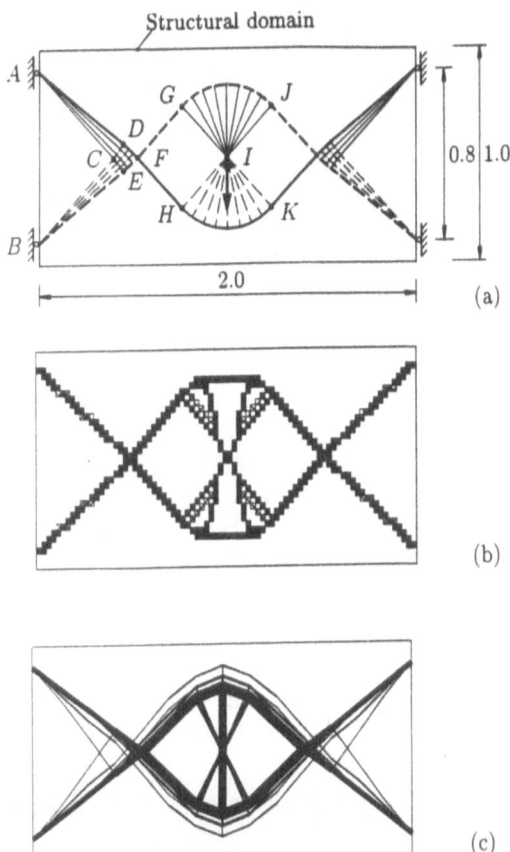

Fig. 13. Rectangular domain with built-in ends: (a) analytical truss solution; (b) numerical plate solution and (c) numerical truss solution.

optimal two-hole solution is about 4.5 times that for the optimal topology.

11.3 Vertical Straight Support with Two Alternative Point Loads

The analytical solution for various load orientations in the above problem is shown in Fig. 15. The numerical truss solution for load orientations of ±5°, using the structural universe in Fig. 16a, is given in Fig. 16b. The weight difference between analytical and numerical solutions is only 0.007%. Finally, solutions for a perforated plate (by DCOC/SIMP) for load orientations of ±10° and ±45° are given in Fig. 17b and c, together with the 2450 square elements used (17a). The weighted combination of the bar directions in Fig. 16b and the results in Figs.

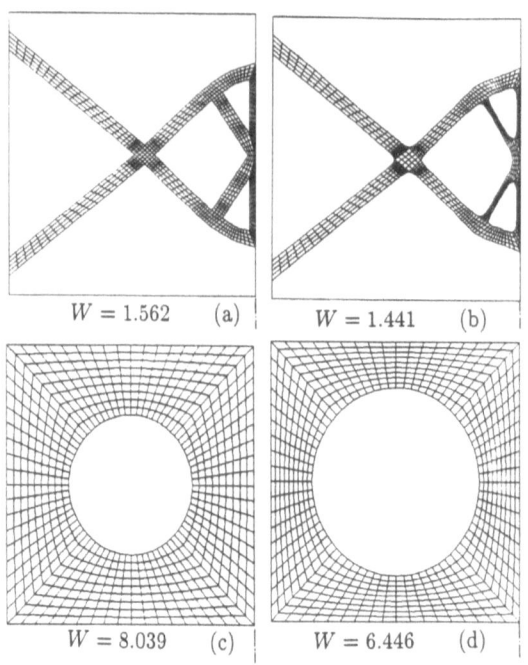

$W = 1.562$ (a) $W = 1.441$ (b)

$W = 8.039$ (c) $W = 6.446$ (d)

Fig. 14. (a,b) Second stage optimization based on optimal topology; (c,d) comparison solution with circular holes of optimized diameter.

17 b and c show an excellent agreement with the analytical solution (Fig. 15).

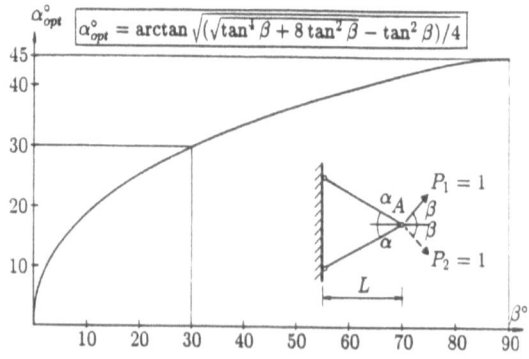

Fig. 15. Straight support with two alternate point loads: analytical solution.

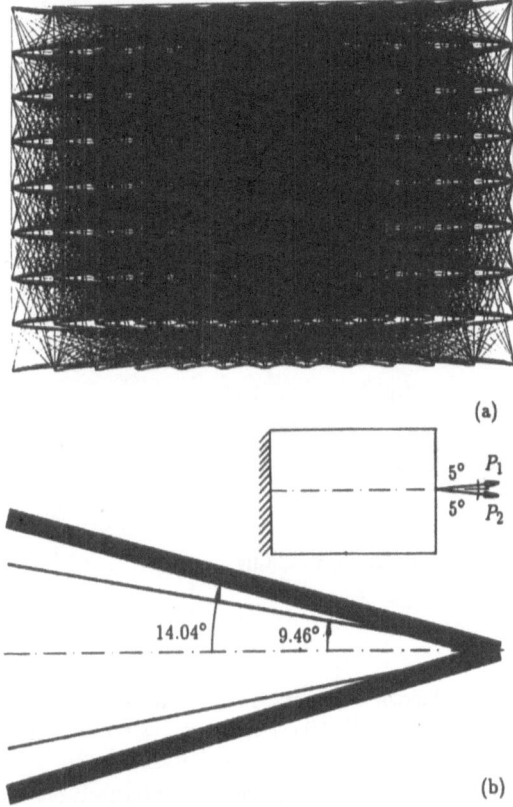

Fig. 16. Numerical truss solution by DCOC for load orientations of ±5°.

12 Concluding Remarks

The results presented in this paper demonstrate the power and versatility of the proposed new optimality criteria methods in both sizing and topology optimization. For the latter, recently developed exact analytical truss solutions, numerical truss solutions by DCOC and generalized shape optimization of perforated plates by the DCOC/SIMP procedure provided a mutual and independent confirmation of the solutions and the correctness of the methods used.

13 Acknowledgements

The authors are indebted to the German Research Foundation (DFG) for financial support (Project No. Ro 744/4-1), to Sabine Liebermann (text processing),

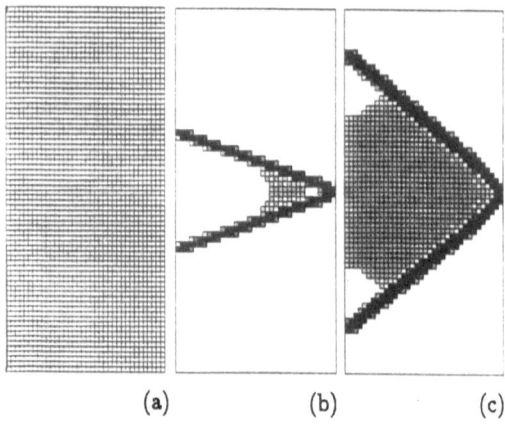

(a)   (b)   (c)

Fig. 17. Numerical perforated plate solutions by DCOC/SIMP for load orientations of $\pm 10°$ and $\pm 45°$.

Peter Moche (drafting) and Susann Rozvany (editing).

14 References

Allaire, G.; Kohn, R.V. 1993: Topology optimization and optimal shape design using homogenization. *Proc. NATO ARW "Topology design of structures"* (held in Sesimbra, Portugal), pp. 207–218. Kluwer, Dordrecht

Avellaneda, M. 1987: Optimal bounds and microgeometries for elastic two-phase composites. *SIAM. J. Appl. Math.* **47**, 1216–1228.

Bendsøe, M.P. 1989: Optimal shape design as a material distribution problem. *Struct. Optim.* **1**, 193–202.

Bendsøe, M.P.; Haber, R.B. 1993: The Michell layout problem as a low volume fraction limit of the perforated plate topology optimization problem: an asymptotic study. *Struct. Optim.* **6**, 263–267.

Bendsøe, M.P.; Kikuchi, N. 1988: Generating optimal topologies in structural design using a homogenization method. *Comp. Meth. Appl. Mech. Eng.* **71**, 197–224.

Bensousson, A.; Lions, J.-L.; Papanicolaou, G. 1978: *Asymptotic Analysis for Periodic Structures*. North Holland, Amsterdam.

Berke, L. 1970: An efficient approach in the minimum weight design of deflection limited structures. *AFFDL-TM-70-FDTR*.

Díaz, A.R.; Bendsøe, M.P. 1992: Shape optimization of multipurpose structures by a homogenization method. *Struct. Optim.* **4**, 17–22.

Hemp, W.S. 1973: *Optimum Structures*. Clarendon, Oxford.

Kirsch, U. 1989: Optimal topologies of structures. *Appl. Mech. Rev.* **42**, 223-238.

Kirsch, U.; Rozvany, G.I.N. 1993: Design considerations in the optimization of structural topologies. In: Rozvany, G.I.N. (Ed.) *Optimization of Large Structural Systems* (Proc. NATO ASI held in Berchtesgaden, 1991), pp. 121-141, Kluwer, Dordrecht.

Kohn, R.V.; Strang, G. 1983: Optimal design for torsional rigidty. In: Atluri, S.N.; Gallagher, R.H.; Zienkiewicz, O.C. (Eds.) *Hybrid and mixed finite element methods*, pp. 281-288. Wiley & Sons, Chichester.

Kohn, R.V.; Strang, G. 1986: Optimal design and relaxation of variational problems, I, II, and III. *Comm. Pure Appl. Math.* **39**, 113-137, 139-182, 353-377.

Lewiński, T.; Zhou, M.; Rozvany, G.I.N. 1993: Exact least-weight truss layouts for rectangular domains with various support conditions. *Struct. Optim.* **6**, 65-67.

Lewiński, T.; Zhou, M.; Rozvany, G.I.N. 1994: Extended exact solutions for least-weight truss layouts - Part I: Cantilever with a horizontal axis of symmetry - Part II: Unsymmetric cantilevers. *Int. J. Mech. Sci.* (accepted).

Lurie, K.A.; Cherkaev, A.V. 1984; G-closure for some particular set of admissible material characteristics for the problem of bending of thin elastic plates. *JOTA*, **42**, 305-316.

Michell, A.G.M. 1904: The limits of economy of material in frame-structures. *Phil. Mag.* **8**, 589-597.

Murat, F.; Tartar, L. 1985: Calcul des variations et homogénéisation. In: *Les méthodes de l'homogénéisation: Théorie et applications en physique*, pp. 319-370. Eyrolles, Paris: Coll. de la Dir. des Etudes et Recherches de Elec. de France.

Olhoff, N. 1981: Optimization of columns against buckling. Optimization of transversely vibrating beams and rotating shafts. In: Haug, E.J.; Cea, J. (Eds.) *Optimization of Distributed Parameter Structures* (Proc. NATO ASI, Iowa 1980), pp. 152-176, 177-199, Sijthoff and Noordhoff, Alphen aan der Rijn.

Olhoff, N.; Bendsøe, M.P.; Rasmussen, J. 1991: On CAD-integrated structural topology and design optimization. *Comp. Meth. Appl. Mech. Eng.* **89**, 259-279.

Ong, T.G.; Rozvany, G.I.N.; Szeto, W.T. 1988: Least-weight design of perforated elastic plates for given compliance: non-zero Poisson's ratio. *Comp. Meth. Appl. Mech. Eng.* **66**, 301-322.

Pedersen, P. 1989: On optimal orientation of orthotropic materials. *Struct. Optim.* **1**, 101-106.

Pedersen, P. 1990: Bounds on elastic energy in solids of orthotropic materials. *Struct. Optim.* **2**, 55-63.

Prager, W. 1974: *Introduction to Structural Optimization.* (Course held in Int. Centre for Mech. Sci. Udine. CISM **212**). Springer-Verlag, Vienna.

Prager, W.; Rozvany, G.I.N. 1977a: Optimal layout of grillages. *J. Struct. Mech.* **5**, 1-18.

Prager, W.; Rozvany, G.I.N. 1977b: Optimization of structural geometry. In: Bednarek, A.R.; Cesari, L. (Eds.) *Dynamical Systems*, pp. 265–293. Academic Press, New York.

Prager, W.; Shield, R.T. 1967: A general theory of optimal plastic design. *J. Appl. Mech.* **34**, 184–186.

Prager, W.; Taylor, J. 1968: Problems of optimal structural design. *J. Appl. Mech.* **5**, 102–106.

Rossow, M.P.; Taylor, J.E. 1973: A finite element method for the optimal design of variable thickness sheets. *AIAA J.* **11**, 1566–1569.

Rozvany, G.I.N. 1972: Grillages of maximum strenght and maximum stiffness. *Int. J. Mech. Sci.* **14**, 651–666.

Rozvany, G.I.N. 1976: *Optimal Design of Flexural Systems*. Pergamon Press, Oxford. Russian translation: Stroiizdat, Moscow, 1980.

Rozvany, G.I.N. 1981: Optimality criteria for grids, shells and arches. In: Haug E.J.; Cea, J. (Eds.) *Optimization of Distributed Parameter Structures* (Proc. NATO ASI held in Iowa City, 1980), pp. 112–151. Sijthoff and Noordhoff, Alphen aan der Rijn, The Netherlands.

Rozvany, G.I.N. 1984: Structural layout theory: the present state of knowledge. In: Atrek, E.; Gallagher, R.H.; Ragsdell, K.M.; Zienkiewicz, O.C. (Eds.) *New Directions in Optimum Structural Design*, pp. 167–195. Wiley & Sons, Chichester, England.

Rozvany, G.I.N. 1989: *Structural Design via Optimality Criteria*, Kluwer, Dordrecht.

Rozvany, G.I.N.; 1992a: Optimal layout theory: analytical solutions for elastic structures with several deflection constraints and load conditions. *Struct. Optim.* **4**, 247–249.

Rozvany, G.I.N. 1992b: Optimal Layout Theory (Chapter 6, see also Chapter 7-10). In: Rozvany, G.I.N. (Ed.) *Shape and Layout Optimization of Structural Systems and Optimality Criteria Methods* (CISM Course held in Udine 1990), pp. 75–163, Springer-Verlag, Vienna.

Rozvany, G.I.N. 1993: Layout theory for grid-type structures. In: Bendsøe, M.P.; Mota Soares, C.A. *Topology Design of Structures* (Proc. NATO ARW. Sesimbra 1992), pp. 251–272. Kluwer, Dordrecht

Rozvany, G.I.N. 1994: Optimal layout theory – allowance for the cost of supports and optimization of support locations. *Mech. Struct. Mach.* (proofs returned).

Rozvany, G.I.N.; Birker, T. 1994: Optimal truss layouts for multiple load conditions via generalized Hencky nets. *Struct. Optim.* (submitted).

Rozvany, G.I.N.; Birker, T.; Lewiński, T. 1994: Some unexpected properties of exact least-weight plane truss layouts with displacement constraints for several alternate loads. *Struct. Optim.* **7**, (scheduled for No. 1)

Rozvany, G.I.N.; Gollub, W. 1990: Michell layouts for various combinations of line supports, Part I. *Int. J. Mech. Sci.* **32**, 12, 1021–1043.

Rozvany, G.I.N.; Gollub, W.; Zhou, M. 1992: Layout optimization in structural design. In: Topping, B.H.V. (Ed.) *Proc. NATO ASI, Optimization and Decision Support Systems in Civil Engeneering*, held 25 June – 7 July 1989, Edinburgh. Kluwer, Dordrecht.

Rozvany, G.I.N.; Lewiński, T.; Gerdes, D.; Birker, T. 1993: Optimal topology of trusses or perforated deep beams with rotational restraints at both ends. *Struct. Optim.* 5, 268–270.

Rozvany, G.I.N.; Olhoff, N.; Bendsøe, M.P.; Ong, T.G.; Sandler, R.; Szeto, W.T. 1987: Least-weight design of perforated elastic plates I, II. *Int. J. Solids Struct.* 23, 521–536, 537–550.

Rozvany, G.I.N.; Olhoff, N.; Cheng, K.-T.; Taylor, J.E. 1982: On the solid plate paradox in structural optimization. *J. Struct. Mech.* 10, 1–32.

Rozvany, G.I.N.; Ong, T.G. 1987: Minimum-weight plate design via Prager's layout theory (Prager memorial lecture). In: Mota Soares (Ed.) *Computer Aided Optimal Design: Structural and Mechanical Systems* (Proc. NATO ASI held in Troia, Portugal, 1986), pp. 165–179. Springer-Verlag, Berlin.

Rozvany, G.I.N.; Sigmund, O.; Lewiński, T.; Gerdes, D.; Birker, T. 1993: Exact optimal structural layouts for non-self-adjoint problems. *Struct. Optim.* 5, 204–206.

Rozvany, G.I.N.; Zhou, M. 1991a: Applications of the COC method in layout optimization. In: Eschenauer, H.; Matteck, C.; Olhoff, N. (Eds.) *Proc. Int. Conf. "Engineering Optimization in Design Processes"* (Karlsruhe 1990), pp. 59–70, Springer, Berlin.

Rozvany, G.I.N.; Zhou, M. 1991b: The COC algorithm, Part I: cross-section optimization or sizing (presented 2nd World Congr. Comp. Mech., Stuttgart, 1990). *Comp. Meth. Appl. Mech. Eng.* 89, 281–308.

Rozvany, G.I.N.; Zhou, M. 1994: Optimality Criteria Methods for Large Structural Systems. Chaper 2 in: Adeli, H. (Ed.) *Advances in Design Optimization.* Chapman and Hall, London.

Rozvany, G.I.N.; Zhou, M., Birker, T. 1992: Generalized shape optimization without homogenization. *Struct. Optim.* 4, 250–252.

Rozvany, G.I.N.; Zhou, M., Birker, T. 1993: Why multi-load design based on orthogonal microstructures are in general non-optimal. *Struct. Optim.* 6, 200–204.

Rozvany, G.I.N.; Zhou, M., Birker, T.; Sigmund, O. 1993: Topology optimization using iterative continuum-type optimality criteria (COC) methods for discretized systems. In: Bendsøe, M.P.; Mota Soares, C.A. *Topology Design of Structures. Proc. NATO ARW.* (held in Sesimbra 1992)

Rozvany, G.I.N.; Zhou, M.; Gollub, 1990: Continuum-type optimality criteria methods for large finite element systems with a displacement constraint – Part II. *Struct. Optim.* 2, 77–104.

Rozvany, G.I.N.; Zhou, M.; Gollub, W. 1993: Layout optimization by COC methods: analytical solutions. In: Rozvany, G.I.N. (Ed.) *Optimization of Large*

Structural Systems (Proc. NATO ASI held in Berchtesgaden, 1991), pp. 77–102, Kluwer, Dordrecht.

Rozvany, G.I.N.; Zhou, M.; Rotthaus, M.; Gollub, W.; Spengemann, F. 1989: Continuum-type optimality criteria methods for large systems with a displacement constraint – Part I. *Struct. Optim.* **1**, 47–72.

Rozvany, G.I.N.; Zhou, M.; Sigmund, O. 1994: Optimization of topology. Chapter 10 in Adeli, H. (Ed.) *Advances in Design Optimization*, Chapman and Hall, London.

Sankaranarayanan, S.; Haftka, R.T.; Kapania, R.K. 1992: Truss topology optimization with simultaneous analysis and design. *Proc. 33rd AIAA/ASME/ ASCE/AHS/ASC Struct. Dyn. Mat. Conf.* (held in Dallas), pp. 2576–2585. AIAA, Washington DC.

Strang, G.; Kohn, R.V. 1983: Hencky-Prandtl nets and constrained Michell trusses. *Comp. Meth. Appl. Mech. Eng.* **36**, 207–222

Suzuki, K.; Kikuchi, N. 1991: A homogenization method for shape and topology optimization. *Comp. Meth. Appl. Mech. Eng.* **93**, 291–318.

Venkayya, V.B.; Khot, N.S.; Berke, L. 1973: Application of optimality criteria approaches to automated design of large practical structures. *AGARD-CP-123*, pp. 3.1–3.9.

Vigdergauz, S. 1992: Two-dimensional grained composites of extreme rigidity. *18th Int. Congr. Theor. Appl. Mech.* (held in Haifa).

Zhou, M.; Haftka, R.T. 1994: A comparison study of optimality criteria methods. *Proc. 35th AIAA/ASME/ASCE/AHS/ASCSDM Conf.* (held in South Carolina, April 1994) (to appear).

Zhou, M.; Rozvany, G.I.N. 1991: The COC algorithm, Part II: topological, geometrical and generalized shape optimization. *Comp. Meth. Appl. Eng.* **89**, 309–336.

Zhou, M.; Rozvany, G.I.N. 1992: DCOC: an optimality critria method for large systems, Part I: theory. *Struct. Optim.* **5**, 12–25.

Zhou, M.; Rozvany, G.I.N. 1993a: Iterative COC methods– Parts I and II. In: Rozvany, G.I.N. (Ed.) *Optimization of Large Structural Systems* (Proc. NATO/DFG ASI, Berchtesgaden 1991), pp. 27–75, Kluwer, Dordrecht.

Zhou, M.; Rozvany, G.I.N. 1993b: DCOC: an optimality critria method for large systems, Part II: algorithm. *Struct. Optim.* **6**, 250–262.

Stochastic Optimization Models for Machining Operations

T. Szántai[1], I. Mészáros[2] and J. Völgyi[3]

[1] Technical University of Budapest, Department of Mathematics

[2] Technical University of Budapest, Department of Production Engineering

[3] Technical University of Budapest, Department of Machine Elements,
Institute of Machine Design

Abstract. In the paper the deterministic optimization models are summarized first. In these models the main decision variables are the *feed, speed* and *depth of cut*. Then we treat tool life as a random parameter and select a probability model that defines the tool life variations. Finally stochastic optimization models are defined for determining machining operations.

Keywords. Machining operations, Taylor's equation, probabilistic tool life models, optimization.

1. A deterministic optimization model

We are dealing with cutting by machine tool (*turnery*) but remark that the models introduced are applicable in the case of other metal workings (*drilling, milling* etc.), too. Cutting by machine tool has three main parameters which can be adjusted independently:

a – *depth of cut* (in mm),

s – *feed rate* (in mm per revolution),

v – *cutting speed* (in m per minute).

The main problem is that when one increases the above parameter values the main machine time and the different costs related to that decrease but the wear of the tool and the different losses related to that increase. The depth of cut used to be fixed in the optimization models and instead of the cutting speed v the revolution per minute n used to be applied.

Two important deterministic formulae:

Taylor formula for the tool life ([7]):

$$T = \frac{C_t K_t}{a^{x_t} s^{y_t} v^{z_t}} \ ,$$

where

C_t constant, depending on the material of the work-piece,

K_t constant, depending on the quality of the tool and on the other conditions of working,

x_t, y_t, z_t exponents depending on the conditions of working ($x_t \approx 3 - 5$, $y_t \approx 1$, $z_t \approx 1$).

Formula for the main cutting power:

$$F = C_F a^{x_F} s^{y_F} v^{z_F},$$

where

C_F constant according to the work-piece and the tool,

x_F, y_F, z_F exponents depending on the work-piece and the conditions of working ($z_F < 0$ usually).

We remark that in the above formulae the cutting speed v can be replaced by the revolution per minute n according to the connection

$$\frac{1}{v} = \frac{1000}{d\pi n} \ ,$$

where d denotes the workpiece diameter.

Figure 1.1 illustrates the feasible domain of the deterministic optimization model in logarithmic scale. The notation of the lines are the following:

    ① – constraints according to revolution per minute,

    ② – constraints according to feed rate,

    ③ – constraint on the main cutting power,

    ④ – constraint on the time life (lower bound).

The objective functions may be the following:

    – find the minimal cost,

- find the maximal productivity,
- find the minimal main operating time,
- find the maximal profit.

The objective function may be nonlinear but the optimization problem can be solved with very simple tools.

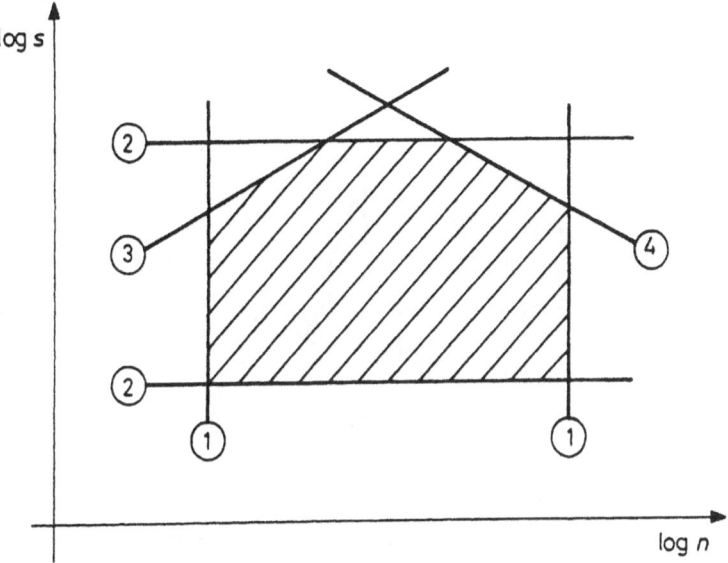

Fig. 1.1. The feasible domain of the deterministic optimization model in logarithmic scale.

We remark that without using the logarithmic scale the problem is a special nonlinear programming problem called geometric programming problem. But as it was shown by J. DUPACOVA in the paper [2] the random variables of the problem may not be statistically independent. Therefore the known results from the field of stochastic geometric programming, cf. M. AVRIEL and D. J. WILDE [1], R. JAGANNATHAN [3] cannot be applied in their original form. J. DUPACOVA formulates the metal cutting problem with the *Taylor's equation* considered as a regression relationship and with the coefficients obtained as statistical estimates of the true values. Accordingly, the optimal cutting conditions and the optimal value of the objective are estimates of the true ones.

2. Probability models of the tool life variations

Normal probability distribution

It is applied when the tool quality is high and stabil and the technological procedure is not disturbed by outer effects.

The Figure 2.1 shows the tool life frequencies of a machining operation whose distribution can be approached by normal probability distribution.

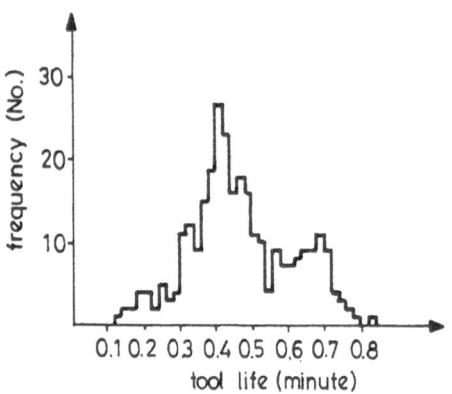

Fig. 2.1. The tool life frequencies of side turnery without cooling ([8]).

Example for the normal probability distribution approach.

Exponential probability distribution

It is applied when the failure occurs suddenly and it is independent of the working time and the actual condition of the element.

The Figure 2.2 shows the tool life frequencies of a machining operation whose distribution can be approached by exponential probability distribution.

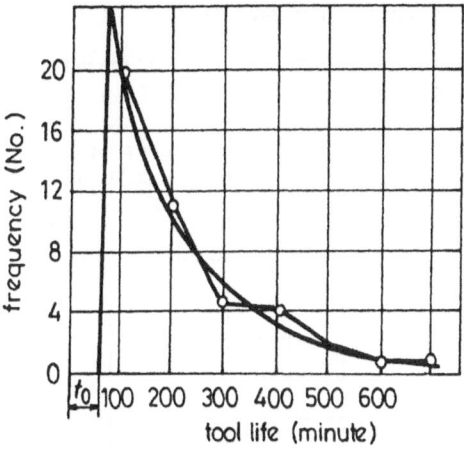

Fig. 2.2. The tool life frequencies of drill with diameter 20 millimetre ([4]).
Example for the exponential probability distribution approach.

Weibull probability distribution

It is applied when in addition of the unexpected, sudden failures some time
dependent effects are also taken into account.

The Figure 2.3 shows the tool life frequencies of a machining operation whose
distribution can be approached by Weibull probability distribution.

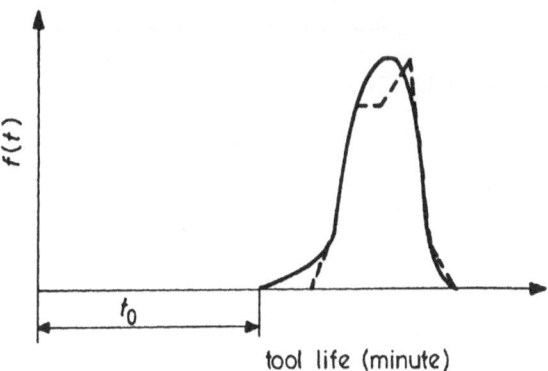

Fig. 2.3. The tool life frequencies of spiral milling ([5]). Example for the
normal probability distribution approach.

Gamma probability distribution

It models the accumulated failures supposing that the initial quality of the tools is homogeneous and the wear is uniform.

The Figure 2.4 shows the tool life frequencies of a machining operation whose distribution can be approached by gamma probability distribution.

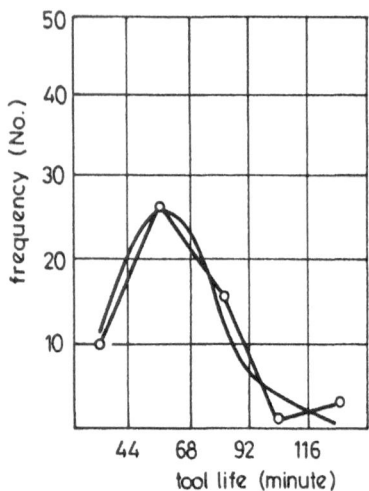

Fig. 2.4. The tool life frequencies of smoothing of the anti-friction bearing circle ([4]). Example for the gamma probability distribution approach.

Lognormal probability distribution

It models the tool life caused by wear only.

$$f(t) = \frac{1}{\sqrt{2\pi}\sigma t} e^{-\frac{(\ln t - m)^2}{2\sigma^2}}, \quad F(t) = \Phi(\frac{\ln t - m}{\sigma}),$$

where

m - the expected value of logarithm of the tool life,

σ - the standard deviation of logarithm of the tool life.

Rossetto-Levi probability distribution

A cutting tool is often subject to the combined effect of progressive wear and several mechanisms of sudden fracture ranging from light chipping to catastrophic breakage of the cutting edge. Mathematical expressions for the cumulative distributions and probability density functions of tool life have been derived by ROSSETTO and LEVI ([6]) for the case of constant hazard function for fracture only combined with lognormal distributions of tool life caused by wear only.

$$f(t) = e^{-\lambda t}\{\lambda - \lambda\Phi(\frac{\ln t - m}{\sigma}) + \frac{1}{\sqrt{2\pi}\sigma t}e^{-\frac{(\ln t - m)^2}{2\sigma^2}}\},$$

$$F(t) = 1 - e^{-\lambda t}\{1 - \Phi(\frac{\ln t - m}{\sigma})\}.$$

where m, λ and σ are three parameters of the distribution.

It is easy to see that this is a superposition of an exponential probability distribution with λ parameter and a lognormal probability distribution with m, σ parameters.

If we take

$$f_1(t) = \lambda e^{-\lambda t}, \ \lambda > 0, \ t > 0, \quad F_1(t) = 1 - e^{-\lambda t}, \ t \geq 0,$$

and

$$f_2(t) = \frac{1}{\sqrt{2\pi}\sigma t}e^{-\frac{(\ln t - m)^2}{2\sigma^2}}, \quad F_2(t) = \Phi(\frac{\ln t - m}{\sigma}),$$

then applying the formula

$$f(t) = f_1(t)[1 - F_2(t)] + f_2(t)[1 - F_1(t)],$$

we get

$$f(t) = \lambda e^{-\lambda t}\{1 - \Phi(\frac{\ln t - m}{\sigma})\} + \frac{1}{\sqrt{2\pi}\sigma t}e^{-\frac{(\ln t - m)^2}{2\sigma^2}}e^{-\lambda t} =$$

$$= e^{-\lambda t}\{\lambda - \lambda\Phi(\frac{\ln t - m}{\sigma}) + \frac{1}{\sqrt{2\pi}\sigma t}e^{-\frac{(\ln t - m)^2}{2\sigma^2}}\},$$

and integrating from minus infinity to plus infinity

$$F(t) = 1 - e^{-\lambda t}\{1 - \Phi(\frac{\ln t - m}{\sigma})\}.$$

According to the Rossetto-Levi probability distribution we have two different cases.

If $\frac{1}{\lambda} \le e^{(m+\sigma^2/2)}$, then the expected time of the failure caused by a sudden fracture is less than that caused by progressive wear. This situation can be seen on the Figure 2.5 where $\lambda = 0.1$, $m = 3$, and $\sigma_1 = 0.5$, $\sigma_2 = 0.3$, $\sigma_3 = 0.15$, $\sigma_4 = 0.03$.
In this case

$$10 = \frac{1}{\lambda} < e^{(m+\sigma_1^2/2)} = 22.76,$$

$$10 = \frac{1}{\lambda} < e^{(m+\sigma_2^2/2)} = 21.01,$$

$$10 = \frac{1}{\lambda} < e^{(m+\sigma_3^2/2)} = 20.31,$$

$$10 = \frac{1}{\lambda} < e^{(m+\sigma_4^2/2)} = 20.09.$$

If $\frac{1}{\lambda} > e^{(m+\sigma^2/2)}$, then the expected time of the failure caused by progressive wear is less than that caused by a sudden fracture. This situation can be seen on the Figure 2.6 where $\lambda = 0.1$, $m = 2$ and $\sigma_1 = 0.6$, $\sigma_2 = 0.4$, $\sigma_3 = 0.2$, $\sigma_4 = 0.1$.
In this case

$$10 = \frac{1}{\lambda} > e^{(m+\sigma_1^2/2)} = 8.85,$$

$$10 = \frac{1}{\lambda} > e^{(m+\sigma_2^2/2)} = 8.01,$$

$$10 = \frac{1}{\lambda} > e^{(m+\sigma_3^2/2)} = 7.54,$$

$$10 = \frac{1}{\lambda} > e^{(m+\sigma_4^2/2)} = 7.43.$$

In both cases the functions drawn are

$f(t)$ − the probability density function,

$F(t)$ − the probability distribution function,

$Z(t) = \frac{f(t)}{1-F(t)}$ − the hazard rate function.

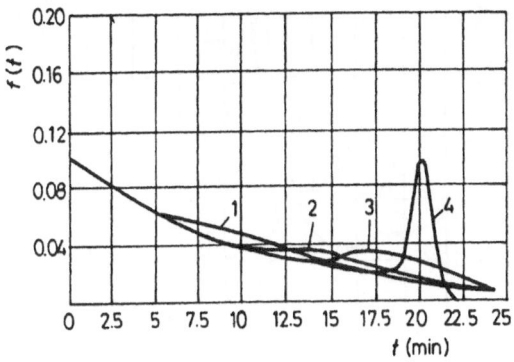

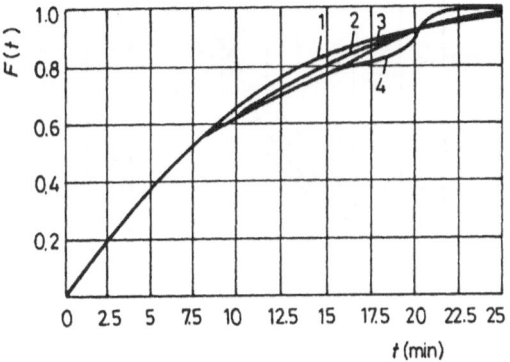

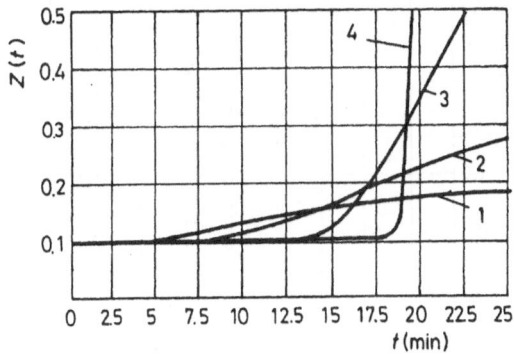

Fig. 2.5. The Rossetto-Levi probability distribution when $\frac{1}{\lambda}$ is less than $e^{(m+\sigma^2/2)}$.

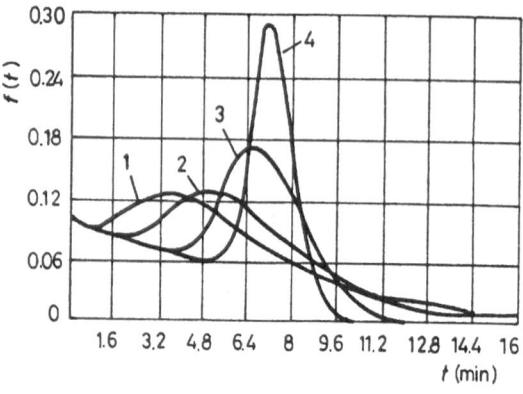

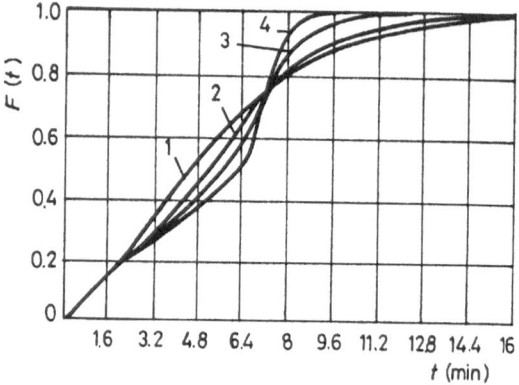

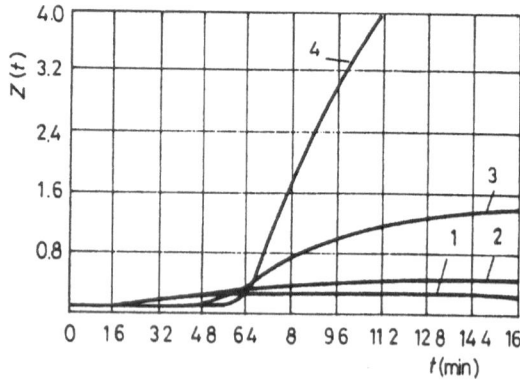

Fig. 2.6. The Rossetto-Levi probability distribution when $\frac{1}{\lambda}$ is greater than $e^{(m+\sigma^2/2)}$.

3. Stochastic models of machining operations based on Rosetto-Levi probability distribution

– Stochastic versions of the Taylor-curves

The Taylor-curve is one of the oldest techniques applied in deterministic machining theory. It gives an exponential connection between the tool life and the machining parameters. If we take the logarithm of this formula as the parameter value m in the Rossetto-Levi probability distribution and take the parameter value σ as $0.1m$ (this means that the relative standard deviation equals 0.1), take the parameter value λ according to the two different cases mentioned before, then by fixing somehow the depth of cut a and the feed rate s we can determine the tool life with different confidence levels as a function of the cutting speed v or alternatively as a function of the revolution per minute n:

$$F(t) = 1 - e^{-\lambda t}\{1 - \Phi(\frac{\ln t - m}{\sigma})\},$$

where

$$m = \ln(\frac{C_t K_t}{a^{x_t} s^{y_t} v^{z_t}}) \ ,$$

and

$$R(t) = 1 - F(t), \ T_\alpha = R^{-1}(\alpha).$$

Using logarithmic scales the deterministic and α-level Taylor- curves are linear. Such Taylor-curves according to different confidence levels α can be seen on the Figures 3.1 and 3.2. In the cases of Figure 3.1 the parameter value λ is determined as

$$\frac{1}{\lambda} = \frac{1}{5}e^{(m+\sigma^2/2)}$$

and in the cases of Figure 3.2 as

$$\frac{1}{\lambda} = 5e^{(m+\sigma^2/2)}.$$

In both cases the feed rate s equals 0.1 according to the figure above and 0.4 according to the figure below.

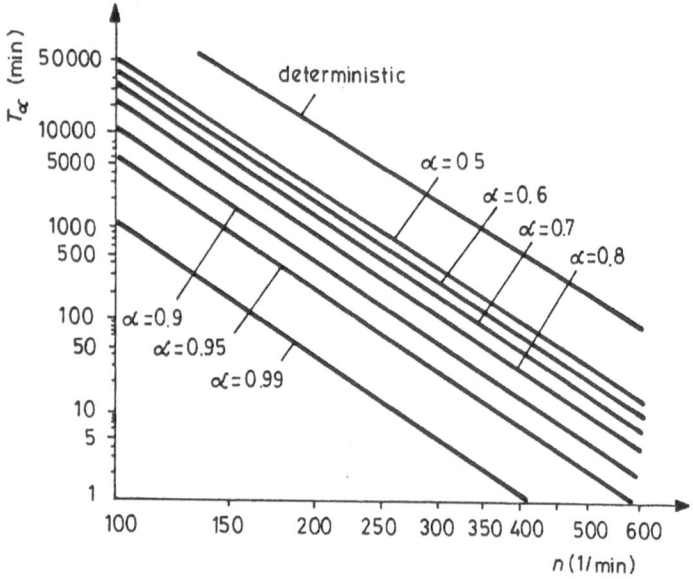

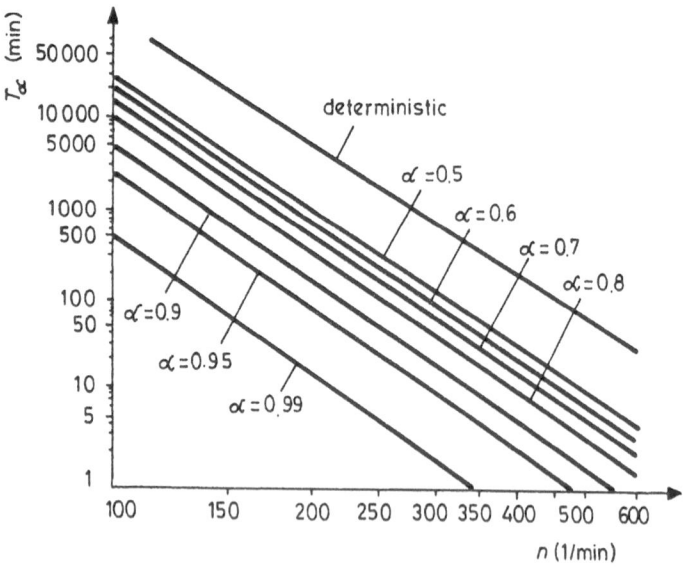

Fig. 3.1. Stochastic Taylor-curves according to the Rossetto-Levi probability distributions when $\frac{1}{\lambda} = \frac{1}{5}e^{(m+\sigma^2/2)}$.

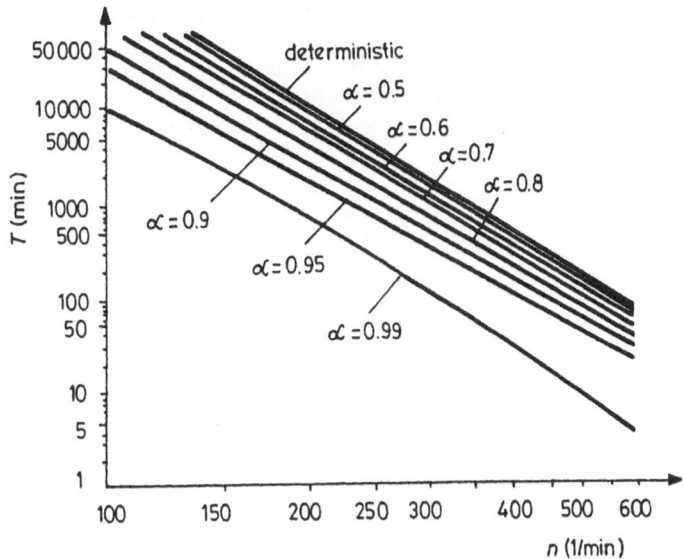

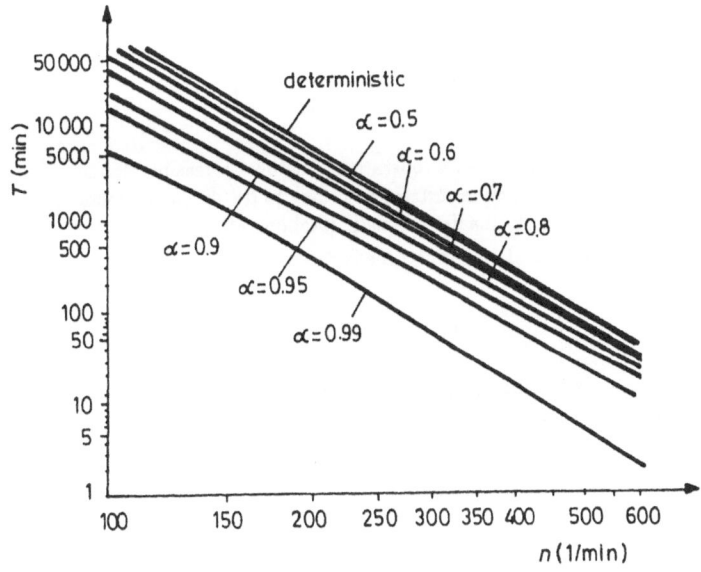

Fig. 3.2. Stochastic Taylor-curves according to the Rossetto-Levi probability distributions when $\frac{1}{\lambda} = 5e^{(m+\sigma^2/2)}$.

– Objective function surfaces defined by the total operating times

The total operating time consists of three main components: the real time of operation T_R, the time of the tool changes T_C, the loss caused by a failure occurred at an α confidence level operation T_L. The definitions of these specific time components are

$$T_R = \frac{1}{vs}, \quad T_C = t_c\frac{T_R}{T_P}, \quad T_L = (1-\alpha)t_l\frac{T_R}{T_P}$$

where

T_P – the productive tool life,

t_c – the tool changing time regarding one tool life,

t_l – the time loss caused by an early tool break down,

α – the confidence level of the operation.

The productive tool life T_P is determined as

$$T_P = T_\alpha - (1-\alpha)t_l, \quad \text{where} \quad T_\alpha = R^{-1}(\alpha) \quad \text{and} \quad R(t) = 1 - F(t)$$

according to the specified Rossetto-Levi probability distribution.
The final form of the total operating time is

$$T_{tot} = T_R + T_C + T_L = \frac{1000}{d\pi ns}\{1 + \frac{t_c + (1-\alpha)t_l}{T_\alpha - (1-\alpha)t_l}\}.$$

Instead of optimizing the total operating time over a feasible domain we give the whole objective function surfaces according to the different confidence levels of the operation and to the deterministic total operating time.
The deterministic total operating time is determined by

$$T_{det} = \frac{1000}{d\pi ns}(1 + \frac{t_c}{T_{Taylor}})$$

where T_{Taylor} is the tool life given by the deterministic Taylor-formula.
Such objective function surfaces according to different confidence levels α can be seen on the Figures 3.3 and 3.4.

In both cases we took the logarithm of the tool life given by the Taylor formula as the parameter value m in the Rossetto-Levi probability distribution and took the parameter value σ as $0.1m$. In the case of Figure 3.3 the parameter value λ was determined as $\frac{1}{\lambda} = \frac{1}{5}e^{(m+\sigma^2/2)}$ and in the case of Figure 3.4 as $\frac{1}{\lambda} = 5e^{(m+\sigma^2/2)}$.

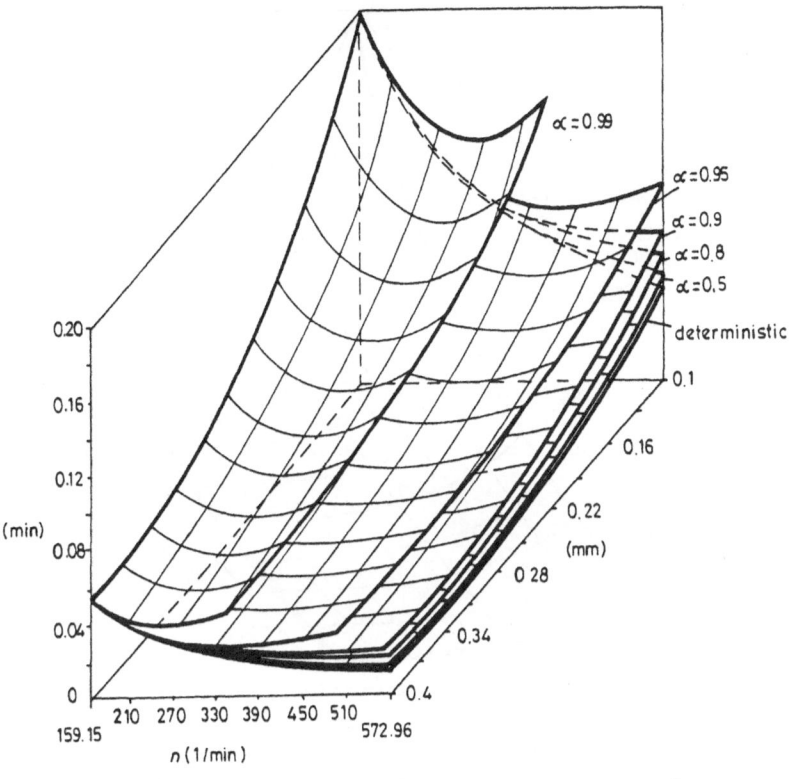

Fig. 3.3. Objective function surfaces defined by the operating times according to the Rossetto-Levi probability distribution when $\frac{1}{\lambda} = \frac{1}{5}e^{(m+\sigma^2/2)}$.

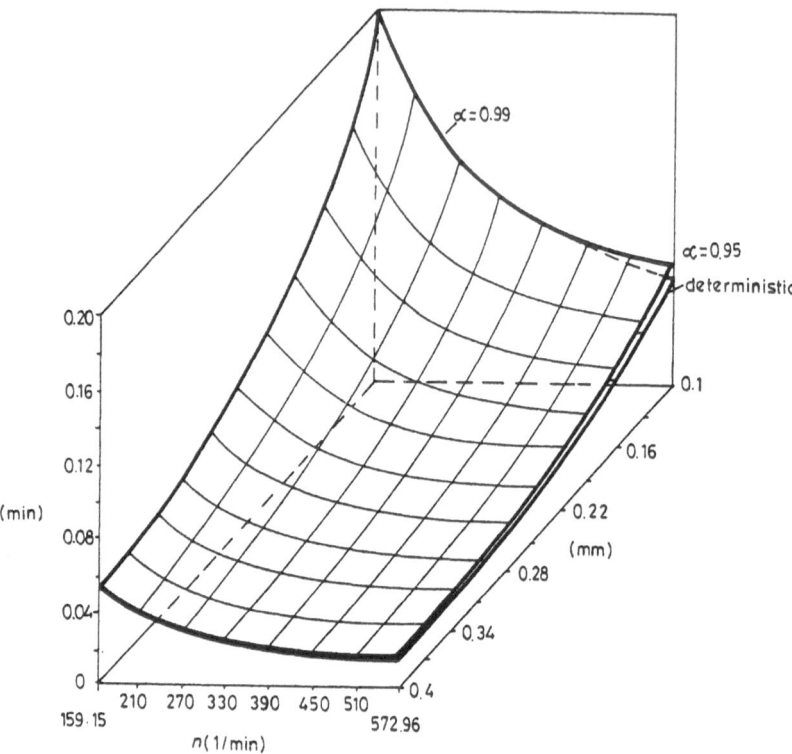

Fig. 3.4. Objective function surfaces defined by the operating times according to the Rossetto-Levi probability distribution when $\frac{1}{\lambda} = 5e^{(m+\sigma^2/2)}$.

- Objective function surfaces defined by the total operating costs

The total operating cost consists of four main components: the cost of the machine C_M, the cost of the tool C_T, the cost of the tool changes C_C and the operating cost at an α confidence level operation C_α. The definitions of these specific cost components are

$$C_M = C_m T_R, \quad C_T = C_t \frac{T_R}{T_P}, \quad C_C = t_c C_m \frac{T_R}{T_P}, \quad C_\alpha = (1-\alpha)(t_l C_m + R_w) \frac{T_R}{T_P}$$

where the time variables t_l, t_c, T_R, T_P were defined before and

C_m – the machine cost for one minute,

C_t – the cost of the tool for one tool life,

R_w – the cost of an early tool break down.

The final form of the total cost is

$$C_{tot} = C_M + C_T + C_C + C_\alpha = \frac{1000 C_m}{d\pi n s} \{ 1 + \frac{\frac{C_t}{C_m} + t_c + (1-\alpha)(t_l + frac R_w C_m}{T_\alpha - (1-\alpha)t_l} \}.$$

Instead of optimizing the total operating cost over a feasible domain we give the whole objective function surfaces according to the different confidence levels of the operation and to the deterministic total operating cost.

The deterministic total operating cost is determined by

$$C_{det} = \frac{1000 C_m}{d\pi n s} (1 + \frac{\frac{C_t}{C_m} + t_c}{T_{Taylor}})$$

where T_{Taylor} is as before.

Such objective function surfaces according to different confidence levels α can be seen on the Figures 3.5 and 3.6.

In both cases we took the logarithm of the tool life given by the Taylor formula as the parameter value m in the Rossetto-Levi probability distribution and took the parameter value σ as $0.1m$. In the case of Figure 3.5 the parameter value λ was determined as $\frac{1}{\lambda} = \frac{1}{5} e^{(m+\sigma^2/2)}$ and in the case of Figure 3.6 as $\frac{1}{\lambda} = 5 e^{(m+\sigma^2/2)}$.

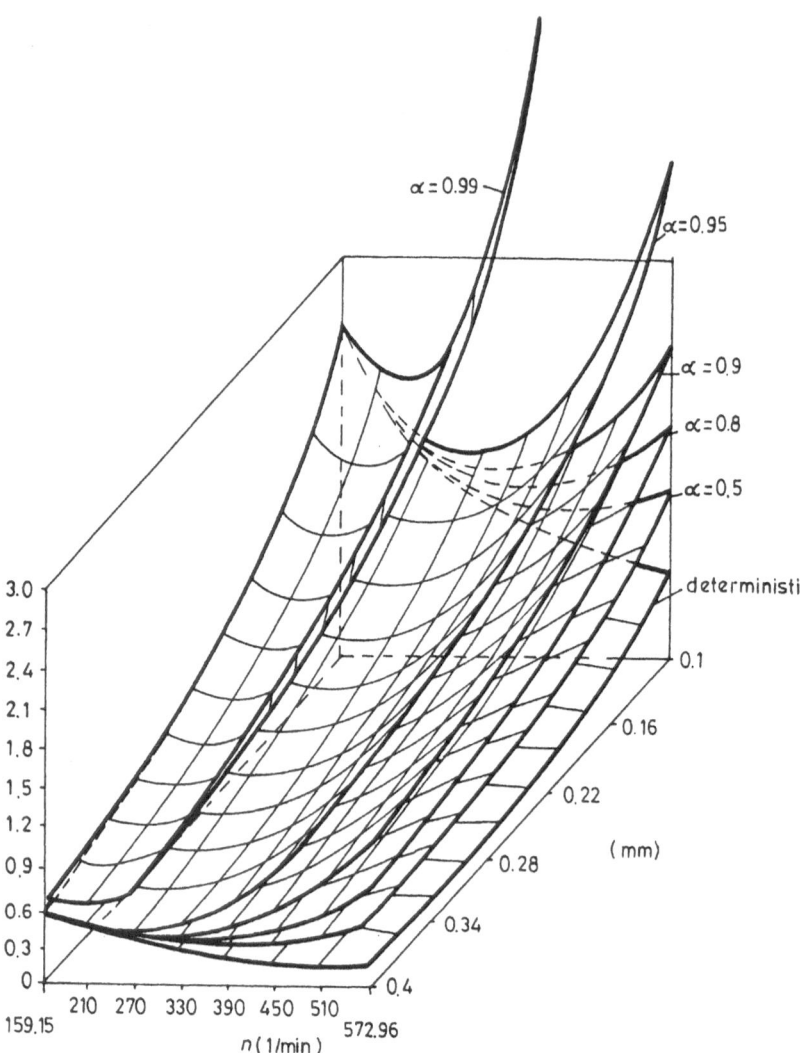

Fig. 3.5. Objective function surfaces defined by the operating costs according to the Rossetto-Levi probability distribution when $\frac{1}{\lambda} = \frac{1}{5}e^{(m+\sigma^2/2)}$.

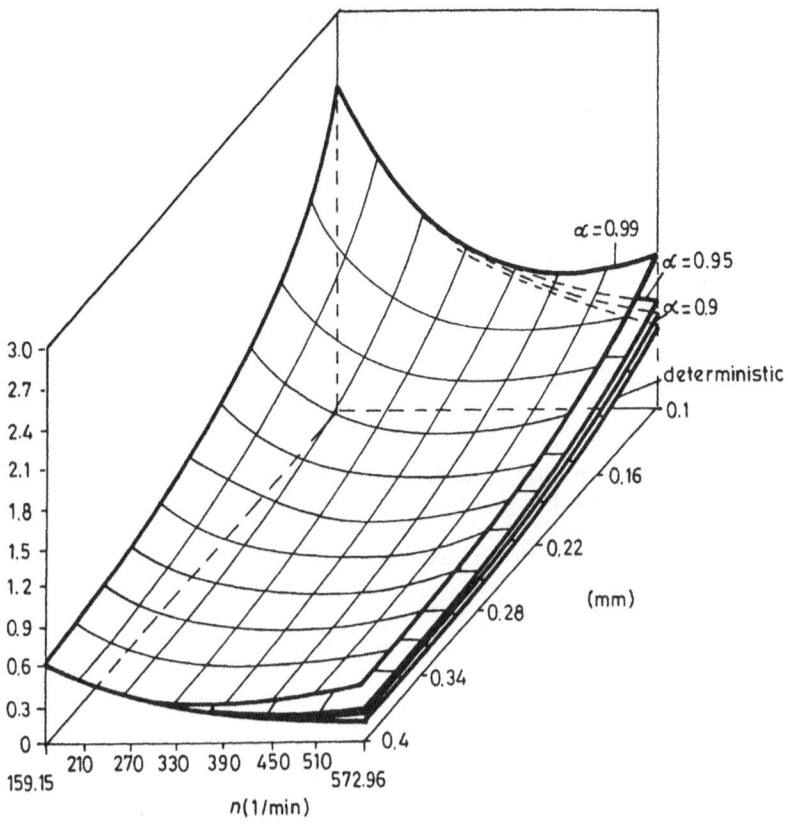

Fig. 3.6. Objective function surfaces defined by the operating costs according to the Rossetto-Levi probability distribution when $\frac{1}{\lambda} = 5e^{(m+\sigma^2/2)}$.

– Probabilistic constraint for lognormally distributed tool life

Let us regard the deterministic Taylor formula in the following form:

$$\tau = \frac{\xi}{a^{x_t} s^{y_t} v^{z_t}} \ ,$$

where

ξ random variable having lognormal probability distribution with parameters $E(\log\xi) = m$ and $D(\log\xi) = \sigma$,

a depth of cut (in mm),

s feed rate (in mm per revolution),

v cutting speed (in m per minute),

x_t, y_t, z_t exponents depending on the conditions of working ($x_t \approx 3 - 5$, $y_t \approx 1$, $z_t \approx 1$).

Then the probabilistic constraint of level $1 - \alpha$ according to the tool life takes the form:

$$P(\tau \geq T_0) \geq 1 - \alpha.$$

Taking the logarithm of the expression inside the probability:

$$P(\log\tau \geq \log T_0) \geq 1 - \alpha$$

$$P(\log\xi \geq \log T_0 + x_t\log a + y_t\log s + z_t\log v) \geq 1 - \alpha$$

$$P\{\frac{\log\xi - m}{\sigma} \geq \frac{\log T_0 + x_t\log a + y_t\log s + z_t\log v - m}{\sigma}\} \geq 1 - \alpha$$

$$1 - \Phi\{\frac{\log T_0 + x_t\log a + y_t\log s + z_t\log v - m}{\sigma}\} \geq 1 - \alpha$$

$$y_t\log s + z_t\log v \leq \sigma\Phi^{-1}(\alpha) - \log T_0 - x_t\log a + m.$$

This is a linear inequality according to the unknown $\log s$ and $\log v$ so taking this constraint the linearity of the deterministic constraint set drawn on the Figure 1.1 remains true. This means that the solution of the probabilistic constrained optimization problem is not more difficult than that of the deterministic one.

As an example let us regard the following constraints according to a single tool machining operation:

$$50 \le v \le 280,$$
$$0.05 \le s \le 0.4,$$
$$s^{0.8}v^{-0.08} \le 0.33,$$
$$s^{0.8}v^{0.92} \le 60.1.$$

The deterministic Taylor-formula let be

$$T = \frac{6.15 \times 10^8}{v^{3.7}s^{1.94}}.$$

Let us suppose that some kind of optimum is reached when

$$v = 190.3, \quad f = 0.4.$$

In this case the deterministic Taylor-formula gives 13.4 tool life as an optimal value.

If T has a lognormally distributed nominator ξ with parameters

$$m = E(\log\xi) = log(6.15 \times 10^8) = 8.789,$$
$$\sigma = D(\log\xi) = 0.1E(\log\xi) = 0.879$$

then one easily can check that the probability of the random tool life to be greater than the deterministic value 13.4 is only about 0.5.

This situation and the proposed probabilistic constraints according to the optimal tool life 13.4 with different $1 - \alpha$ levels are illustrated on the Figure 3.7.

We remark that after taking the logarithm of the expressions and applying some simple linear transformations the horizontal axis of the figure corresponds to

$$x_1 = 10\log v$$

and the vertical axis of the figure corresponds to

$$x_2 = 10\log s + 13.$$

Such a way all of the constraints including probabilistic ones became linear.

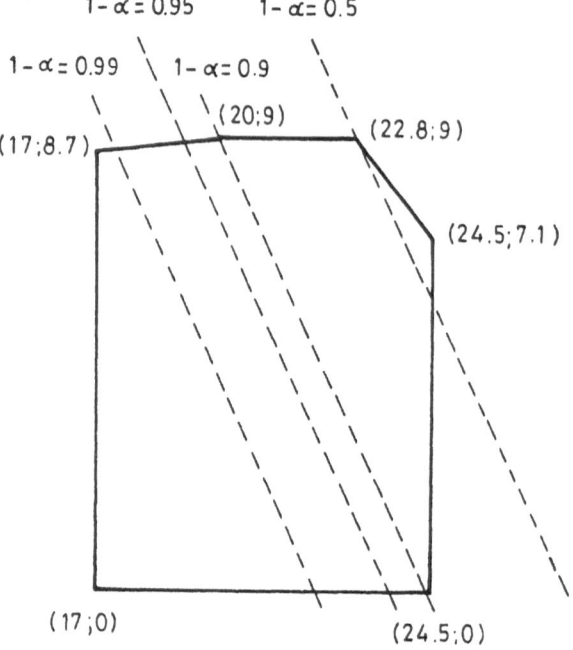

Fig. 3.7. Illustration of the proposed probabilistic constraints.

Acknowledgement. The authors would like to express their thanks to the referees for their invaluable comments and advices.

References

[1] AVRIEL, M. and D. J. WILDE, Stochastic geometric programming, in: H. W. Kuhn (ed.) *Proceedings of the Princeton Symposium of Mathematical Programming* (Princeton University Press, Princeton, 1970).

[2] DUPAČOVA, J., On stochastic aspects of a metal cutting problem. Lecture presented at 2nd GAMM/IFIP Workshop on Stochastic Optimization: Numerical Methods and Technical Applications, UniBw Munich-Neubiberg, June 15-17, 1993.

[3] JAGANNATHAN, R., A stochastic geometric programming problem with multiplicative recourse, *Operations Research Letters* 9 (1990) 99-104.

[4] КАЦЕВ, П. Г. , Статистические методы исследования режущего инструмента, (Москва, Машиностроение, 1974).

[5] KOMOROWSKI, R., Rozkładu Weibulla jako ogólny model rozkładu truvatości narçedzi skrawajaçych, *Mechanik* 44 (1971) 136-139. (in Polish).

[6] ROSSETTO, S. and LEVI, R., Fracture and wear as factors affecting stochastic tool-life models and machining economics, *Journal of Engineering for Industry, Transactions of ASME, Series B* 99 (1977) 281-286.

[7] TAYLOR, F. W., On the art of cutting metals, *Transactions of ASME* 28 (1907) 31-35.

[8] WAGER, J. G. and BARASH, M. M., Study of the distribution of the life of HSS tools, *Journal of Engineering for Industry, Transactions of ASME, Series B* 93 (1971) 1044-1050.